Manual of Cross-Connection Control

Ninth Edition

Published by

Foundation for Cross-Connection Control
and Hydraulic Research
University of Southern California
KAP-200 University Park MC-2531
Los Angeles, CA 90089-2531

December, 1993

Foundation for Cross-Connection Control
and Hydraulic Research
University of Southern California
KAP-200 University Park MC-2531
Los Angeles, CA 90089-2531

First Edition, 1960
Second Edition, 1965
Third Edition, 1966
Fourth Edition, 1969
Fifth Edition, 1974
Sixth Edition, 1979
Revised Sixth Edition, 1982
Seventh Edition, 1985
Eighth Edition, 1988
Ninth Edition, 1993

ISBN 0-9638912-0-0

Preface

to the
Ninth Edition

This Ninth Edition of the *Manual of Cross-Connection Control* is the most extensive revision of this Manual to date. The Manual Review Committee has spent almost three years considering modifications, additions and deletions. Some of the most significant changes are in Section 9, *Backflow Prevention Assembly Field Test Procedures and Gage Accuracy Verification.* The test procedure for the double check valve assembly has been changed to a direction of flow test. This change came after months of discussion. It was finally decided that the direction of flow test can give more accurate results and more information can be derived from the test when shutoff valves are leaking. The most noticeable change in Section 9 is the change to illustrated testing procedures. Almost every step of the procedures is accompanied by an illustration making all the steps necessary to carry out the field test of the backflow preventers clear to the tester.

Many new illustrations were included in Section 7 covering installation guidelines and typical uses of the various backflow preventers. Section 10, *Specifications of Backflow Prevention Assemblies,* now contains a life cycle test for the backflow preventers to help discover problems which would have been discovered in the field evaluation before the time and effort is expended on the field evaluation. There is also a new specification for a Spill-Resistant Vacuum Breaker, a new type of pressure vacuum breaker which will minimize spillage during testing and use.

The Manual Review Committee was comprised of representatives from water agencies, health agencies, certified testers and Foundation staff members who volunteered a considerable amount of time and effort over the past three years considering many suggestions for the Ninth Edition. As with previous editions, manufacturers of backflow prevention assemblies were not included in the Manual Review Committee. However, manufacturers were asked to comment on Section 10, *Specifications of Backflow Prevention Assemblies,* and a general concurrence was received. The Committee is assured these changes will result in significantly better backflow prevention assemblies. We salute the Members of the Manual Review Committee for their tireless efforts and contributions in making this Manual even more useful to those involved in cross-connection control. We appreciate, very much, the comments and feedback received from the many people who contributed information and made suggestions for improvements in this Ninth Edition of the *Manual of Cross-Connection Control.*

We trust that you will find this Manual invaluable to you in your efforts in cross-connection control. We believe this new edition will be used much more effectively than previous editions by the many people involved in cross-connection control. As always, should you have any questions or comments, please feel free to contact the Foundation office.

Paul H. Schwartz,
Chief Engineer

J. J. Lee,
Director

Dedication

Walter O. Weight

Walter Weight was born on October 29, 1896. He was raised in St. Paul Minnesota and attended the University of Minnesota. Because of his interest in agriculture at the University of Minnesota, Walter took on a position in sales, with an agricultural equipment company, in Alberta, Canada. He moved to Los Angeles in 1924 after passing through Los Angeles.

Walter began selling water meters in 1927 and by 1932 was meeting with managers of several water agencies in Southern California on a monthly basis. This was the beginning of the Southern California Water Utilities Association which now, because of Walter Weight, has over 650 members.

Walter Weight was the first person in the American Water Works Association to be recognized for recruiting 1000 new members and he received the *Platinum Award* for this achievement. Walter also received the highly esteemed *Fuller Award* from AWWA for his many contributions to the industry. These are just two of the many awards which Walter has received.

Walter first became involved in cross-connection control in the late 1940's through the work the Los Angeles Department of Water and Power was doing in conjunction with the Foundation for Cross-Connection Control and Hydraulic Research at the University of Southern California. Others had begun to look to USC for guidance in selecting appropriate backflow preventers and in developing programs to detect and prevent cross-connections. This resulted in the publication of the first edition of the *Manual of Cross-Connection Control* in 1960. When USC's new building program caused the Foundation to seek a new facility to conduct its laboratory work, Walter was instrumental in working out an arrangement with the Los Angeles Department of Water and Power to secure the Foundation's current laboratory. Walter worked with various manufacturers and water entities to provide the necessary funds and equipment to reestablish the Foundation's laboratory at its current location in 1968.

Walter, as chairman of the Southern California Water Utilities Association, encouraged many of its members in providing funds to the Foundation on an annual basis to assure USC that the work of the Foundation was worthwhile and an essential service for the water industry. Walter presented the first check for $10,000 from the Southern California Water Utilities Association to Dr. Norman Topping, President of the University. This initiated the Foundation's Membership Program in 1967.

Walter Weight has contributed greatly to the success of the Foundation. Walter, approaching his 100th year, is still actively getting more people interested in providing a safer water supply to everyone.

Dedication

William Minder Whiteside

1919-1991

William M. Whiteside graduated from Huntington Park High School in 1937. As a youth he achieved the rank of Eagle Scout. He entered the Army Air Corps in 1942 and served as a flight instructor until 1945. Thereafter he entered the School of Engineering at the University of Southern California. He graduated in 1948 with a Bachelor of Science Degree in Mechanical Engineering. Upon graduation, he took a position in sales with a water meter manufacturing company. In 1955 he became vice-president and General Manager of the California Domestic Water Company, serving until 1982. At this time he became the General Manager of the San Gabriel Valley Municipal Water District until 1989 when he retired.

In 1950 Mr. Whiteside became active in the American Water Works Association. In 1976 he was elected as the Chairman of the California/Nevada Section of the American Water Works Association. He served in this capacity for one year and then served as a Director to the National Board of the American Water Works Association until 1981. In 1972 he received the *Fuller Award* by the American Water Works Association for his outstanding contributions to the water works field.

In 1966 Mr. Whiteside was elected as President of the Southern California Water Utilities Association. He served in this capacity until 1983. He was instrumental in having the members of the Southern California Water Utilities Association contribute to the ongoing support of the Foundation for Cross-Connection Control and Hydraulic Research at the University of Southern California which led to the establishment of the Foundation's Membership Program. Working along side with Mr. Walter Weight and many pioneers in the cross-connection control field, Mr. Whiteside contributed tirelessly time, wisdom, and experience to assure the continued success of the Foundation. William Whiteside served as a member of the Foundation's Board of Directors from 1966 until 1991.

In 1991, in memory of William Whiteside and his great contributions to the water industry and the field of cross-connection control, his friends established the William Whiteside Scholarship Fund at the University of Southern California. Each year a scholarship is awarded to outstanding USC students involved in water related fields.

The Foundation is deeply indebted to Mr. William Whiteside for the generous support he has given throughout the years.

This Ninth Edition of the *Manual of Cross-Connection Control* was prepared by the following Manual Review Committee which spent many, many hours in considering the changes incorporated herein. It is the sincere hope of this Committee that the water agencies, health agencies, plumbing officials, and all who enjoy the use of clean water will understand better the problems created by cross-connections and will thereby work toward the control or elimination of all cross-connections.

Mr. Mike Ahlee
Ahlee Backflow Service
El Cajon, CA

Mr. E. D. Alterton
Hermosa Beach, CA

Mr. Kenneth Anderson
Riverside, CA

Mr. Richard Bird
Sweetwater Authority
Spring Valley, CA

Mr. Henry W. Chang
Mechanical Engineer
Foundation for Cross-Connection Control
and Hydraulic Research
University of Southern California
Los Angeles, CA

Mr. Martin Friebert
Environmental Health Division
Orange County Health Department
Santa Ana, CA

Mr. William Gedney
State of California
Department of Health Services
San Bernardino, CA

Mr. Ernest J. Havlina
Los Angeles, CA

Mr. Orvil Kirk
Cross-Connection Control Specialist
County of Ventura
Ventura, CA

Dr. J. J. Lee
Director
Foundation for Cross-Connection Control
and Hydraulic Research
University of Southern California
Los Angeles, CA

Mr. Charles L. Nena
Water Quality Supervisor
California Water Service Company
Los Angeles, CA

Mr. Robert Purzycki
Backflow Apparatus and Valve Company
Gardena, CA

Mr. Paul H. Schwartz
Chief Engineer
Foundation for Cross-Connection Control
and Hydraulic Research
University of Southern California
Los Angeles, CA

Mr. Robert Smith
Chief, Cross-Connection Control
Department of Health Services
County of Los Angeles
Los Angeles, CA

Prof. E. Kent Springer
Emeritus Director
Foundation for Cross-Connection Control
and Hydraulic Research
University of Southern California
Los Angeles, CA

Mr. Patrick A. Sylvester
Communications Coordinator
Foundation for Cross-Connection Control
and Hydraulic Research
University of Southern California
Los Angeles, CA

This Manual is prepared for the benefit of the *General Public* who have the implicit faith that the water is always *safe to drink*

Table of Contents

List of Figures

List of Tables

SECTION 1

Objectives

This **MANUAL** is designed to provide the water purveyor, health agency, plumbing official and other persons responsible for the public and in-plant potable water supply with information, procedures for the development, implementation and enforcement of backflow prevention practices which will meet the Federal, State and local regulations governing cross-connection control. It is promulgated for the purpose of:

1. Providing minimum standards for the protection of the public water supply - both as system protection and as internal protection. It is recognized that water purveyors and agencies having cognizance may establish requirements more rigid than those set forth herein.

2. Protecting the public potable water system at the service connection by containing within the consumer's premises the actual or potential pollution or contamination which may result from backflow through cross-connections.

3. Providing means whereby the consumer may segregate the domestic and industrial water uses into separate systems to prevent possible pollution or contamination of the private potable water system which may result from backflow through cross-connections.

More specifically, this **MANUAL** contains chapters dealing with the following:

1. **OBJECTIVES**
   To provide for uniform cross-connection control practices.

2. **POLICY: CROSS-CONNECTION CONTROL**
   A statement of policy.

3. **RESPONSIBILITIES: HEALTH AGENCY, WATER PURVEYOR, PLUMBING OFFICIAL, CONSUMER AND CERTIFIED BACKFLOW PREVENTION ASSEMBLY TESTER**
   Defining the responsibilities of the owner, consumer, water user, certified tester, plumbing official, water purveyor and the health authorities having jurisdiction.

4. **DEFINITIONS**
   Definitions of words and phrases which need clarification of meaning in order that uniform interpretation of this **MANUAL** is possible.

5. **TYPICAL FACILITIES: CROSS-CONNECTIONS OR WATER USES WHICH MAY ENDANGER THE PUBLIC WATER SYSTEM**

6. **RESULTS OF NON-COMPLIANCE**

7. **CROSS-CONNECTION CONTROL PRACTICES**
   Water using plants or facilities where protection is recommended. The type of assembly or method of protection involves the principle that the degree of protection should be commensurate with the degree of hazard.

8. **SAMPLE LETTERS, FORMS, INSTALLATION GUIDELINES, MODEL ORDINANCE, ETC.**
   Installation, testing, maintenance and reporting procedures and requirements for backflow prevention assembly testers.

9. **BACKFLOW PREVENTION ASSEMBLY FIELD TEST PROCEDURES AND GAGE ACCURACY VERIFICATION**

10. **SPECIFICATIONS OF BACKFLOW PREVENTION ASSEMBLIES**
    Specifications for backflow prevention assemblies, design, operation and construction.

11. **SUMMARY OF CASE HISTORIES**
    Summary of documented backflow incidents.

SECTION 2

Policy: Cross-Connection Control

Under Public Law 99-339 — the Safe Drinking Water Act Amendments of 1986 — and regulations of most states the water purveyor has the primary responsibility for preventing water from unapproved sources, or any other substances, from entering the public potable water system. The health agency has the overall responsibility for preventing water from unapproved sources entering either the potable water system within the water consumer's premises or the public water supply directly.

To the water purveyors this means that they must plan and diligently execute a program of cross-connection control which either eliminates all cross-connections or requires the installation and maintenance of a proper type of approved backflow prevention assembly at the water service connection whenever a potential hazard is determined to exist in the consumer's system.

To the health agencies this means that they must plan and diligently execute a program of cross-connection control specifically within the consumer's system. Or, where the health agency and water purveyor develop a joint coordinated program, the cross-connection control program may go all the way to the last outlet in the consumer's system.

To the plumbing inspection agencies this means that they must plan and diligently execute a program whereby plumbing type cross-connections will not be permitted to be built into a building; or, where a cross-connection to the consumer's potable water system can not be avoided, an approved backflow prevention assembly will be required.

In view of the above noted responsibilities, this *Manual of Cross-Connection Control* is intended to be of assistance to not only the water purveyors, the health agencies and the plumbing officials, but to the manufacturers of backflow prevention assemblies and to the public in their joint effort in developing a cross-connection control program which will satisfy the minimum responsibility factors as set forth in Public Law 99-339 — the Safe Drinking Water Act Amendments of 1986 — and/or the regulations of the states and territories.

Where it is probable that a pollutional, contaminant, system or plumbing hazard may be created by a water user; i.e., where materials dangerous to health or toxic substances in toxic concentrations are handled in tanks, piping systems or other vessels on the premises; or, where the water system is unstable and cross-connections may be installed or reinstalled, an approved backflow prevention assembly of the proper type should be considered at each service connection on the premises. In addition, to protect the occupants of the premises the administrative authority shall require that any unprotected cross-connection be eliminated or that the water system be protected with an approved backflow prevention assembly at the point of use. Provisions of most regulations allow the administrative authority to recognize that there are varying degrees of hazard and to allow the administrative authority to exercise its judgement as to the type of backflow protection to be used and its location.

Illustrative examples of types of conditions and protective assemblies are given in the following sections in an effort to assist each of the several parties that must coordinate their efforts in a program to understand better the various facets of cross-connection control.

SECTION 3

Responsibilities: Health Agency, Water Purveyor, Plumbing Official, Consumer, and Certified Backflow Prevention Assembly Tester

The implementation of regulations for the effective control of cross-connections requires the full cooperation of the water purveyor, the health agency, the plumbing official, the consumer and the certified backflow prevention assembly tester. Each has its responsibilities and each must carry out its phase of a coordinated cross-connection control program in order to prevent pollution or contamination of the potable water supplies. The responsibilities of each are outlined hereafter.

3.1 Responsibility: Health Agency

3.1.1 General

The health agency has the responsibility for promulgating and enforcing laws, rules, regulations and policies to be followed in carrying out an effective cross-connection control program.

3.1.2 Public Potable Water System

The health agency has the primary responsibility of insuring that the water purveyor operates the public potable water system free of actual or potential sanitary hazards, including unprotected cross-connections. This agency has the further responsibility of insuring that the water purveyor provides an approved water supply at the point of delivery to the consumer's water system and, further, that he requires the consumer to install, test and properly maintain an approved backflow prevention assembly(s) on the service connection(s) when required.

3.1.3 Consumer's Water System(s)

The health agency has the primary responsibility of insuring that the consumer's potable water system is provided with an approved water supply and that its potable water system(s) is maintained free of sanitary hazards, including unprotected cross-connections.

3.2 Responsibility: Water Purveyor

Under the cross-connection control regulations or rules of most states and territories the water purveyor has primary responsibility to prevent water from unapproved sources, or any other substance, entering the public water supply system. The water purveyor is prohibited by these regulations or rules from installing or maintaining a water service connection to a consumer's water system within its jurisdiction where a health, system, plumbing or pollutional hazard exists, or will probably exist, unless the public potable water supply is protected against backflow by an approved backflow prevention assembly(s) installed at the service connection(s) (i.e., point of delivery).

The water purveyor's responsibility begins at the source and includes all of the public water distribution system, including the service connection, and ends at the point of delivery to the consumer's water system(s). In addition, the water purveyor shall exercise reasonable vigilance to insure that the consumer has taken the proper steps to protect the public potable water system. To insure that the proper precautions are taken the water purveyor is required to determine the degree of hazard to the public potable water system. When it is determined that a backflow prevention assembly is required for the protection of the public system the water purveyor shall require the consumer, at the consumer's expense, to install an approved backflow prevention assembly at each service connection, to test immediately upon installation and annually, or more often, to properly repair and maintain such assembly or assemblies and to keep adequate records of each test and subsequent maintenance and repair, including materials or replacement parts.

3.3 Responsibility: Plumbing Official

The building and safety department of the appropriate political jurisdiction has the responsibility to not only review building plans and inspect plumbing as it is installed; but, it has the explicit responsibility of preventing cross-connections from being designed and built into the structures within its jurisdiction. Where the review of building plans suggests or detects the potential for cross-connections being made as an integral part of the plumbing system the plumbing official has the responsibility under most building codes for requiring that such cross-connection practices be either eliminated or provided with approved backflow prevention equipment.

The plumbing official's responsibility begins at the point of service (i.e., the downstream side of the water meter or service connection) and carries throughout the entire length of the consumer's water system. The plan inspector should inquire about the intended use of water at any point where it is suspect that a cross-connection might be made or where one is actually called for by the plans. When such is discovered, a suitable approved backflow prevention assembly shall be required and be properly installed.

3.4 Responsibility: Consumer

The consumer has the responsibility of preventing pollutants and contaminants from entering his/her potable water system(s) or the public potable water system. The consumer's responsibility starts at the point of delivery from the public potable water system and includes all of his/her water systems. The consumer, at his/her own expense, shall install, operate, test and maintain approved backflow prevention assemblies as directed by the authority having jurisdiction. The consumer shall maintain accurate records of tests and repairs made to backflow prevention assemblies and provide the administrative authority having jurisdiction with copies of such records. The records shall be on forms approved by the administrative authority having jurisdiction and shall include the list of materials or replacement parts used. Following any repair, overhaul, re-piping or relocation of an assembly the consumer shall have it tested to insure that it is in good operating condition and will prevent backflow. Tests, maintenance and repairs of backflow prevention assemblies shall be made by a certified backflow prevention assembly tester.

When requested by the water purveyor or health agency, the consumer shall appoint a water supervisor who shall be responsible for conformance with all applicable laws, rules and regulations pertaining to cross-connection control; for the installation, operation and use of all water piping systems, backflow prevention assemblies and water-using equipment on the premises; and for the avoidance of unprotected cross-connections. The water supervisor should be the consumer or any full-time employee appointed by the consumer who has a thorough knowledge of the installation, operation and maintenance of all water systems and backflow prevention assemblies on the consumer's premises. In the event of pollution or contamination of the public or the consumer's potable water system due to backflow on or from the consumer's premises, the owner or water supervisor shall promptly take steps to confine further spread of the pollution or contamination within the system and should notify the local health officer and the water purveyor of the condition. The person responsible for the consumer's water system(s) (consumer or water supervisor) shall take appropriate measures to free the water system(s) of any pollutants or contaminants.

3.5 Responsibility: Certified Backflow Prevention Assembly Tester

(where there is a Certification Program in force)

When directed to test, repair, overhaul or maintain backflow prevention assemblies, a certified backflow prevention assembly tester will have the following responsibilities:

The tester will be responsible for performing accurate field tests and for repairing or overhauling backflow prevention assemblies and making reports of such repair to the consumer and responsible authorities on forms approved by the administrative authority having jurisdiction. The tester shall include the list of materials or replacement parts used. The tester shall be equipped with and be capable of using all the necessary tools, gages, and other equipment necessary to properly test, repair and maintain backflow prevention assemblies. It will be the tester's responsibility to insure that original manufactured replacement parts are used in the repair of or replacement of parts in a backflow prevention assembly. It will be the tester's further responsibility not to change the design, material or operational characteristics of an assembly during repair or maintenance without prior approval of the approving authority. A certified tester shall perform the work and be responsible for the accuracy of all tests and reports.

SECTION 4

Definitions

4.1 Accessible

The term "accessible," when referring to a backflow prevention assembly, shall mean capable of being reached for testing and/or maintenance, but which first may require the removal of an access panel, door, or similar obstruction. (See Readily Accessible.)

4.2 Administrative Authority

The term "administrative authority" shall mean the individual official, board, department, or agency established and authorized by a state, county, city, or other political subdivision created by law to administer and enforce the provisions of the cross-connection control program.

4.3 Air gap

The term "air gap" shall mean a physical separation between the free flowing discharge end of a potable water supply pipeline and an open or non-pressure receiving vessel. An "approved air gap" shall be at least double the diameter of the supply pipe measured vertically above the overflow rim of the vessel — in no case less than 1 inch (2.54 cm). (Additional reference: ASME A112.1.2 - 1991 *Air Gaps in Plumbing Systems* [1].) See Fig. 4.1.

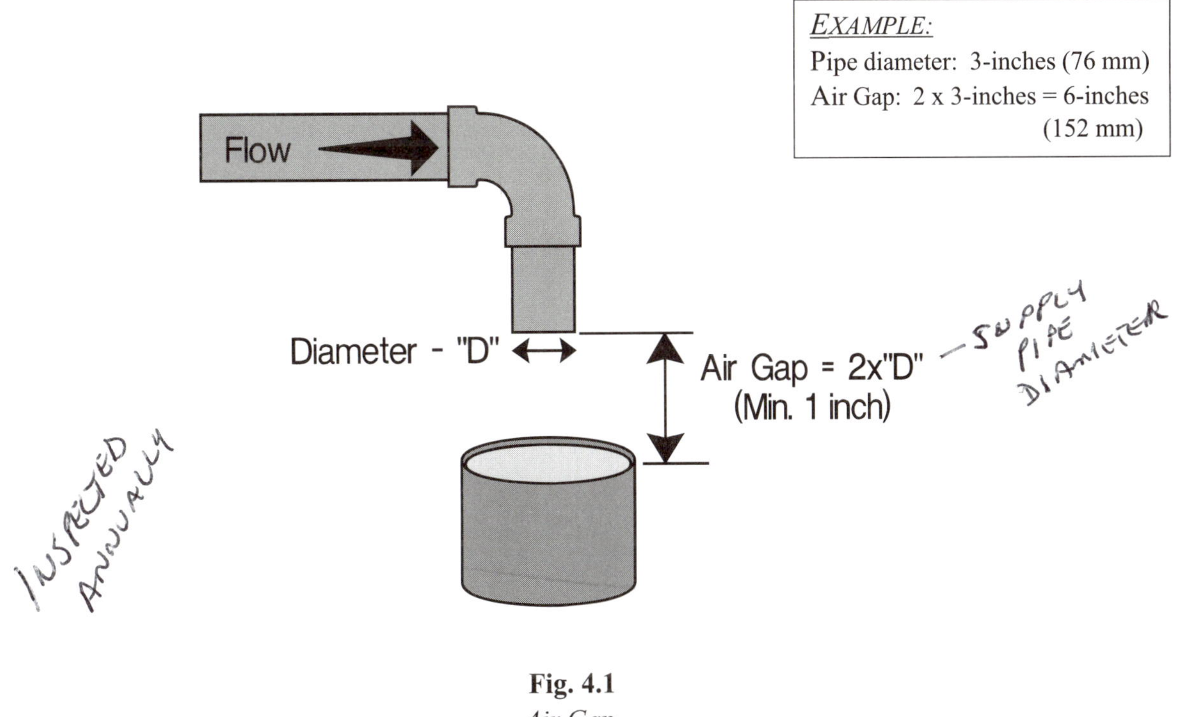

Fig. 4.1
Air Gap

[1] American Society of Mechanical Engineers, 345 East 47th Street, New York, NY 10017 (212) 705-7736

4.4 Approved

a. The term "approved" as herein used in reference to a water supply shall mean a water supply that has been approved by the health agency having jurisdiction.

b. The term "approved" as herein used in reference to an air gap, a double check valve assembly, a reduced pressure principle backflow prevention assembly or other backflow prevention assemblies or methods shall mean an approval by the administrative authority having jurisdiction.

4.5 Atmospheric Vacuum Breaker Backsiphonage Prevention Assembly (AVB)

The term "atmospheric vacuum breaker backsiphonage prevention assembly" (also known as the non-pressure type vacuum breaker) shall mean an assembly containing an air inlet valve, a check seat and an air inlet port(s). The flow of water into the body causes the air inlet valve to close the air inlet port(s). When the flow of water stops the air inlet valve falls and forms a check valve against backsiphonage. At the same time it opens the air inlet port(s) allowing air to enter and satisfy the vacuum. A shutoff valve immediately upstream may be an integral part of the assembly, but the assembly shall not be subjected to operating pressure for more than twelve (12) hours in any twenty-four (24) hour period. An atmospheric vacuum breaker is designed to protect against a non-health hazard (i.e., pollutant) or a health hazard (i.e., contaminant) under a *backsiphonage* condition only. See Fig. 4.2.

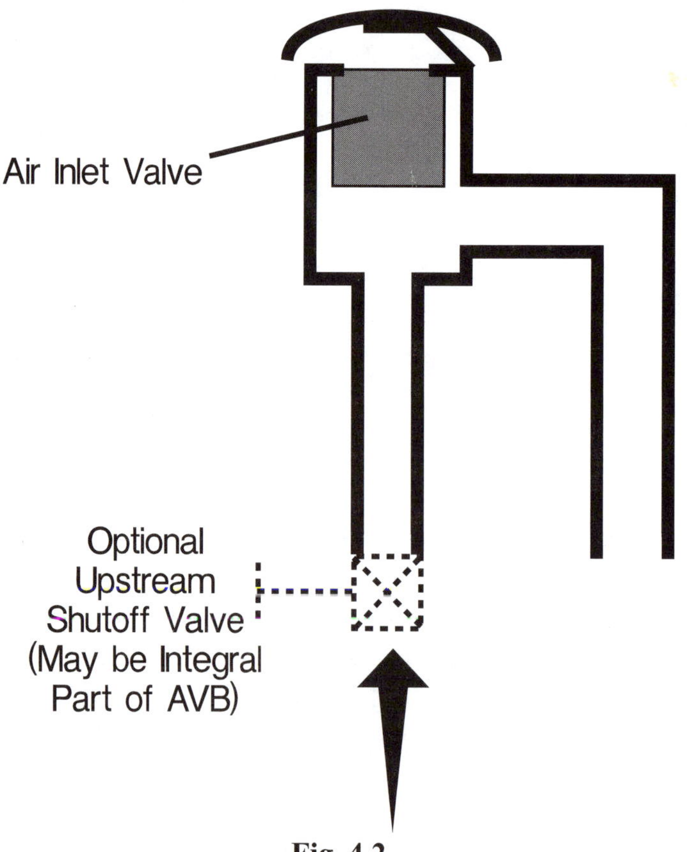

Fig. 4.2
Atmospheric Vacuum Breaker Backsiphonage Prevention Assembly (AVB)

4.6 Backflow

The term "backflow" shall mean the undesirable reversal of flow of water or mixtures of water and other liquids, gases or other substances into the distribution pipes of the potable supply of water from any source or sources. See terms Backsiphonage (4.11) and Backpressure (4.10).

4.7 Backflow Prevention Assembly — Approved

The term "approved backflow prevention assembly" shall mean an assembly that has been investigated and approved by the administrative authority having jurisdiction. The approval of backflow prevention assemblies by the administrative authority shall be on the basis of a favorable laboratory and field evaluation report by an approved testing laboratory (see Sec. 4.32) recommending such approval.

4.8 Backflow Prevention Assembly — Type

The term "backflow prevention assembly" shall mean any effective assembly used to prevent backflow into a potable water system. The type of assembly used shall be based on the existing or potential degree of hazard, and backflow condition. The types of backflow prevention assemblies are:

a. Atmospheric Vacuum Breaker Backsiphonage Prevention Assembly — see Sec. 4.5.

b. Double Check Valve Backflow Prevention Assembly — see Sec. 4.19.

c. Double Check - Detector Backflow Prevention Assembly — see Sec. 4.20.

d. Pressure Vacuum Breaker Backsiphonage Prevention Assembly — see Sec. 4.35.

e. Reduced Pressure Principle Backflow Prevention Assembly — see Sec. 4.38.

f. Reduced Pressure Principle-Detector Backflow Prevention Assembly — see Sec. 4.39.

g. Spill-Resistant Pressure Vacuum Breaker Backsiphonage Prevention Assembly — see Sec. 4.43

4.9 Backflow Prevention Assembly Tester — Certified

The term "certified backflow prevention assembly tester" shall mean a person who has proven his/her ability to the satisfaction of the administrative authority having jurisdiction. Each person who is certified to make field tests and make reports on backflow prevention assemblies shall be conversant with applicable laws, rules and regulations and have had experience in plumbing or pipe fitting or have other equivalent qualifications in the opinion of the administrative authority having jurisdiction.

4.10 Backpressure

The term "backpressure" shall mean any elevation of pressure in the downstream piping system (by pump, elevation of piping, or steam and/or air pressure) above the supply pressure at the point of consideration which would cause, or tend to cause, a reversal of the normal direction of flow.

4.11 Backsiphonage

The term "backsiphonage" shall mean a form of backflow due to a reduction in system pressure which causes a subatmospheric pressure to exist at a site in the water system.

4.12 Check Valve — Approved

The term "approved check valve" shall mean a check valve that is drip-tight in the normal direction of flow when the inlet pressure is at least one (1) psi (pound per square inch) and the outlet pressure is zero. The check valve shall permit no leakage in a direction reverse to the normal flow. The closure element (e.g., clapper or poppet) shall be internally loaded to promote rapid and positive closure. An approved check valve is only one component of an approved backflow prevention assembly — i.e., pressure vacuum breaker (PVB and SVB), double check valve assembly (DC) or reduced pressure principle assembly (RP). (See Specifications, Section 10, for more details.)

4.13 Consumer

The term "consumer" shall mean the owner or operator of an on-site water system(s) having a service from a public potable water system.

4.14 Containment

See Section 4.42, Service Protection.

4.15 Contamination

The term "contamination" shall mean an impairment of the quality of the water which creates an actual hazard to the public health through poisoning or through the spread of disease by sewage, industrial fluids, waste, etc.

4.16 Critical Level

The term "critical level" shall mean the marking (C-L or C/L) on atmospheric vacuum breakers and pressure vacuum breakers that determines the minimum elevation above the flood level rim of the fixture or receptacle served, as well as downstream piping and water uses, at which the unit may be installed. When an AVB, PVB, or SVB does not bear a critical level marking, the bottom of the assembly shall constitute the critical level.

4.17 Cross-Connection

The term "cross-connection" shall mean any unprotected actual or potential connection or structural arrangement between a public or a consumer's potable water system and any other source or system through which it is possible to introduce into any part of the potable system any used water, industrial fluid, gas, or substance other than the intended potable water with which the system is supplied. Bypass arrangements, jumper connections, removable sections, swivel or change-over devices and other temporary or permanent devices through which or because of which backflow can occur are considered to be cross-connections.

a. The term "direct cross-connection" shall mean a cross-connection which is subject to both backsiphonage and backpressure.

b. The term "indirect cross-connection" shall mean a cross-connection which is subject to backsiphonage only.

4.18 Cross-Connection — Point of

The term "point of cross-connection" shall mean the specific point or location in a public or a consumer's potable water system where a cross-connection exists.

4.19 Double Check Valve Backflow Prevention Assembly (DC)

The term "double check valve backflow prevention assembly" shall mean an assembly composed of two independently acting, approved check valves, including tightly closing resilient seated shutoff valves attached at each end of the assembly and fitted with properly located resilient seated test cocks. (See Specifications, Section 10 for additional details.) This assembly shall only be used to protect against a non-health hazard (i.e., pollutant). See Fig. 4.3.

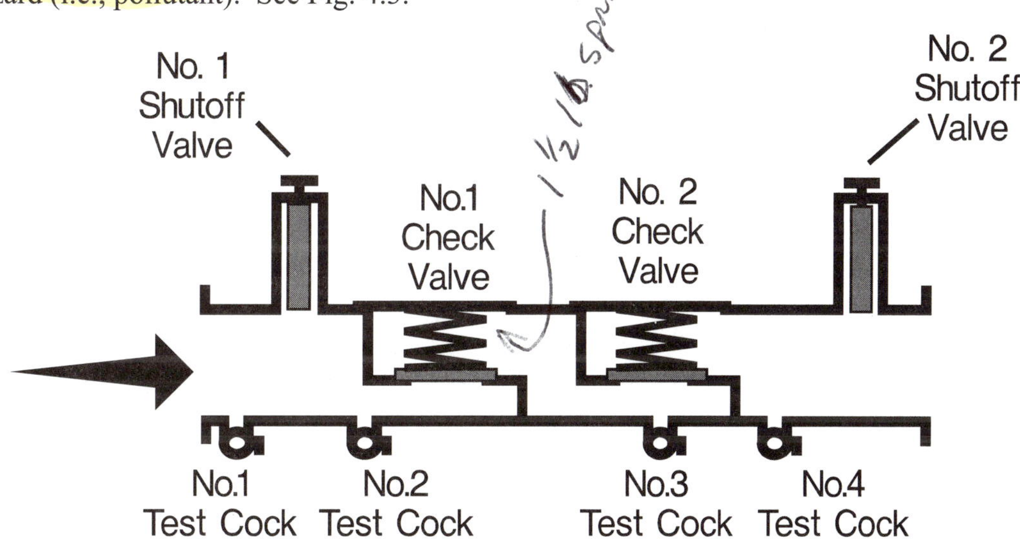

Fig. 4.3
Double Check Valve Backflow Prevention Assembly (DC)

4.20 Double Check-Detector Backflow Prevention Assembly (DCDA)

The term "double check-detector backflow prevention assembly" shall mean a specially designed assembly composed of a line-size approved double check valve assembly with a bypass containing a specific water meter and an approved double check valve assembly. The meter shall register accurately for only very low rates of flow up to 3 gpm (gallons per minute) and shall show a registration for all rates of flow. (See Specifications, Section 10, for additional details.) This assembly shall only be used to protect against a non-health hazard (i.e., pollutant). The DCDA is primarily used on fire sprinkler systems (see Section 7.2.3.14). See Fig. 4.4.

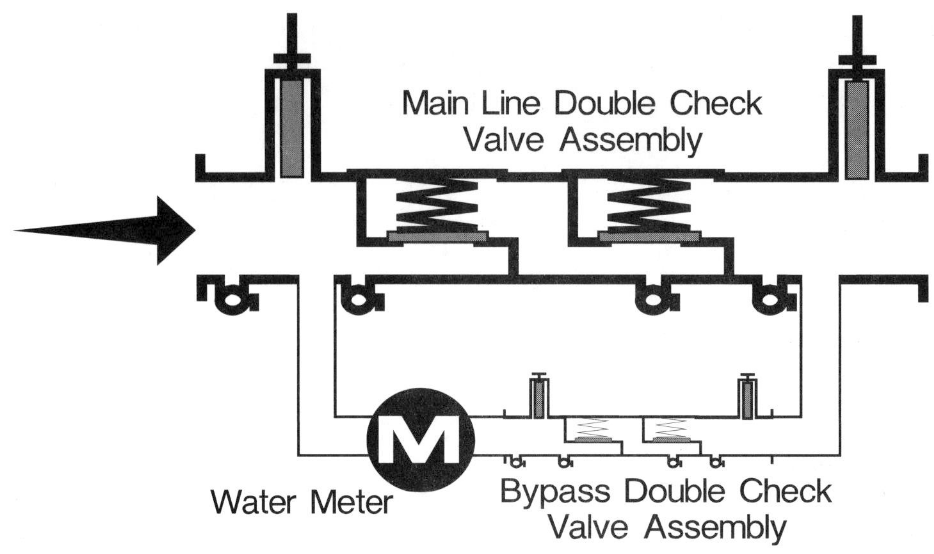

Fig. 4.4

Double Check-Detector Backflow Prevention Assembly (DCDA)

4.21 Hazard — Degree of

The term "degree of hazard" shall mean either a pollutional (non-health) or contamination (health) hazard and is derived from the evaluation of conditions within a system.

4.22 Hazard — Health

See Contamination — Section 4.15.

4.23 Hazard — Plumbing

The term "plumbing hazard" shall mean an internal or plumbing type cross-connection in a consumer's potable water system that may be either a pollutional or a contamination type hazard. Plumbing type cross-connections can be located in many types of structures including homes, apartment houses, hotels and commercial or industrial establishments. Such a connection, if permitted to exist, must be properly protected by an appropriate type of backflow prevention assembly.

4.24 Hazard — Non-health

See Pollution — Section 4.34.

4.25 Hazard — System

The term "system hazard" shall mean an actual or potential threat of severe danger to the physical properties of the public or the consumer's potable water system or of a pollution or contamination which would have a protracted effect on the quality of the potable water in the system.

4.26 Health Agency

The term "health agency" shall mean the health authority having jurisdiction.

4.27 Hospital

The term "hospital" shall mean any institution, place, building, or agency which maintains and operates facilities for one or more persons for the diagnosis, care and treatment of human illness, including convalescence and care during and after pregnancy or which maintains and operates organized facilities for any such purpose, and to which persons may be admitted for overnight stay or longer. The term "hospital" includes sanitarium, nursing home and maternity home.

4.28 Industrial Fluids

The term "industrial fluids" shall mean any fluid or solution which may be chemically, biologically or otherwise contaminated or polluted in a form or concentration which would constitute a health, system, pollutional or plumbing hazard if introduced into an approved water supply. This may include, but not be limited to: polluted or contaminated used waters; all types of process waters and used waters originating from the public potable water system which may deteriorate in sanitary quality; chemicals in fluid form; plating acids and alkalies; circulated cooling waters connected to an open cooling tower and/or cooling waters that are chemically or biologically treated or stabilized with toxic substances; contaminated natural waters such as from wells, springs, streams, rivers, bays, harbors, seas, irrigation canals or systems, etc.; oils, gases, glycerine, paraffins, caustic and acid solutions and other liquid and gaseous fluids used industrially, for other processes, or for fire fighting purposes.

4.29 Industrial Piping System — Consumer's

The term "consumer's industrial piping system" shall mean any system used by the consumer for transmission of or to confine or store any fluid, solid or gaseous substance other than an approved water supply. Such a system would include all pipes, conduits, tanks, receptacles, fixtures, equipment and appurtenances used to produce, convey or store substances which are or may be polluted or contaminated.

4.30 Internal Protection

The term "internal protection" shall mean the appropriate type or method of backflow prevention within the consumer's potable water system at the point of use, commensurate with the degree of hazard.

4.31 Isolation

See Section 4.30, Internal Protection.

4.32 Laboratory — Approved Testing

The term "approved testing laboratory" shall mean the Foundation for Cross-Connection Control and Hydraulic Research of the University of Southern California or any other laboratory having equivalent capabilities for both the laboratory and field evaluation of backflow prevention assemblies.

4.33 Point of Delivery

See Service Connection — Sec. 4.41

4.34 Pollution

The term "pollution" shall mean an impairment of the quality of the water to a degree which does not create a hazard to the public health but which does adversely and unreasonably affect the aesthetic qualities of such waters for domestic use.

4.35 Pressure Vacuum Breaker Backsiphonage Prevention Assembly (PVB)

The term "pressure vacuum breaker backsiphonage prevention assembly" shall mean an assembly containing an independently operating internally loaded check valve and an independently operating loaded air inlet valve located on the discharge side of the check valve. The assembly is to be equipped with properly located resilient seated test cocks and tightly closing resilient seated shutoff valves attached at each end of the assembly. (See Specifications, Section 10, for additional details.) This assembly is designed to protect against a non-health hazard (i.e., pollutant) or a health hazard (i.e., contaminant) under a backsiphonage condition only. See Fig. 4.5.

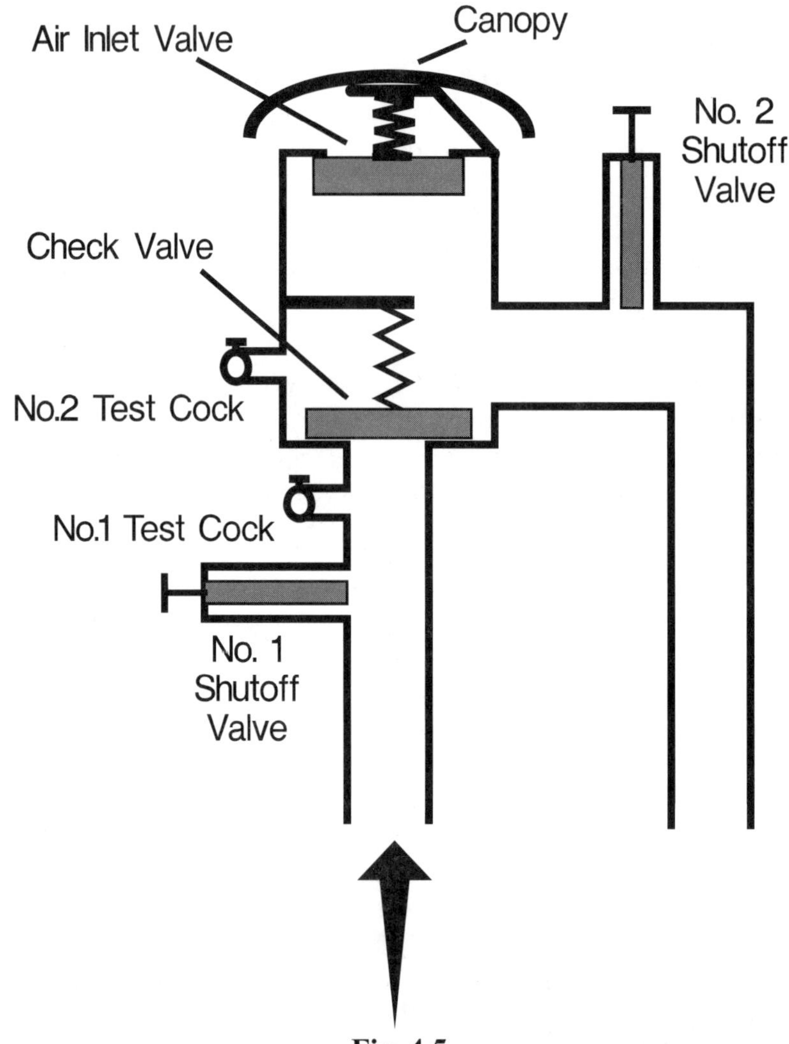

Fig. 4.5
Pressure Vacuum Breaker Backsiphonage Prevention Assembly (PVB)

4.36 Readily Accessible

The term "readily accessible," when referring to a backflow prevention assembly, shall mean capable of being reached for testing and/or maintenance, *without* the need of removing any access panel, door, or similar obstruction.

4.37 Reclaimed Water

The term "reclaimed water" shall mean water which, as a result of treatment of wastewater, is suitable for a direct beneficial use or a controlled use that would not otherwise occur, and is not safe for human consumption.

4.38 Reduced Pressure Principle Backflow Prevention Assembly (RP)

The term "reduced pressure principle backflow prevention assembly" shall mean an assembly containing two independently acting approved check valves together with a hydraulically operating, mechanically independent pressure differential relief valve located between the check valves and at the same time below the first check valve. The unit shall include properly located resilient seated test cocks and tightly closing resilient seated shutoff valves at each end of the assembly. (See Specifications, Section 10, for additional details.) This assembly is designed to protect against a non-health (i.e., pollutant) or a health hazard (i.e., contaminant). This assembly shall not be used for backflow protection of sewage or reclaimed water. (Note: Check with local administrative authority for acceptable uses.) See Fig. 4.6.

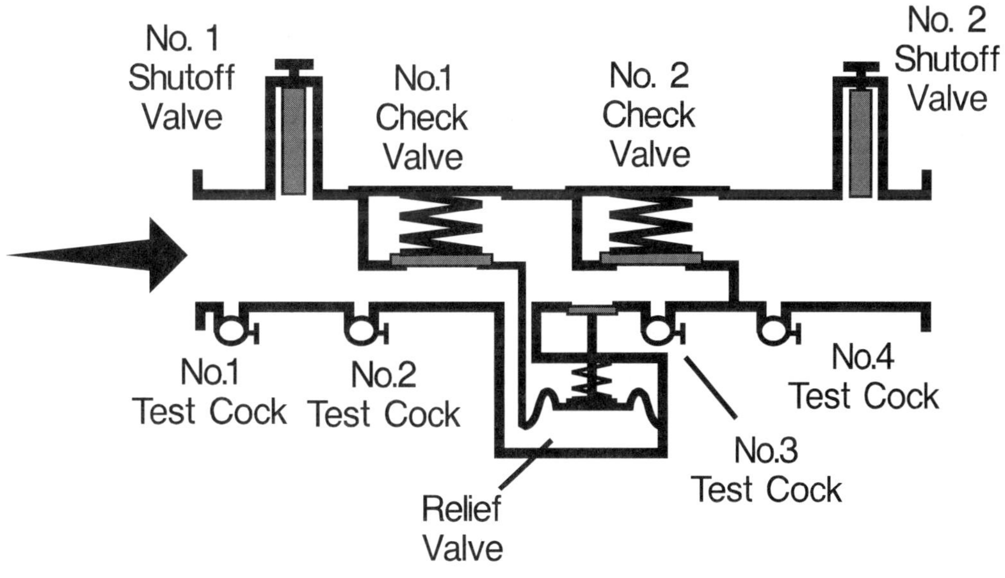

Fig. 4.6
Reduced Pressure Principle Backflow Prevention Assembly (RP)

4.39 Reduced Pressure Principle-Detector Backflow Prevention Assembly (RPDA)

The term "reduced pressure principle-detector backflow prevention assembly" shall mean a specially designed assembly composed of a line-size approved reduced pressure principle backflow prevention assembly with a bypass containing a specific water meter and an approved reduced pressure principle backflow prevention assembly. The meter shall register accurately for only very low rates of flow up to 3 gpm and shall show a registration for all rates of flow. (See Specifications, Section 10, for additional details.) This assembly shall be used to protect against a non-health hazard (i.e., pollutant) or a health hazard (i.e., contaminant). The RPDA is primarily used on fire sprinkler systems (see Section 7.2.3.14). See Fig. 4.7.

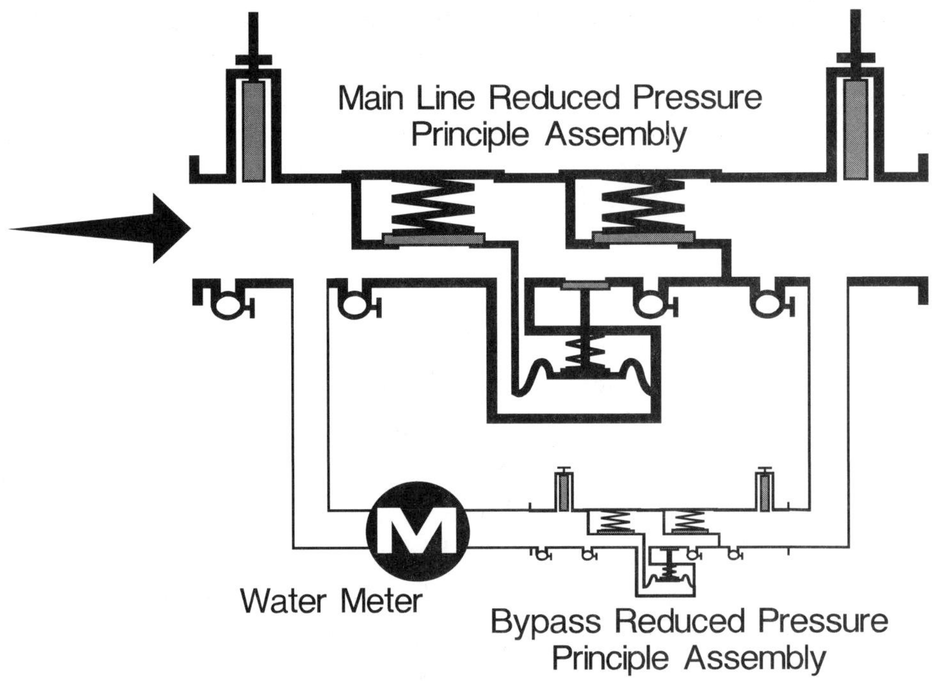

Fig. 4.7
Reduced Pressure Principle-Detector Backflow Prevention Assembly (RPDA)

4.40 Sanitary Sewer

The term "sanitary sewer" shall mean the pipe that carries sewage.

4.41 Service Connection

The term "service connection" shall mean the terminal end of a service connection from the public potable water system, (i.e., where the water purveyor may lose jurisdiction and sanitary control of the water at its point of delivery to the consumer's water system). If a water meter is installed at the end of the service connection, then the service connection shall mean the downstream end of the water meter.

4. 42 Service Protection

The term "service protection" shall mean the appropriate type or method of backflow protection at the service connection, commensurate with the degree of hazard of the consumer's potable water system.

4.43 Spill-Resistant Pressure Vacuum Breaker Backsiphonage Prevention Assembly (SVB)

The term "spill-resistant pressure vacuum breaker backsiphonage prevention assembly" shall mean an assembly containing an independently operating internally loaded check valve and independently operating loaded air inlet valve located on the discharge side of the check valve. The assembly is to be equipped with a properly located resilient seated test cock, a properly located bleed/vent valve, and tightly closing resilient seated shutoff valves attached at each end of the assembly. (See Specifications Section 10, for additional details.) This assembly is designed to protect against a a non-health hazard (i.e., pollutant) or a health hazard (i.e., contaminant) under a backsiphonage condition only. See Fig. 4.8.

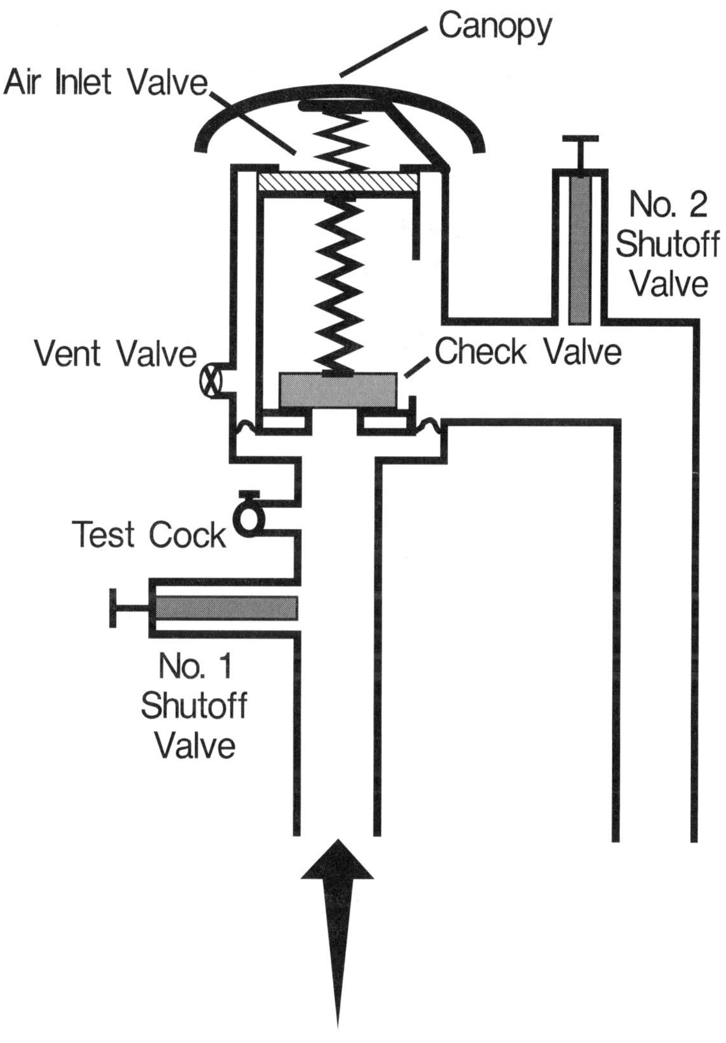

Fig. 4.8
Spill-Resistant Pressure Vacuum Breaker Backsiphonage Prevention Assembly (SVB)

4.44 Water — Potable

The term "potable water" shall mean water from any source which has been investigated by the health agency having jurisdiction, and which has been approved for human consumption.

4.45 Water Purveyor

The term "water purveyor" shall mean the public or private owner or operator of the potable water system supplying an approved water supply to the public.

4.46 Water Supply — Approved

The term "approved water supply" shall mean any public potable water supply which has been investigated and approved by the health agency. The system must be operating under a valid health permit. In determining what constitutes an approved water supply, the health agency has final judgment as to its safety and potability.

4.47 Water Supply — Auxiliary

The term "auxiliary water supply" shall mean any water supply on or available to the premises other than the water purveyor's approved public potable water supply. These auxiliary waters may include water from another purveyor's public potable water supply or any natural source such as a well, spring, river, stream, harbor, etc. They may be polluted or contaminated or they may be objectionable and constitute an unacceptable water source over which the water purveyor does not have sanitary control.

4.48 Water Supply — Unapproved

The term "unapproved water supply" shall mean a water supply which has not been approved for human consumption by the health agency having jurisdiction.

4.49 Water System(s) — Consumer's

The term "consumer's water system(s)" shall include any water system located on the consumer's premises whether supplied by a public potable water system or an auxiliary water supply. The system or systems may be either a potable water system or an industrial piping system.

4.50 Water System — Consumer's Potable

The term "consumer's potable water system" shall mean that portion of the privately owned potable water system lying between the point of delivery and the point of use. This system includes all pipes, conduits, tanks, receptacles, fixtures, equipment and appurtenances used to produce, convey, store or utilize the potable water.

4.51 Water System - Public Potable

The term "public potable water system" shall mean any publicly or privately owned water system operated as a public utility under a valid health permit to supply water for domestic purposes. This system will include all sources, facilities and appurtenances between the source and the point of delivery such as valves, pumps, pipes, conduits, tanks, receptacles, fixtures, equipment and appurtenances used to produce, convey, treat or store potable water for public consumption or use.

4.52 Water - Used

The term "used water" shall mean any water supplied by a water purveyor from a public potable water system to a consumer's water system after it has passed through the service connection and is no longer under the control of the water purveyor. See Section 7.2.3.33.

4.53 Water Supervisor

The term "water supervisor" shall mean the consumer or a person on the premises appointed by the consumer charged with the responsibility of maintaining the consumer's water system(s) on the property free from cross-connections and other sanitary defects, as required by regulations and laws. A certified backflow prevention assembly tester may act as a water supervisor if he/she is a full-time employee of the consumer having the day-to-day responsibility for the installation and use of pipelines and equipment on the premises and for avoidance of cross-connections.

BACKFLOW PREVENTION ASSEMBLY GENERAL APPLICATIONS *

	Non-Health Hazard *(POLLUTANT)*		Health Hazard *(CONTAMINANT)*		Sewage	
	BACKSIPHONAGE	BACKPRESSURE	BACKSIPHONAGE	BACKPRESSURE	BACKSIPHONAGE	BACKPRESSURE
AIR GAP	X	X	X	X	X	X
RP	X	X	X	X		
DC	X	X				
PVB	X		X			
SVB	X		X			
AVB	X		X			

* *NOTE: These are general guidelines. Check with local administrative authority(s) for local applications.*

TABLE 4-1
BACKFLOW PREVENTION ASSEMBLY GENERAL APPLICATIONS

SECTION 5

Typical Facilities: Cross-Connections or Water Uses Which May Endanger the Public Water System

The following conditions and problems should be reviewed in all cases where the regulatory agency adopts a policy of requiring that the degree of hazard be based upon a complete inspection of the consumer's water-using facilities. For convenience, these conditions have been divided into three groups. Group I includes cross-connections typical to certain industries or uses. Group II includes potential cross-connections involving water-using fixtures, equipment, facilities, etc., grouped in associated categories. Group III lists chemical and chemical compounds used in water treatment, hydraulically distributed in the consumer's water system or used in conjunction with water by industry. It should be noted that the following are intended to be representative, they are not all inclusive.

5.1 Group I: Includes Cross-Connections Typical to Certain Industries or Uses

5.1.1 Sewage Systems

Cross-connections to sewage or surface water pumps for priming, cleaning, flushing or unclogging purposes.

Water-operated sewage sump ejectors for operational purposes.

Sewers for the purpose of disposing of filter or softener backwash water or water from cooling systems or for the purpose of providing for a quick drain for the building water lines or of flushing or blowing out obstructions in the sewer lines, etc. (NOTE: Most state regulations require backflow protection at the service connection to any premise on which there is located a sewage or pumping station, even though there are no cross-connections.)

5.1.2 Reservoirs, Cooling Towers, etc.

Reservoirs, cooling towers and circulating systems which may be heavily contaminated either with bird droppings, vermin, algae, bacterial slimes or with toxic water treatment compounds such as pentachlorophenol, copper, chromates, metallic glucosides, compounds of mercury, quaternary ammonium compounds, etc. (See Fig. 7.2)

5.1.3 Industrial Fluid Systems

Industrial fluid systems and lines containing cutting and hydraulic fluids, coolants, hydrocarbon products, glycerine, paraffin, caustic and acid solutions, etc.

5.1.4 Fire Fighting Systems

Fire fighting systems, including storage reservoirs which may be treated for prevention of scale formation, corrosion, algae, slime growths, etc.

Fire systems which may be subject to contamination with antifreeze solutions, Foamite or other chemicals or chemical compounds used in fighting fire.

Fire systems which are subject to contamination with auxiliary or used water supplies or industrial fluids. (See Figs. 7.6 through 7.12.)

5.1.5 Plating Facilities

Plating facilities involving the use of highly toxic cyanides, heavy metals in solution (such as copper, cadmium, chrome, nickel, etc.), acids and caustic solutions.

Plating solution filtering equipment with pumps and circulating lines.

Tanks, vats or other vessels used in painting, descaling, anodizing, cleaning, stripping, oxidizing, etching, passivating, pickling, dipping, rinsing operations.

Other lines or facilities needed in the preparation or finishing of the products.

5.1.6 Steam Generating Facilities

Steam generating facilities and lines which may be contaminated with boiler compounds such as pentachlorophenol, hydrazine, cyclohexylamine, etc. (NOTE: A particular hazard is the possibility of steam getting back into the domestic system, causing either a system or a health hazard.)

5.1.7 Plumbing Hazards

Inadequately protected (improperly installed, improperly maintained or without vacuum breakers) flush valve toilets, urinals, aspirators, retorts, pipet tube washers and similar contaminated and/or sewer-connected facilities.

Laboratory equipment which may be chemically or bacteriologically contaminated such as steam sterilizers, autoclaves, specimen tanks, autopsy and mortuary equipment. (NOTE: These hazards are critical because little or no attention is given to the maintenance of the vacuum breakers often supplied with such equipment.)

5.1.8 Cooling Systems — Single Pass

Compressors, heat exchangers, air-conditioning equipment and other water cooled equipment which may be sewer-connected.

5.1.9 Irrigation Systems

Irrigation systems which may be equipped with pumps, injectors, pressurized tanks or vessels, or other facilities for injecting into the irrigation system agricultural chemicals such as fungicides, pesticides, soil conditioning and other similar noxious, toxic or objectionable substances.

Irrigation systems subject to contamination from submerged inlets, auxiliary water supplies, ponds, reservoirs, swimming pools and other sources of stagnant, polluted or contaminated waters.

5.1.10 Plumbing — Hospitals

Contaminated or sewer-connected equipment such as bed pan washers, flush valve toilets and urinals, autoclaves, specimen tanks, sterilizers, pipet tube washers, cuspidors, aspirators, autopsy and mortuary equipment, etc. (NOTE: It has been found that in this type of facility little or no attention is given to the maintenance of air gaps or vacuum breakers.)

5.1.11 Plumbing — Multi-storied Buildings

Where the upper floors of multi-storied buildings are above the reach of the water purveyor's system pressure it will be necessary to use booster pumps. Considerable care must be exercised to prevent the use of the suction side line to these pumps from also being used as the takeoff for domestic, sanitary, laboratory or industrial uses on the lower floors. Pollutants or contaminants from equipment supplied by takeoffs from the suction side line may be easily pumped throughout the upper floors. (See Fig. 7.14)

5.1.12 Industrial Systems — Chemical Contamination

Tanks, can and bottle washing machines and lines where caustics, acids, detergents and other compounds are used in cleaning, sterilizing and flushing.

5.1.13 Photo Processing Equipment

Tanks, automatic film processing machines or other facilities used in processing films, which may be contaminated with chemicals such as acetic acid, potassium ferricyanide and/or one of the many different types of aromatic series of organic chemicals.

5.1.14 Laundries and Dye Works

Laundry machines having submerged or low inlets.

Dye vats in which toxic chemicals and dyes are used.

Wash water storage tanks equipped with pumps and recirculating systems.

Retention and mixing tanks.

Shrinking, bluing and dyeing machines with direct connections to circulating systems.

(NOTE: Some of these machines are equipped with pumps capable of forcing contaminated fluids into the public water supply through cross-connections.)

5.1.15 Industrial Facilities

Tanks, lines, valves, fittings, and other equipment being subjected to hydraulic tests.

Hydraulically operated equipment where the public water pressure is used directly and may be subject to backpressure.

Equipment under hydraulic tests where pumps, rams, pressure cylinders or other hydraulic principles are used to provide pressures for testing purposes.

(NOTE: In such cases, air, gas or hydraulic fluids may be forced back into the public system.)

5.1.16 Motion Picture Studios

Open reservoirs, lagoons, tanks or similar facilities, used as props in the making of motion pictures. (NOTE: These facilities may be heavily contaminated with body wastes, dyes, biological or chemical contaminates used in the prevention of algae and slime growths and to color the waters for filming purposes.)

Automatic film processing machines, tanks, vats and other facilities, used in processing films. (NOTE: Toxic chemicals such as acetic acid, potassium ferricyanide and different types of organic chemicals may be used in these facilities.)

Special effects equipment in which chemicals and other materials may be injected into the water supply for special effects.

5.1.17 Petroleum Processing

Steam boilers, steam lines, mud pumps and mud tanks, hydraulically operated Tretolite tanks, oil well casings (for dampening gas pressures) dehydration tanks, outlet lines from storage and dehydration tanks (for purging purposes), oil and gas tanks (to create hydraulic pressures and to hydraulically raise the oil and gas levels), gas and oil lines (for testing, evacuating and slugging purposes).

5.1.18 Paper Processing

Pulp, bleaching, dyeing and other processing equipment which may be contaminated with toxic chemicals.

5.1.19 Cannery Equipment

Pressure cookers, autoclaves, retorts and other similar steam-connected facilities, washers, cookers, tanks, lines, flumes, and other equipment used for storing, washing, cleaning, blanching, cooking, flushing, fluming, or for transmission of foods, fertilizers or wastes.

5.1.20 Auxiliary Water Systems

Most state regulations require the service connection from an approved water supply be protected by a suitable backflow prevention assembly where there is an auxiliary water supply system on the premises even though there are no overt cross-connections. For more specific information regarding an auxiliary water supply and recommended protection see Section 7.2.3.3 "Auxiliary Water Systems." (See Figs. 7.3 and 7.4)

5.1.21 Solar Energy Systems

Solar energy systems for domestic hot water heating, space heating or cooling, industrial process water heating, swimming pool heating which may have cross-connections with the domestic water system. The solar energy system may employ antifreeze solutions or chemical corrosion inhibitors. (See Figs. 7.19 and 7.20)

5.2 Group II: Includes Potential Cross-Connections Involving Water-Using Fixtures and Equipment

5.2.1 Auxiliary Water Supplies, Fire Fighting, Irrigation, Swimming Pools, etc.

Fire Fighting systems — booster pumps to tank systems, storage facilities and pumper connections
Fish ponds — pump connected
Hot water systems — drainage and flushing facilities
Irrigation systems — parks, golf courses, playgrounds, schools, etc.
Jumper connections
Lawn sprinklers under pressure
Ocean water for fire protection
Fountains — display, public and private
Public and private water companies
Private wells for domestic, commercial, irrigation, and industrial use
Swimming pool inlets, recirculation systems, chlorinators, and drains

5.2.2 Process Waters — Recirculated

Air conditioning — refrigerated, air wash, makeup and drains
Ball mills
Cooling systems — refrigeration, diesel engines, compressors
Any industries practicing water conservation
Ink mills
Paint mills

5.2.3 Water Treatment Facilities

Addition of chemicals
Boiler feed treatment
Compound feeders
Scale, corrosion, slime control
Water filtration and water softening

5.2.4 Situations Where Toxic or Objectionable Chemicals Are or May Be Transmitted, Stored or Used in a Manner Which May Endanger the Water System

Brine lines
Foamite lines
Glycerine lines
Laboratory equipment
Mixing tanks
Oil systems
Photo processing and washing
Pickling tanks
Plating works
Refrigerants

5.2.5 Priming Lines — Connected to:

Acid pumps
Air conditioner pumps
Air pumps
Booster pumps
Cadmium solution pumps
Caustic pumps
Chromic acid pumps
Cyanide pumps
Gasoline lifts
Glycerine pumps
Hydraulic elevator pumps
Sewer pumps
Sump ejectors
Venturi float lines

5.2.6 Direct Water Connections to Steam Systems and Hydraulic Elevators and Air Lines, etc.

Boilers — high and low pressure
Cold and hot water return to steam systems
Compressors
Direct-connected hydraulic elevators
Elevator air lines
Return and surge tank hydraulic elevator systems
Steam ejectors
Steam lines
Suction tees
Turbo burners
Vacuum systems

5.2.7 Industrialized Lines

All types of industries
Laboratories

5.2.8 Interstreet Services — Low pressure and fringe areas

Elevation and pressure conditions
More than one service to a premise

5.2.9 Industrial Water-use Connections

Box plants — glue pots, soaking vats, steaming processes
Canneries — pressure cookers, retorts, wash lines, salt wash lines

Creameries — distilled water, ice water, tap water, hot water, steam, milk, other products
Laundries — caustic soap solutions, hot and cold water, softened hot and cold water, chlorinated water, and boiler room equipment
Metal works — testing lines, cooling systems, plating solutions, metal processing lines, cutting oil, lubricant lines, and welding machines
Oil Companies — flushing oil lines, tanks and systems, heating and cooling systems
Packing houses — rendering vats, pressure reduction vats, and hide soaking and pickling vats
Rubber and rubber goods plants — roll cooling machines, cookers, water transmission systems, brine and styrene solutions
Shipyards — salt water systems, tank testing facilities, ship line testing, pier head outlets, fire systems, prestolite systems
Tanneries — chemical solution and dye lines, lanolin lines and soaking tanks
Hospitals — all types

5.2.10 Cross-Connections Involving Sewage or Sewage Disposal Facilities

Baptismal founts
Brewery vats
Brine tanks
Cheese tanks
Compressors — cooling systems with direct connection
Culture vats
Diesel engines — cooling systems with direct connection
Dipper vats
Direct water lines to sewers for drains or flushing
Dye tanks
Fire sprinkler drain lines
Flush manholes — water supply to
Flush tanks
Food mixing tanks
Holding tanks — camper or trailer toilet flushing facilities
Kitchen equipment
Morticians' aspirators
Photographic tanks
Pickling tanks
Plating tanks
Potato peelers
Priming lines
Reservoir bypasses and drains to sewer or storm drains
Reverse Osmosis Purifiers — brineline
Sewage chlorinators — direct injection
Sewage sump ejectors
Sewage sump pumps and ejectors — water-operated
Sewer flushing equipment — water connection
Shrinking tanks
Sinks
Soaking tanks
Spring-loaded glass washers

Steam soap washing devices
Steam table connections
Swimming pool gutter drains
Therapeutic baths
Various blowoffs or drains to sewers
Water jacketed tanks, vats, and pots
Water softeners
Water street mains draining to sewer or storm drains
Water-operated pumps

5.2.11 Special Uses Where Cross-connections Are Usually Found

Baptismal tanks
Blood plasma equipment
Blueprint machines
Car washing equipment — caustic and soap guns, mixers, and boiler equipment
Chillers
Commercial vacuum cleaning equipment
Construction equipment lines
Deaerators
Garbage washing with steam and cold water connections
Humidity controls
Hydraulic fertilizer applications
Mortuaries
Oil well leases
Pest control equipment
Pressure and steam cookers
Roof and house tanks
Soap mixing layouts
Solar heating systems
Steamer supply equipment
Storage reservoirs
Veterinary hospitals
Water-operated siphons — all types
Weed control equipment
X-ray equipment

5.2.12 Plumbing and Water Piping Cross-connections

Aspirators
Autoclaves
Auto shampoo
Basins
Bathtubs
Bedpan washers
Bidets
Blueprint machines
Bottle washers

Carbonators
California washers (below flood level)
Can washers
Coffee urns
Colonic irrigators
Cooking kettles
Cuspidors — water operated
Dental cuspidors — water operated
Dishwashers
Drains, tanks, vats
Drinking fountains
Fish ponds
Frostproof toilets
Garbage grinding devices
Grease traps
Hoppers (utility)
Hose bibs
Hydraulic vacuum cleaners
Instrument sterilizers
Insecticide sprayers water operated
Integral tank and closet bowls
Laboratory operated vacuum pumps
Lawn sprinklers
Laundry trays
Laundry washers
Overflow tanks
Pasteurizers
Plumbers enemy
Plumbers friend — indentical gadgets: removable hose connection between bib and lavatory or sink drain
Pressure cookers
Refrigeration units
Shampoo units
Soda fountains
Turbo burner drains
Toilets — flush valves low tanks
Urinals
Washing machines
Water Purifiers — brine line to sewer
Watering troughs
Yard outlets — submerged
Yard sprinkling nozzles

5.3 Group III: Chemicals and Chemical Compounds Used in Water Treatment

5.3.1 Agriculture

Solutions of chemicals are used by agriculture for many purposes. The following are some of the chemical compounds which may be injected into irrigation systems for application purposes. All of them are toxic in concentrated solutions.

Fertilizers	Ammonium salts Ammonium gas	Phosphates Potassium salts
Weedicides	2.4.D Dinitrophenol Karmex 2.4.5.T Pentachlorophenol	Sodium chlorate Borax Sodium arsenite Methyl bromide
Pesticides	D D T T D E B H C Lindane TEPP	Parathion Malathion Nicotine M H

5.3.2 Cooling Systems — Open or Closed

Cooling systems — including cooling towers — usually require some treatment of the water for algae, slime or corrosion control.

Chemicals frequently used for this purpose may include the following highly toxic chemicals:

Quaternary ammonium compounds	Mercury
Pentachlorophenol	Chromium

OR the following chemicals which are toxic in higher concentrations:

Chlorine	Permanganate
Bromine	Glucosides
Copper	

5.3.3 Plating Plants

In plating work, materials are first cleaned in acid or caustic solutions at concentrations that are highly toxic, after which they are immersed in plating solutions which are highly toxic. Such solutions may contain:

Cyanides
Fluorides

Or metals in solution such as:

Copper	Cadmium
Chromium	Antimony
Nickel	Silver salts, etc.

5.3.4 Steam Boiler Plants

Most boiler plants will use some form of boiler feed water treatment. The chemicals normally used for this purpose include:

Highly toxic compounds such as:

Cyclohexylamine
Hydrazine
Morpholine
Benzylamine

OR the less toxic compounds such as:

Acids
Sodium hydroxide
Sodium sulphite
Sodium phosphate
Sodium nitrate
Sodium aluminate
Sodium alginate

5.3.5 Dye Plants

Most solutions used in dyeing are highly toxic. The toxicity depends on the chemicals used and their concentrations. The following types or chemical groups of dyes are generally used:

Vat Dye
Mordant Dye
Chrome Dye
Nitro Dye
Metallized Dye
Thiazol Dye

SECTION 6

Results of Non-Compliance

6.1 Consumer

Service of water to any premise should be discontinued by the water purveyor if a backflow prevention assembly required by law, rules or regulations is not installed, tested and maintained; or, if it is found that a backflow prevention assembly has been removed or bypassed; or, if unprotected cross-connections exist on the premises and there is inadequate backflow protection at the service connection. Water service should not be restored until such conditions or defects are corrected.

Water service to any premise may also be ordered discontinued by the state or local health agency having jurisdiction where the health agency finds that an actual or potential cross-connection exists which provides a hazard to either the public water supply system or the potable water supplied within the premises.

6.2 Water Purveyor

Certification of the public water supply by the health agency and the United States Environmental Protection Agency (EPA) is predicated on the maintenance of a defensible cross-connection control program. Failure to implement such a program may result in the withdrawal of the certification of the public water supply for use on interstate carriers and recision of the permit to supply public potable water. The Safe Drinking Water Act Amendments of 1986 place the full responsibility of the protection of the public water supply upon the water purveyor.

6.3 Backflow Prevention Assembly Tester — Certified

A certification issued to a backflow prevention assembly tester by the administrative authority having jurisdiction may be revoked or suspended for the improper testing, maintenance, reporting or other unethical practices connected with these assemblies.

SECTION 7

Cross-Connection Control Practice

7.1 Operational Requirements

7.1.1 General

It is a primary responsibility of the water purveyor and/or the administrative authority to evaluate the hazards inherent in supplying a consumer's water system, i.e., determine whether solid, liquid or gaseous pollutants or contaminants are, or may be, handled on the consumer's premises in such a manner as to possibly permit pollution or contamination of the public potable water system. When a hazard or potential hazard to the public water system is found on the consumer's premises the consumer shall, under the rules or regulations of the administrative authority, be required to install an approved and appropriate backflow prevention assembly at each public service connection to the premises and/or at key locations within the premises. The type and location of assemblies to be installed should depend upon the nature of the hazard involved.

When an approved backflow prevention assembly has been provided at the point of delivery, the water purveyor may no longer be legally responsible for the consumer's internal water system. The responsibility to evaluate internal hazards and the separation of domestic and industrial water uses devolves to the health agency and/or the administrative authority having jurisdiction. Should there be a change in use of water on the premises that would affect the type of hazard to the public water system the health agency or administrative authority having jurisdiction should inform the water purveyor.

In making a determination of the backflow potential (i.e., whether the consumer's water system(s) is operated under pressure, as set forth in the rules or regulations of the administrative authority), consideration should be given to the fact that backflow may occur under any conceivable pressure differentials, varying from vacuums to very high pressures. This means that backflow can occur from the consumer's premises into the public potable water system when the street main pressure is lower than the consumer's pressure, is at atmospheric pressure or is under a vacuum condition. Therefore, no specific positive or negative pressure range in the public system may be established that could be used as a yardstick for predicting the differentials where backflow may occur. It should be assumed that backflow into the public system can occur when a higher pressure can be attained in the consumer's system.

Plumbing cross-connections can be a serious health hazard, particularly when they are located in multi-storied buildings such as hotels, apartment houses, hospitals, medical and dental buildings, and other buildings having laboratories or other critical water uses, particularly when such buildings have upfeed systems. In such buildings the water purveyor may find that there is a special danger to health or to the public water system because of the failure to install suitable backflow protection, or because of the removal or malfunctioning of such internally located backflow prevention assemblies. These assemblies may be on cross-connections to sewers, such as flush valve toilets, bedpan washers, etc. It may also be found there is no adequate installation or testing and maintenance program, or at best a poor testing and maintenance program, and failure of such internal assemblies can therefore be anticipated. When this is found to be the case, the water purveyor may find it necessary to require protection at the point of delivery even though the individual internal cross-connections are presently protected in accordance with plumbing and health codes.

7.1.2 New Construction

Where possible, a plan check should be made prior to construction to determine the degree of hazard and the type of backflow prevention assembly that may be required at the point of delivery from the public potable water system and/or the type or types that may be required within the consumer's water system. Where adequate plans and specifications are not available and no realistic evaluation of the proposed water uses can be determined, the consumer, architect, engineer or other authorized person should be advised that eventually circumstances may require the installation of maximum backflow protection at the water service connection. Therefore, the responsible person should anticipate the possible need for a backflow prevention assembly. The designers, in planning the service connection to an industrial plant, should divide the system into domestic and industrial uses at the point of delivery and provide for a backflow prevention assembly on each leg or branch.

7.1.3 Existing Systems

A survey should be made of the consumer's presently installed water system in order to determine the degree of hazard to both the consumer's system and the public potable water system. This survey for service protection need not be a detailed inspection of the location or disposition of the water lines; but, can be confined to establishing the water uses on the premises such as: the existence of cross-connections, the availability of auxiliary or used water supplies, the use of or availability of pollutants, contaminants and other liquid, solid or gaseous substances, etc.

7.1.4 Typical Methods of Backflow Prevention

When a degree of hazard has been determined and properly classified, effective steps should be taken to require correction of the condition or the installation of backflow prevention assembly(s) at the service connection(s) and/or internally to prevent the possible pollution or contamination of the consumer's and/or public potable water system. The rules or regulations of the administrative authorities advise this in order to isolate within the consumer's water system(s) any pollution or contamination that may occur which could adversely affect the potability of the public or consumer's water system. Proper backflow protection should be maintained on each service connection or separation of internal systems, at the point of delivery and ahead of any outlet. Where the consumer's water line is divided at the point of delivery, an assembly must be installed on each leg or branch (i.e., industrial and domestic). The type and kind of backflow prevention assembly needed on each leg or branch is to be in accordance with the highest degree of hazard found on the premises. In evaluating the necessity for backflow protection on fire services, refer to Section 7.2.3.14.

It may be expedient or necessary under certain circumstances for the water purveyor to approve the installation of a lead-in line into the premises a considerable distance beyond the service connection (point of delivery) before installing the backflow protection. In using this alternate the water purveyor continues to have jurisdiction over the consumer's lead-in line as far as the backflow prevention assembly(s). It is the purveyor's responsibility to see the consumer does not install any unprotected takeoffs between the point of delivery from the public potable water supply system and the backflow prevention assembly(s). See Fig. 7.1.

A reduced pressure principle backflow prevention assembly must always be installed above grade and provided with freeze protection, where necessary. This assembly shall never be installed in a meter box, pit, or vault. If installed within a consumer's building there must be adequate floor drainage (not a dry well) beneath or near the assembly. No backflow prevention assembly shall be installed in a vertical run of pipe unless it was evaluated and approved in this orientation.

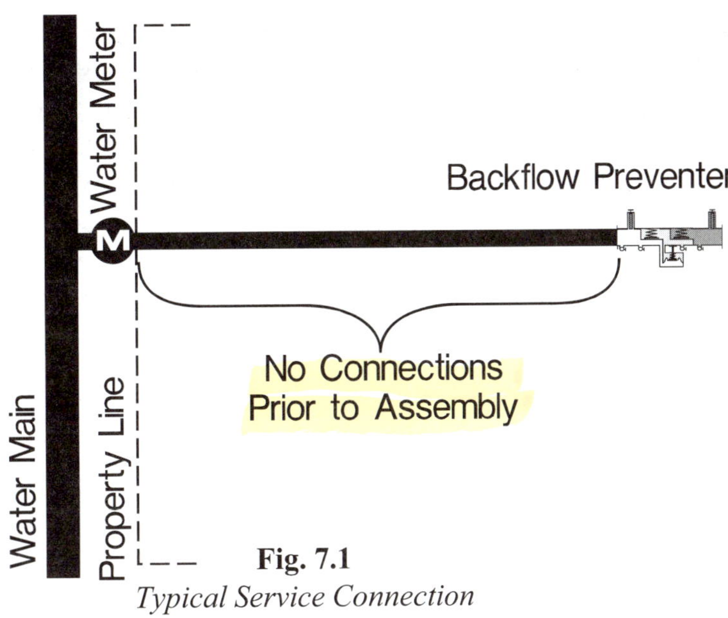

Fig. 7.1
Typical Service Connection

7.2 Plants or Facilities Where Backflow Protection Will Typically be Required

7.2.1 General

The rules or regulations of the administrative authorities are typically predicated upon the policy that an approved backflow prevention assembly is required at the water purveyor's point of delivery where an actual or potential health, system, plumbing or pollutional hazard exists. That is, when, in the opinion of the administrative authority having jurisdiction, there is a substantial hazard to the public system, no water connection should be installed or maintained between the public potable water supply and a consumer's premises without a backflow prevention assembly at the point of delivery. (See Section 5 for lists of typical cross-connections, chemicals or water uses which may endanger the public water system.)

Where the administrative authority chooses to delegate its responsibility for the protection of all internal cross-connections to the water purveyor said water purveyor may then decide whether internal protection will be sufficient of itself; or, whether, due to the general conditions within the premises, there is need for system protection at the meter plus the segregation of system and internal protection.

Where the administrative authorities do not delegate their responsibilities over the consumer's internal systems, the water purveyor may choose to rely only on the control of the administrative authority, or, as is more usual, the water purveyor will act separately to protect its water distribution system in accordance with the law.

Contamination or pollution of a water system is usually brought about by a cross-connection to: sewage and storm drain lines, tanks, vessels and other sewage-connected facilities; systems containing auxiliary water supplies which may be polluted or contaminated; fire systems which are or may be polluted or contaminated; irrigation systems which may be polluted or contaminated with fertilizers, pesticides or other objectionable materials; sewage-connected plumbing fixtures which are not adequately protected; oil wells, tanks, pipelines, storage or refinery equipment containing hydrocarbons in liquid, gaseous or solid form; chemical systems used in plating and manufacturing processes; reservoirs, cooling towers, transmission or circulating systems; industrial fluid or used water systems; and facilities utilizing radioactive substances.

In evaluating the degree of hazard and the type of protection required, consideration must be given to all of the foregoing possibilities of pollution or contamination of the public or the consumer system. In order to simplify this evaluation, it is recommended that one of the following alternate plans be adopted.

7.2.2 Alternate Plans

7.2.2.1 First Alternate

This alternative may be used where manpower is available for making initial inspections and for maintaining an adequate inspection and reinspection program on the consumer's premises. In this case the water purveyor and/or administrative authority should make an evaluation of the degree of hazard based on a meticulous inspection of the plant's water system, water uses and in-plant backflow protection afforded (see Section 5 for typical cross-connections, chemicals and dangerous water uses) and, where deemed necessary, require the installation of an approved backflow prevention assembly at an internal point(s) and/or at the service connection(s).

7.2.2.2 Second Alternate

This alternative may be used where it is not practical or economically feasible to maintain an extensive program of inspection and reinspection of the piping and water-using facilities on premises where cross-connections are known, or suspected, to exist. In using this alternative the water purveyor or administrative authority can determine the need for backflow protection at one or more points internally or at the service connection by comparing the plant under consideration with one or more comparable plants described in Table 7-1. The hazards involved have been partially described and the recommended types of protection have been based on extensive field experience for each of the facilities or conditions listed. With each of the industries listed, a partial description of the types of hazards that may be expected and the recommendations for types of protection have been given. Where a facility has not been listed it will be necessary for the water purveyor and/or the administrative authority to make an investigation of the plant's existing water system and water using facilities. When considering a new or proposed plant, an evaluation should be made of the facility's proposed piping plans, specifications and anticipated water-uses.

Table 7-1

Table 7-1 gives a list of plants or facilities where backflow protection will typically be required. Following the list is a partial description of the hazards involved and the protection recommended at the service connection.

Plants or Facilities Where Backflow Protection Will Typically Be Required

7.2.3 Hazards Involved and Protection Recommended

7.2.3.1 Aircraft and Missile Plants

An approved backflow prevention assembly shall be installed at the service connection to any premise on which there is an aircraft or missile plant. Excluded from this category are plants that are used only for management, mock-up, engineering, assembly and other like purposes where the water supply is not used for industrial or other purposes which would create an actual or potential hazard. (IMPORTANT: Due to frequent changes in models or components, methods of manufacturing and additions to the plants, the in-plant water system may undergo continual change in water-use requirements. As a result, new cross-connections may be installed and existing in-plant protection may be bypassed, removed or be made otherwise ineffective.)

The hazards normally found in a plant of this type include cross-connections between the consumer's water system and: ***reservoirs, cooling towers and circulating systems*** which may be heavily contaminated with bird droppings, vermin, algae, bacterial slimes or toxic water treatment compounds such as copper sulfate, pentachlorophenol, chromates, metallic glucosides, compounds of mercury, quaternary ammonium compounds, etc.; ***steam generating facilities and lines*** which may be contaminated with boiler compounds such as those chemicals listed above (NOTE: a particular hazard is the possibility of steam getting back into the domestic system, causing either a system or a health hazard); ***plating facilities*** involving the use of highly toxic cyanides, heavy metals in solution (such as copper, cadmium, chrome, nickel, etc.), acids and caustic solutions; ***plating solution filtering equipment*** with pumps and circulating lines; ***tanks, vats, and other vessels*** used in painting, descaling, anodizing, cleaning, stripping, oxidizing, etching, passivating, pickling, dipping, rinsing operations, or other lines or facilities needed in the preparation or finishing of the product; ***water-cooled equipment which may be sewer-connected*** such as heat exchangers, compressors, air conditioning equipment, etc.; ***industrial fluid systems and lines*** containing cutting and hydraulic fluids, coolants, hydrocarbon products, glycerine, paraffin, caustic and acid solutions, etc.; ***sewer-connected plumbing fixtures*** such as flush valve toilets and urinals without atmospheric vacuum breakers or with improperly maintained atmospheric vacuum breakers (NOTE: this hazard is critical because little or no attention is given to the maintenance of an atmospheric vacuum breaker and frequently the working parts or the entire assembly may be removed from the line); ***fire fighting systems***, including storage reservoirs which may be treated for the prevention of scale formation, corrosion, algae, slime growths, etc., or fire systems which may be subject to contamination with antifreeze solutions, Foamite or other chemicals or chemical compounds used in fighting fire, or fire systems which are subject to contamination with auxiliary or used water supplies or industrial fluids; ***hydraulically-operated equipment*** where the public potable water pressure is used directly and may be subject to backpressure; ***equipment under hydraulic tests*** such as tanks, lines, valves, fittings and also pumps, pressure cylinders or other hydraulic equipment which may be used to provide pressures for testing purposes (NOTE: in such cases, air, gas or hydraulic fluids may be forced back into the public system); ***a system supplied from an auxiliary water supply.*** (NOTE: The rules or regulations of the administrative authority or the water purveyor require a backflow prevention assembly at the service connection to premises where there is an auxiliary water supply, even though there are no existing cross-connections. For more specific information regarding an auxiliary water supply and recommended protection, see Section 7.2.3.3 "Auxiliary Water Systems".)

Protection Recommended

For service protection an air gap or a reduced pressure principle backflow prevention assembly is recommended. For internal protection it may be possible to utilize a pressure vacuum breaker type of assembly on some points of application; otherwise, the air gap or the reduced pressure principle assembly would be used.

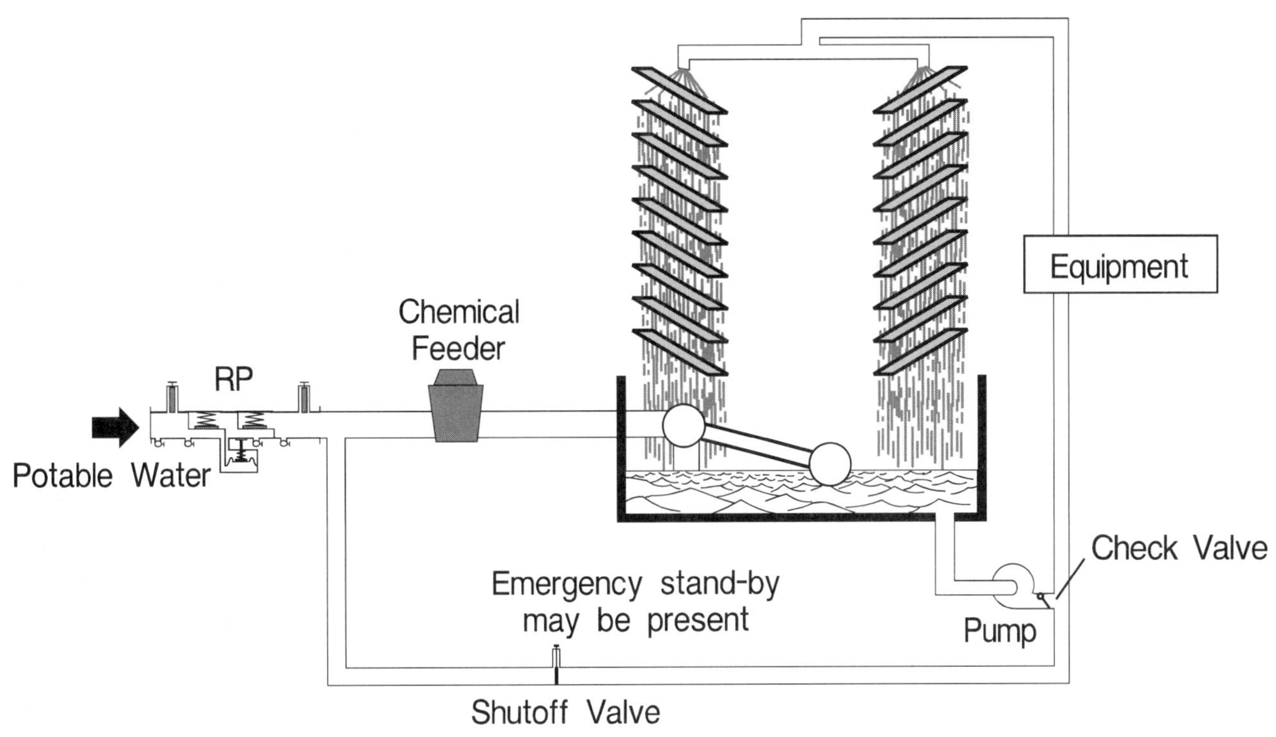

Fig. 7.2
Example of Cooling Tower

7.2.3.2 Automotive Plants

An approved backflow prevention assembly shall be installed at the service connection to any premise where an automotive manufacturing plant is operated or maintained. (IMPORTANT: Due to frequent changes in models or components, methods of manufacturing and additions to the plants, the in-plant water system undergoes continual change in water-use requirements. As a result, new cross-connections may be installed and existing in-plant protection may be bypassed, removed, or otherwise made ineffective.)

The hazards normally found in a plant of this type include cross-connections between the consumer's water system and: ***sewage pumps*** for priming, cleaning, flushing or unclogging purposes; ***water-operated sewage sump ejectors*** for operational purposes; ***sewer lines*** for the purpose of disposing of filter or softener backwash water or water from cooling systems or of providing for a quick drain for the building lines or of flushing or blowing out obstructions, etc. (NOTE: administrative authorities require backflow protection at the service connection to any premise on which there is located a sewage ejector or pumping station, even though there are no cross-connections); ***reservoirs, cooling towers and circulating systems*** which may be heavily contaminated with bird droppings, vermin, algae, bacterial slimes or with toxic water treatment compounds such as pentachlorophenol, copper sulfate, chromates, metallic glucosides, compounds of mercury, quaternary ammonium compounds, etc.; ***steam generating facilities and lines*** which may be contaminated with boiler compounds such as those listed above (NOTE: a particular hazard is the possibility of steam getting back into the domestic system, causing either a system or a health hazard); ***industrial fluid systems and line*** containing cutting and hydraulic fluids, coolants, hydrocarbon products, glycerine, paraffin, caustic and acid solutions, etc.; ***fire fighting systems***, including storage reservoirs which may be treated for prevention of scale formation, corrosion, algae, slime growths, etc., or fire systems which may be subject to contamination with antifreeze solutions, Foamite or other chemicals or chemical compounds used in fire fighting, or fire systems which are subject to contamination with auxiliary or used water supplies or industrial fluids; ***plating facilities*** involving the use of highly toxic cyanides, heavy metals in solution (such as copper, cadmium, chrome, nickel, etc.), acids and caustic solutions; ***plating solution filtering equipment*** with pumps and circulating lines; ***tanks, vats, or other vessels*** used in painting, descaling, anodizing, cleaning, stripping, oxidizing, etching, passivating, pickling, dipping or rinsing operations, or other lines or facilities needed in preparation or finishing of the products; ***sewer-connected plumbing fixtures*** such as flush valve toilets or urinals without atmospheric vacuum breakers or with improperly maintained atmospheric vacuum breakers (NOTE: this hazard is critical because little or no attention is given to the maintenance of atmospheric vacuum breakers and frequently the working parts or the entire assembly may be removed from the line); ***water-cooled equipment which may be sewer-connected*** such as compressors, heat exchangers, air conditioning equipment, etc.; ***hydraulically-operated equipment*** where the public potable water pressure is used directly and may be subject to backpressure; ***equipment under hydraulic tests*** such as tanks, lines, valves, fittings and also pumps, pressure cylinders or other hydraulic facilities which may be used to provide pressures for testing purposes (NOTE: in such cases air, gas or hydraulic fluids may be forced back to the public system); ***a system supplied from an auxiliary water supply.*** (NOTE: The administrative authority normally will require a backflow prevention assembly at the service connection to the premises where there is an auxiliary water supply, even though there are no existing cross-connections. For more specific information regarding auxiliary water supply and recommended protection, see Section 7.2.3.3 "Auxiliary Water Systems".)

Protection Recommended:

An air gap or a reduced pressure principle backflow prevention assembly.

7.2.3.3 Auxiliary Water Systems

An approved backflow prevention assembly shall be installed at the service connection to any premise where there is an auxiliary water supply or system even though there is no connection or cross-connection. NOTE: In order that the water purveyor may more clearly understand the meaning of the term "auxiliary water supply" it is necessary to define its meaning. The following discussion will partially clarify this phrase.

The term "auxiliary water supply" (briefly defined in Section 4.47) describes water supplies or sources not under the control or direct supervision of the water purveyor. Typical of such water supplies are natural waters derived from wells, springs, streams, rivers, harbors, bays and oceans. Also, there may be public potable water supplies furnished by some other water purveyor which may or may not be under good sanitary control or may be otherwise unacceptable to the water purveyor.

In the event that backflow occurs because of inadequate protection at the service connection, polluted or contaminated used waters or industrial fluids may be discharged into the public water system; therefore, it is necessary for the water purveyor to evaluate the potential hazard and to take steps necessary to protect the public water system in accordance with the degree of hazard found. In making such an evaluation it is not deemed necessary that the auxiliary water sources be developed and interconnected with the potable water system through cross-connections. It is deemed necessary only to determine that the water or fluids are available to the premises and of quantity sufficient to be desirable and feasible for the consumer to develop and use the supply.

For administrative purposes, in the use of this Manual the water purveyor may subdivide auxiliary water supplies into two general classifications:

(A) An approved public potable water supply over which the water purveyor does not have sanitary control;

(B) Any private water supply other than the water purveyor's approved public potable water supply on or available to the premises;

In the following recommendations the degree of hazard is classified and the type of assembly is recommended:

Protection Recommended:

(A) Public Potable Water System — Interconnection To

An approved backflow prevention assembly shall be installed on any direct interconnection, except as noted hereafter, between the water purveyor's approved public potable water supply and another approved public potable water supply over which the purveyor does not have sanitary control. This may be accomplished in the following manner:

(1) An air gap or a reduced pressure principle backflow prevention assembly is recommended at the service connection when the auxiliary water supply is or may be contaminated to a degree that would constitute a health hazard;

(2) A double check valve assembly is recommended at the service connection when the auxiliary water supply is being operated under a public health permit but is not acceptable to the water purveyor as a source;

(3) No backflow protection at the service connection is recommended if the auxiliary water system has a properly conducted sanitary control program in force and the auxiliary water supply is acceptable to the water purveyor as a source.

(B) Private Water Supply

An approved backflow prevention assembly shall be installed at the service connection to any premise on which there is any available water supply other than the water purveyor's public potable water supply. This may be accomplished in the following manner:

(1) An air gap or a reduced pressure principle backflow prevention assembly is recommended at each service connection when the auxiliary water supply is or may be contaminated to a degree that it would constitute a health or system hazard;

(2) A double check valve assembly is recommended at each service connection when the auxiliary water supply is or may be subject to pollution (i.e., when there is no health or system hazard).

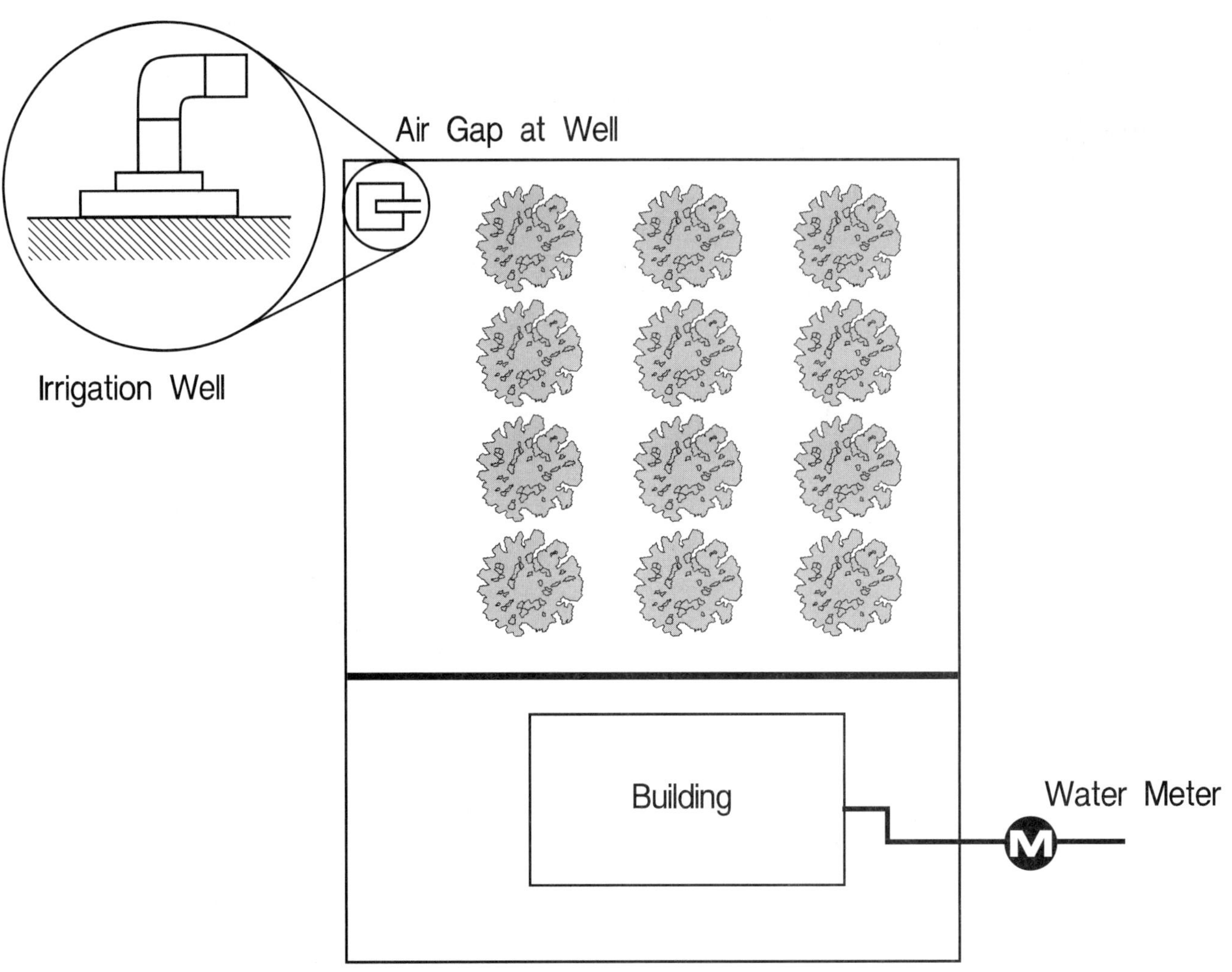

Fig. 7.3
Example of Agricultural Auxiliary Water Supply
Typically Would Not Require Backflow Protection at User Connection

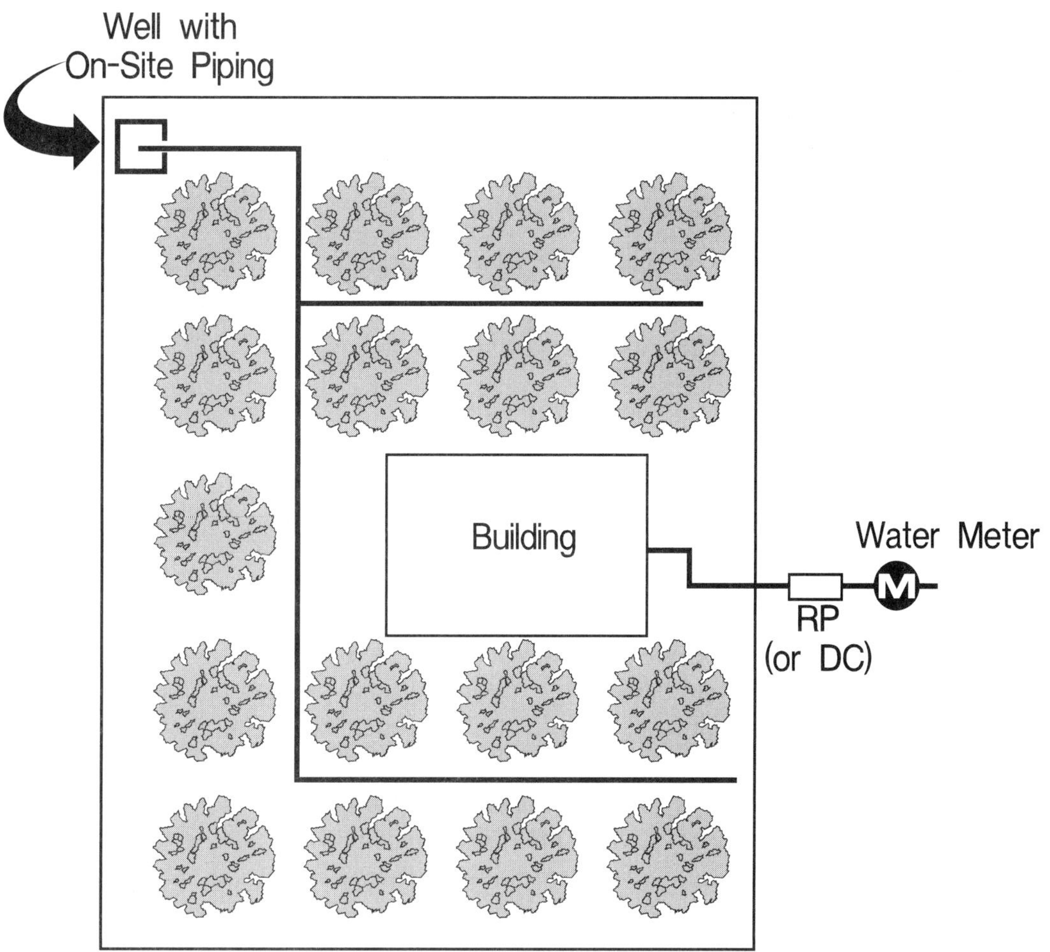

Fig. 7.4
Example of Auxiliary Water Supply
Minimum of DC required at User Connection

7.2.3.4 Beverage Bottling Plants

An approved backflow prevention assembly shall be installed on the service connection to any premise where a beverage bottling plant is operated or maintained and water is used for industrial purposes.

The hazards normally found in a plant of this type include cross-connections between the consumer's water system and: ***steam connected facilities*** such as pressure ***cookers, autoclaves, retorts, etc.; washers, cookers, tanks, lines, flumes and other equipment*** used for storing, washing, cleaning, blanching, cooking, flushing or fluming or for transportation of foods, fertilizers or wastes; ***can and bottle washing machines and lines*** where caustics, acids, detergents, and other compounds are used in cleaning, sterilizing and flushing; reservoirs, cooling towers and circulating systems which may be heavily contaminated with bird droppings, vermin, algae, bacterial slimes or with toxic water treatment compounds such as pentachlorophenol, copper sulfate, chromates, metallic glucosides, compounds of mercury, quaternary ammonium compounds, etc.; steam generating facilities and lines which may be contaminated with boiler compounds such as those chemicals listed above (NOTE: A particular hazard is the possibility of steam getting back into the domestic system, causing either a system or a health hazard); ***industrial fluid systems and lines*** containing cutting and hydraulic fluids, coolants, hydrocarbon products, glycerine, paraffin, caustic and acid solutions, etc.; ***water cooled equipment which may be sewer-connected*** such as compressors, heat exchangers, air conditioning equipment, etc.; ***a system supplied from an auxiliary water supply.*** (NOTE: The administrative authority will normally require a backflow prevention assembly at the service connection to premises where there is auxiliary supply, even though there is no existing cross-connections. For more specific information regarding an auxiliary water supply and recommended protection, see Section 7.2.3.3 "Auxiliary Water Systems".)

Protection Recommended:

(a) An air gap or a reduced pressure principle backflow prevention assembly where there is a potential health or system hazard;

(b) A double check valve assembly where there is only a pollutional hazard.

7.2.3.5 Breweries

An approved backflow prevention assembly shall be installed on each service connection to any premise where a brewery is operated or maintained.

The hazards normally found in a plant of this type include a cross-connection between the consumer's water system and: ***industrial fluid systems and lines*** containing cutting and hydraulic fluids, coolants, hydrocarbon products, glycerine, paraffin, caustic and acid solutions, etc.; ***tanks, can and bottle washing machine lines*** where caustics, acids, detergents and other compounds are used in cleaning, sterilizing and flushing; ***water cooled equipment which may be sewer-connected*** such as compressors, heat exchangers, air conditioning equipment, etc.; ***reservoirs, cooling towers and circulating systems*** which may be heavily contaminated with bird droppings, vermin, algae, bacterial slimes or with toxic water treatment compounds such as pentachlorophenol, copper sulfate chromates, metallic glucosides, compounds of mercury, quaternary ammonium compounds, etc.; ***steam generating facilities and lines*** which may be contaminated with boiler compounds such as those chemicals listed above (NOTE: A particular hazard is the possibility of steam getting back into the domestic water system, causing either a system or a health hazard); ***fire fighting systems***, including storage reservoirs which may be treated for prevention of scale formation, corrosion, algae, slime growths, etc., or fire systems which may be subject to contamination with antifreeze solutions, of Foamite or other chemicals or chemical compounds used in fire fighting, or fire systems which are subject to contamination with auxiliary or used water supplies or industrial fluids.

Protection Recommended:

An air gap or a reduced pressure principle backflow prevention assembly.

7.2.3.6 Buildings — Hotels, Apartment Houses, Public and Private Buildings, or any Other Structures Having Unprotected Cross-Connections

An approved backflow prevention assembly shall be installed at the service connection to any premise where multistoried buildings such as a hotel, apartment house, office or loft building are operated or maintained, if the buildings have unprotected or inadequately protected cross-connections, sewage pumping facilities, auxiliary water supplies or other like sources of contamination which would create a potential hazard to the public water system.

Also, an approved backflow prevention assembly shall be installed at the service connection to any premise where there are existing cross-connections or where potential cross-connections exist or where it is expected that the consumer may make piping or equipment changes which would result in the installation of cross-connections.

The hazards normally found in buildings of this type include cross-connections between the consumer's water system and: ***sewage pumps*** for priming, cleaning, flushing or unclogging purposes; ***water-operated sewage sump ejectors*** for operational purposes; sewer lines for the purpose of disposing of filter or softener backwash water or water from cooling systems or of providing for a quick drain for the building water lines or of flushing or blowing out obstructions, etc. (NOTE: administrative authorities will require backflow protection at the service connection to any premise on which there is located a sewage ejector or pumping station, even though there are no cross-connections); ***reservoirs, cooling towers and circulating systems*** which may be heavily contaminated with bird droppings, vermin, algae, bacterial slimes or with toxic water treatments compounds such as pentachlorophenol, copper sulfate, chromates, metallic glucosides, compounds of mercury, quaternary ammonium compounds, etc.; ***steam generating facilities and lines*** which may be contaminated with boiler compounds such as those chemicals listed above (NOTE: a particular hazard is the possibility of steam getting back into the domestic system, causing either a system or a health hazard); ***fire fighting systems***, including storage reservoirs which may be treated for prevention of scale formation, corrosion, algae, slime growths, etc., or fire systems which may be subject to contamination with antifreeze solutions, Foamite or other chemicals or chemical compounds used in fighting fire, or fire systems which are subject to contamination with auxiliary or used water supplies or industrial fluids; ***contaminated and/or sewer-connected facilities*** such as inadequately protected flush valve toilets, urinals, aspirators, retorts, pipet tube washers, etc.; ***laboratory equipment*** which may be chemically or bacteriologically contaminated; ***steam sterilizers, autoclaves, specimen tanks, autopsy and mortuary equipment*** (NOTE: this hazard is critical because little or no attention is given to the maintenance of atmospheric vacuum breakers and frequently they are removed from the line); ***tanks, automatic film processing machines*** or other facilities used in processing films, which may be contaminated with chemicals such as acetic acid, potassium ferricyanide and/or one of the many different types of the aromatic series of organic chemicals; ***water cooled equipment which may be sewer-connected*** such as compressors, heat exchangers, air conditioning equipment, etc. (NOTE: in multi-storied buildings of this type the supply line to the toilets, urinals, lavatories, laboratory sinks, tanks, etc., on the lower floors may be taken off the suction side of the house pump and, as a result, sewage or other contaminated substances may be drawn into the house supply line); ***a system supplied from an auxiliary water supply.*** (NOTE: The administrative authority will normally require a backflow prevention assembly at the service connection to premises where there is auxiliary supply, even though there are no existing cross-connections. For more specific information regarding an auxiliary water supply and recommended protection, see Section 7.2.3.3 "Auxiliary Water Systems".)

Protection Recommended:

(a) An air gap or a reduced pressure principle backflow prevention assembly where there is a potential health or system hazard;

(b) A double check valve assembly where there is only a pollutional hazard.

7.2.3.7 Canneries, Packing Houses and Rendering Plants

An approved backflow prevention assembly shall be installed at the service connection to any premise where vegetable or animal matter is canned, concentrated or processed. A cannery, packing house or rendering plant as used herein does not include the small plants where water is not used industrially (i.e., where there are no cross-connections or auxiliary water supplies including used waters or industrial fluids).

The hazards normally found in a plant of this type include cross-connections between the consumer's water system and: ***steam-connected facilities*** such as pressure cookers, autoclaves, retorts, etc.; ***washers, cookers, tanks, lines, flumes, and other equipment*** used for storing, washing, cleaning, blanching, cooking, flushing, fluming or for transmission of food, fertilizers or wastes; ***reservoirs, cooling towers and circulating systems*** which may be heavily contaminated with bird droppings, vermin, algae, bacterial slimes or with toxic water treatments compounds such as pentachlorophenol, copper sulfate, chromates, metallic glucosides, compounds of mercury, quaternary ammonium compounds, etc.; ***steam generating facilities and lines*** which may be contaminated with boiler compounds such as those chemicals listed above (NOTE: a particular hazard is the possibility of steam getting back into the domestic system, causing either a system or a health hazard); ***industrial fluid systems and lines*** containing cutting and hydraulic fluids, coolants, hydrocarbon products, glycerine, paraffin, caustic and acid solutions, etc.; ***fire fighting systems***, including storage reservoirs which may be treated for prevention of scale formation, corrosion, algae, slime growths, etc., or fire systems which may be subject to contamination with antifreeze solutions, Foamite or other chemicals or chemical compounds used in fighting fire, or fire systems which are subject to contamination with auxiliary or used water supplies or industrial fluids; ***water cooled equipment which may be sewer-connected*** such as compressors, heat exchangers, air conditioning equipment, etc.; ***tanks, can and bottle washing machines and lines*** where caustic, acids, detergents and other compounds are used in cleaning, sterilizing and flushing; ***a system supplied from an auxiliary water supply.*** (NOTE: The administrative authority will normally require a backflow prevention assembly at the service connection to premises where there is auxiliary supply, even though there are no existing cross-connections. For more specific information regarding an auxiliary water supply and recommended protection, see Section 7.2.3.3 "Auxiliary Water Systems".)

Protection Recommended:

An air gap or a reduced pressure principle backflow prevention assembly.

7.2.3.8 Car Wash Facilities

An approved air gap or a reduced pressure principle backflow prevention assembly shall be installed at the service connection to any premise where a car wash facility is operated or maintained. In addition, where a domestic water system is on the property and supplied by a single service connection an approved air gap or a reduced pressure principle backflow prevention assembly shall be installed on the industrial line at the point of takeoff from the domestic line.

Most automatic car washes utilize a recirculating system as well as soap or wax aspirating systems.

Coin operated car washes shall provide backflow protection with installation of a reduced pressure principle backflow prevention assembly or air gap.

Protection Recommended:

An air gap or reduced pressure principle backflow prevention assembly :

(a) At the point of service where the industrial water is separated from the domestic water; or,

(b) At the point of service and at the separation of the services internally.

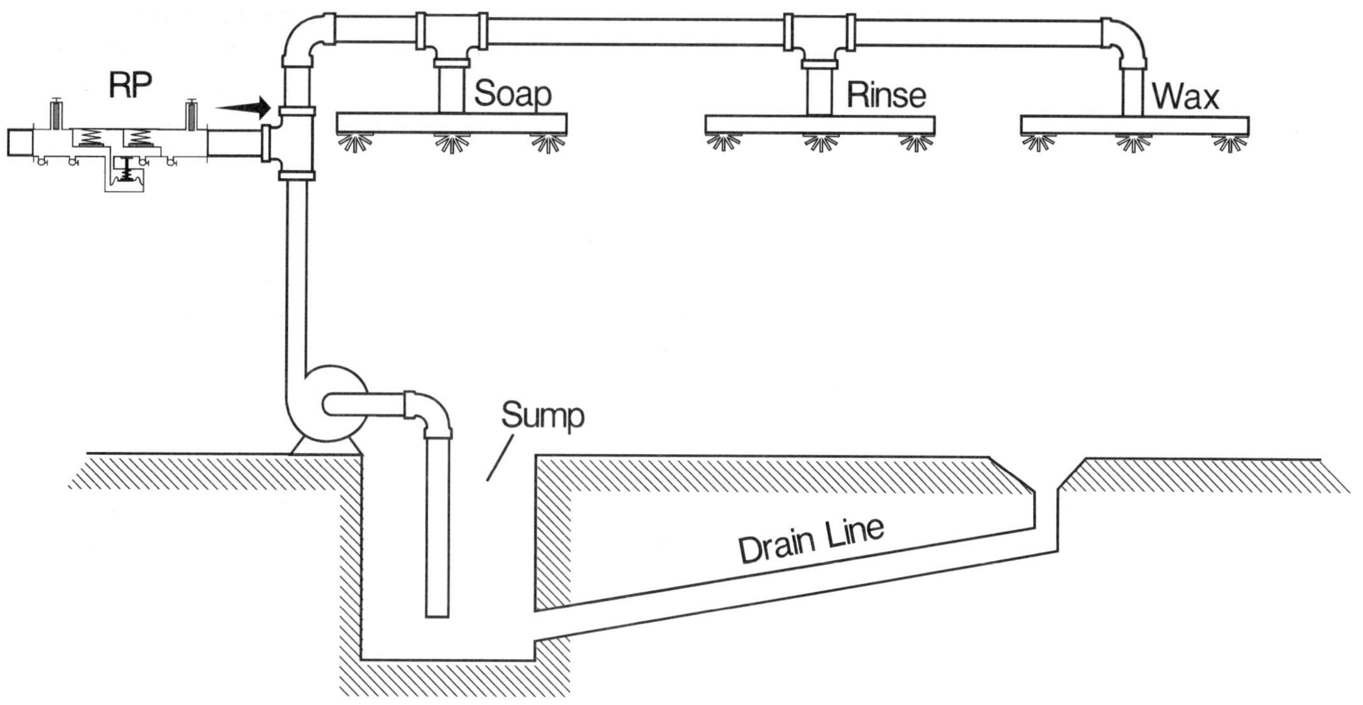

Fig. 7.5
Typical Car Wash Facilities

7.2.3.9 Chemical Plants — Manufacturing, Processing, Compounding or Treatment

An approved backflow prevention assembly shall be installed on the service connection to any premise where there is a facility requiring the use of water in the industrial process of manufacturing, storing, compounding or processing chemicals. This will include facilities where chemicals are used as an additive to the water supply or in processing products.

This is a very broad category and will require careful consideration of the processes involved in the plants. Water for manufacturing purposes is a prime requisite in most chemical plants. Cross-connections between the public water supplies and the chemical tanks, lines and other facilities are needed for purging the lines, cleaning the tanks and vats, and in compounding processes. Toxicity of the chemicals will vary. It is customary for the manufacturer to change processes and water uses as needed.

The hazards normally found in a plant of this type include cross-connections between the consumer's water system and: ***reservoirs, cooling towers and circulating systems*** which may be heavily contaminated with bird droppings, vermin, algae, bacterial slimes or with toxic water treatments compounds such as pentachlorophenol, copper sulfate, chromates, metallic glucosides, compounds of mercury, quaternary ammonium compounds, etc.; ***steam generating facilities and lines*** which may be contaminated with boiler compounds such as those chemicals listed above (NOTE: a particular hazard is the possibility of steam getting back into the domestic system, causing either a system or a health hazard); ***industrial fluid systems and lines*** containing cutting and hydraulic fluids, coolants, hydrocarbon products, glycerine, paraffin, caustic and acid solutions, etc.; ***fire fighting systems***, including storage reservoirs which may be treated for prevention of scale formation, corrosion, algae, slime growths, etc., or fire systems which may be subject to contamination with antifreeze solutions, Foamite or other chemicals or chemical compounds used in fighting fire, or fire systems which are subject to contamination with auxiliary or used water supplies or industrial fluids; ***water cooled equipment which may be sewer-connected*** such as compressors, heat exchangers, air conditioning equipment, etc.; ***hydraulically-operated equipment*** where the public potable water

pressure is used directly and may be subject to back pressure; ***equipment under hydraulic tests*** such as tanks, lines, valves, fittings and also pumps, pressure cylinders or other hydraulic facilities which may be used to provide pressures for testing purposes (NOTE: in such cases, air, gas or hydraulic fluids may be forced back into the public system); ***pressure cookers, autoclaves, retorts, and other similar steam-connected facilities; washers, cookers, tanks, flumes, and other equipment*** used for storing, washing, cleaning, blanching, cooking, flushing, fluming, or for transmission of foods, fertilizers or wastes; ***a system supplied from an auxiliary water supply.*** (NOTE: The administrative authority will normally require a backflow prevention assembly at the service connection to premises where there is auxiliary supply, even though there are no existing cross-connections. For more specific information regarding an auxiliary water supply and recommended
protection, see Section 7.2.3.3 "Auxiliary Water Systems".)

Protection Recommended:

An air gap or a reduced pressure principle backflow prevention assembly.

7.2.3.10 Chemically Contaminated Water Systems

An approved backflow prevention assembly shall be installed on the service connection to any premise where chemicals are used as an additive to the water supply for prevention of scale formation, corrosion, algae, slime growths, etc., or where the water supply is used for transmission and distribution of chemicals, or where the chemicals are used with water in compounding or processing products.

Protection Recommended:

An air gap or reduced pressure principle backflow prevention assembly.

7.2.3.11 Civil Works

An approved backflow prevention assembly shall be installed at the service connection to any premise on which there is a civil work facility having unprotected cross-connections, sewage and storm drain facilities, auxiliary water supplies, fire systems, irrigation systems, oil and gas handling facilities, plating facilities, industrial fluids, containers or transmission lines which, if cross-connected to, would create a potential hazard to the public water system. The term "civil works" includes federal, state, city, county and district yards; docks and facilities; or military camps, posts, stations, public buildings and facilities. The reason for differentiating between civil works facilities and other similar privately-owned facilities is, that under some plumbing codes, civil works owned or controlled by governmental agencies are exempt from the local inspection provisions of the code. This has resulted in the installation of internal cross-connections which are frequently left unprotected. In such cases the water purveyor must assume full responsibility for preventing backflow of used waters, industrial fluids and other contaminated or polluted materials from civil works installations into the public potable water system.

Protection Recommended:

(a) An air gap or a reduced pressure principle backflow prevention assembly where there is an existing or potential health or system hazard;

(b) A double check valve assembly where there is a potential pollutional hazard.

7.2.3.12 Dairies and Cold Storage Plants

An approved backflow prevention assembly shall be installed on the service connection to any premise where a dairy, creamery, ice cream plant, cold storage or ice manufacturing plant is operated or maintained, provided such a plant has on the premises an auxiliary water supply, industrial fluid system, sewage handling facilities or other similar sources of contamination which, if cross-connected to, would create a hazard to the public system.

The hazards normally found in a plant of this type include cross-connections between the consumer's water system and: ***reservoirs, cooling towers and circulating systems*** which may be heavily contaminated with bird droppings, vermin, algae, bacterial slimes or with toxic water treatments compounds such as pentachlorophenol, copper sulfate, chromates, metallic glucosides, compounds of mercury, quaternary ammonium compounds, etc.; ***steam generating facilities and lines*** which may be contaminated with boiler compounds such as those chemicals listed above (NOTE: a particular hazard is the possibility of steam getting back into the domestic system, causing either a system or health hazard); ***water cooled equipment which may be sewer-connected*** such as compressors, heat exchangers, air conditioning equipment, etc.; ***tanks, can and bottle washing machines and lines*** where caustics, acids, detergents, and other compounds are used in cleaning, sterilizing and flushing.

Protection Recommended:

(a) An air gap or a reduced pressure principle backflow prevention assembly where there is an existing or potential health or system hazard;

(b) A double check valve assembly where there is an existing or potential pollutional hazard.

7.2.3.13 Film Laboratories

An approved backflow prevention assembly shall be installed on each service connection to any premise where a film laboratory or processing or manufacturing plant is operated or maintained. This may not include the small dark room facilities normally found in X-ray laboratories, drug stores, small shops, etc.

The hazards normally found in a plant of this type include cross-connections between the consumer's water system and: ***tanks, automatic film processing machines*** or other facilities used in processing films, which may be contaminated with chemicals such as acetic acid, potassium ferricyanide and/or one of the many different types of the aromatic series of organic chemicals; ***water cooled equipment which may be sewer-connected*** such as compressors, heat exchangers, air conditioning equipment, etc.

Protection Recommended:

An air gap or a reduced pressure principle backflow prevention assembly.

7.2.3.14 Fire Systems[1]

Industrial fire protection systems consist of sprinklers, hose connections, and hydrants. Sprinkler systems may be dry or wet, open or closed. Systems of fixed-spray nozzles may be used indoors or outdoors for protection of flammable-liquid and other hazardous processes. It is standard practice, especially in cities, to equip automatic sprinkler systems with fire department pumper connections.

---

[1]Additional information regarding fire systems may be found in AWWA Manual M-14

For cross-connection control, fire protection systems may be classified on the basis of water source and arrangement of supplies as follows:

CLASS 1 — direct connections from public water mains only; no pumps, tanks, or reservoirs; no physical connection from other water supplies; no antifreeze or other additives of any kind; all sprinkler drains discharging to atmosphere, dry wells or other safe outlets.

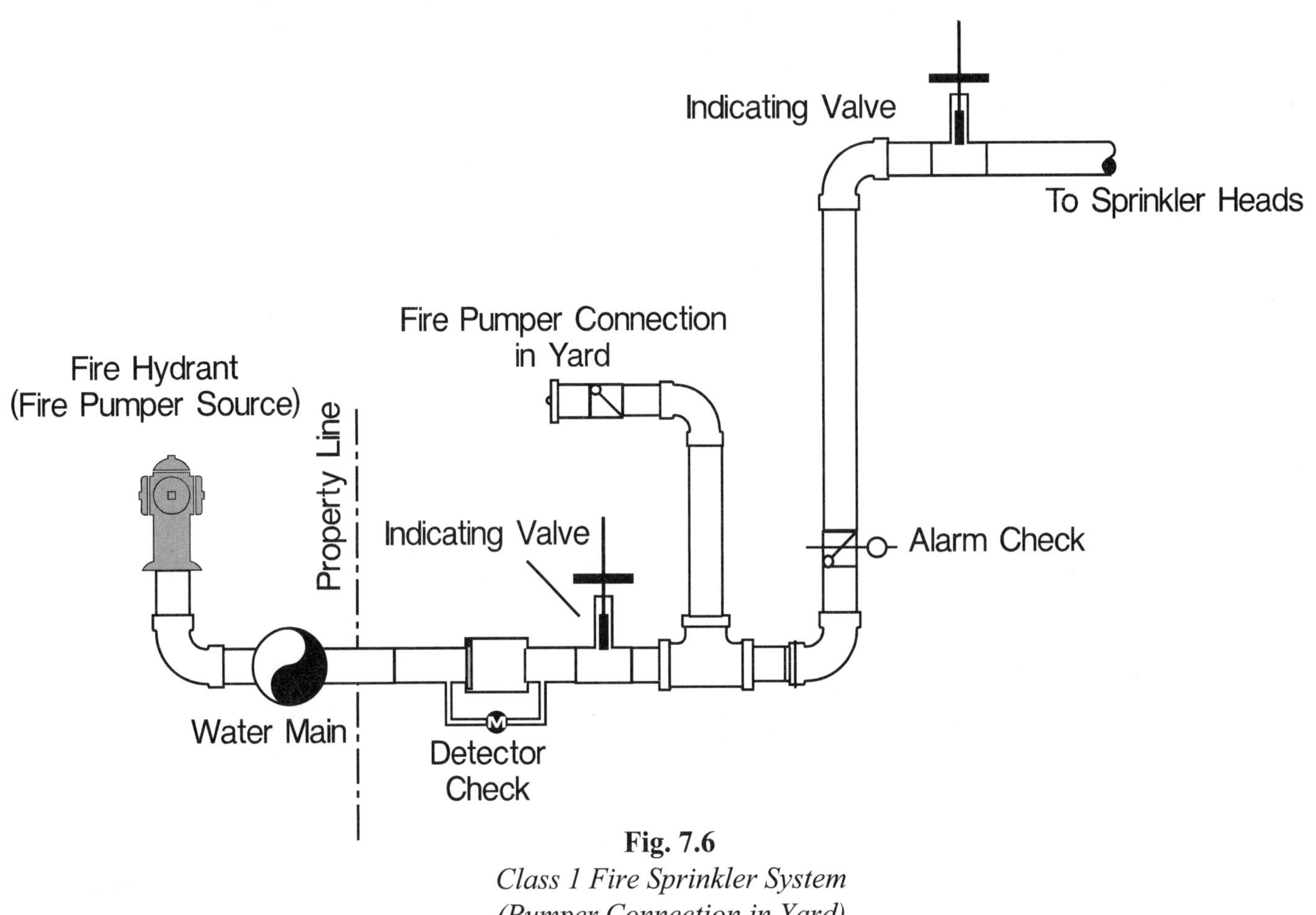

Fig. 7.6
Class 1 Fire Sprinkler System
(Pumper Connection in Yard)

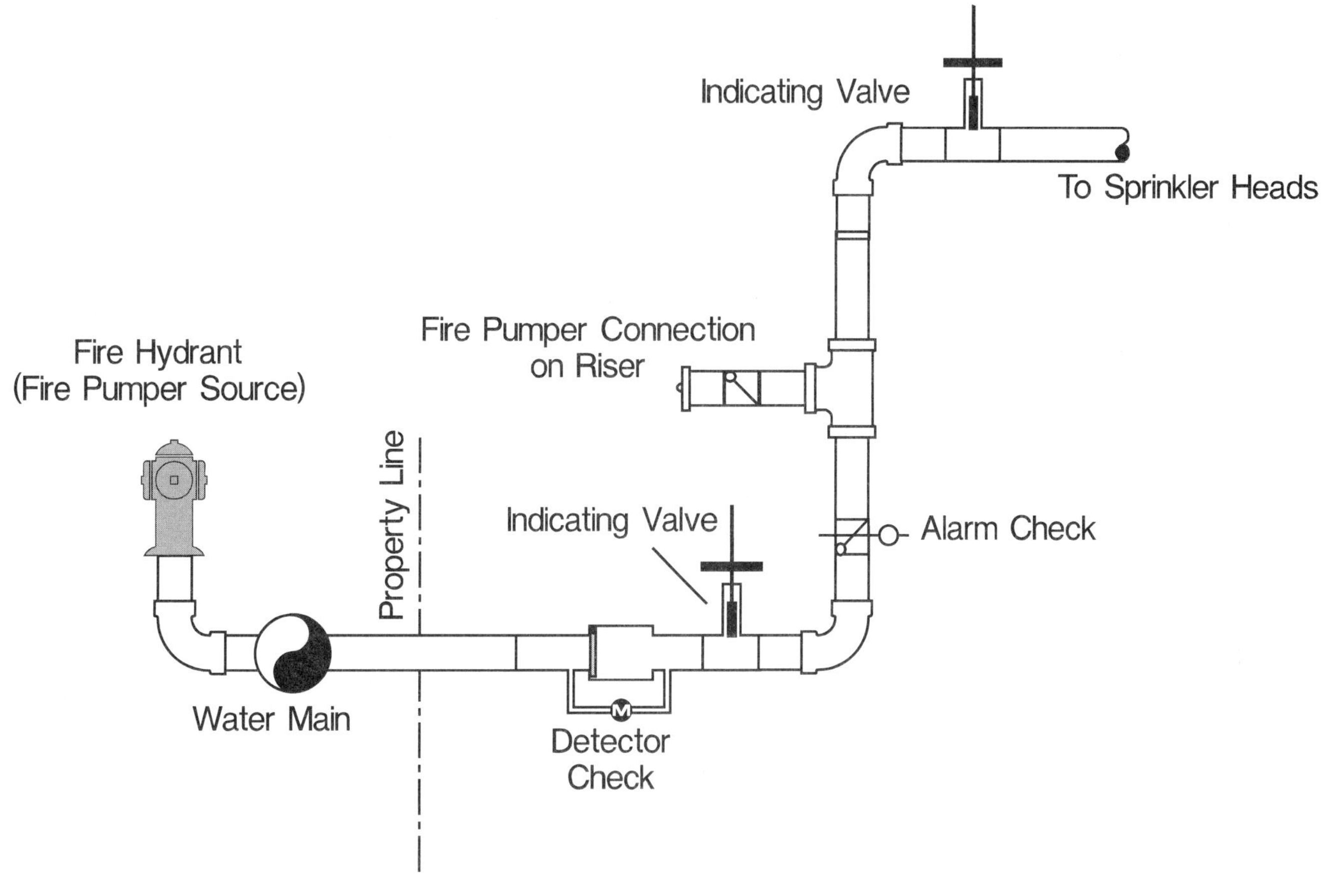

Fig. 7.7
Class 1 Fire Sprinkler System
(Pumper Connection on Riser)

CLASS 2 — same as Class 1, except booster pumps may be installed in the connections from the street mains. (Booster pumps do not affect the potability of the system; it is necessary, however, to avoid drafting so much water that pressure in the water main is reduced below 10 psi.)

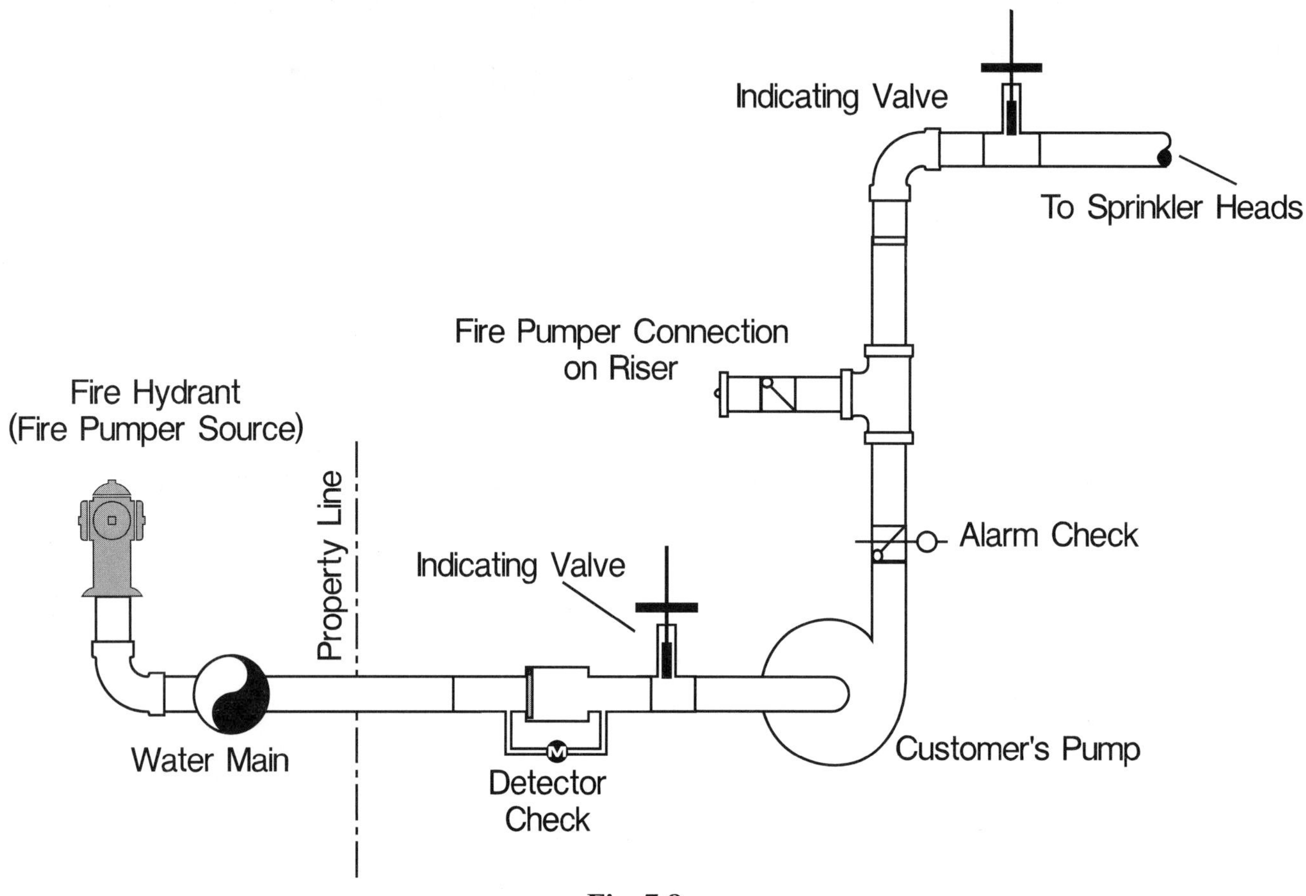

Fig. 7.8
Class 2 Fire Sprinkler System

CLASS 3 — direct connection from public water supply main plus one or more of the following; elevated storage tanks; fire pumps taking suction from above-ground covered reservoirs or tanks; and pressure tanks. (All storage facilities are filled by or connected to public water only, the water in the tanks to be maintained in a potable condition. Otherwise, Class 3 systems are the same as Class 1.)

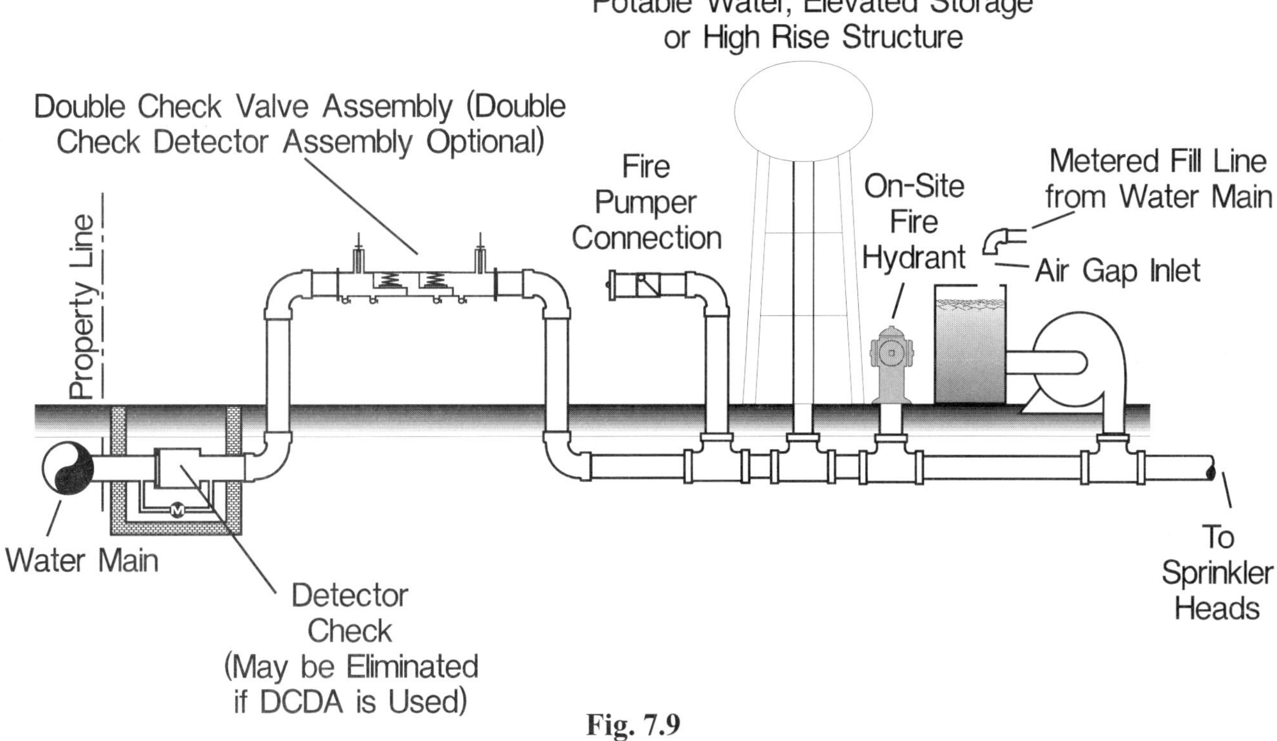

Fig. 7.9
Class 3 Fire Sprinkler System

CLASS 4 — directly supplied from public mains similar to Classes 1 and 2, with an auxiliary water supply on or available to the premises; or an auxiliary supply located within 1,700 feet of the pumper connection.

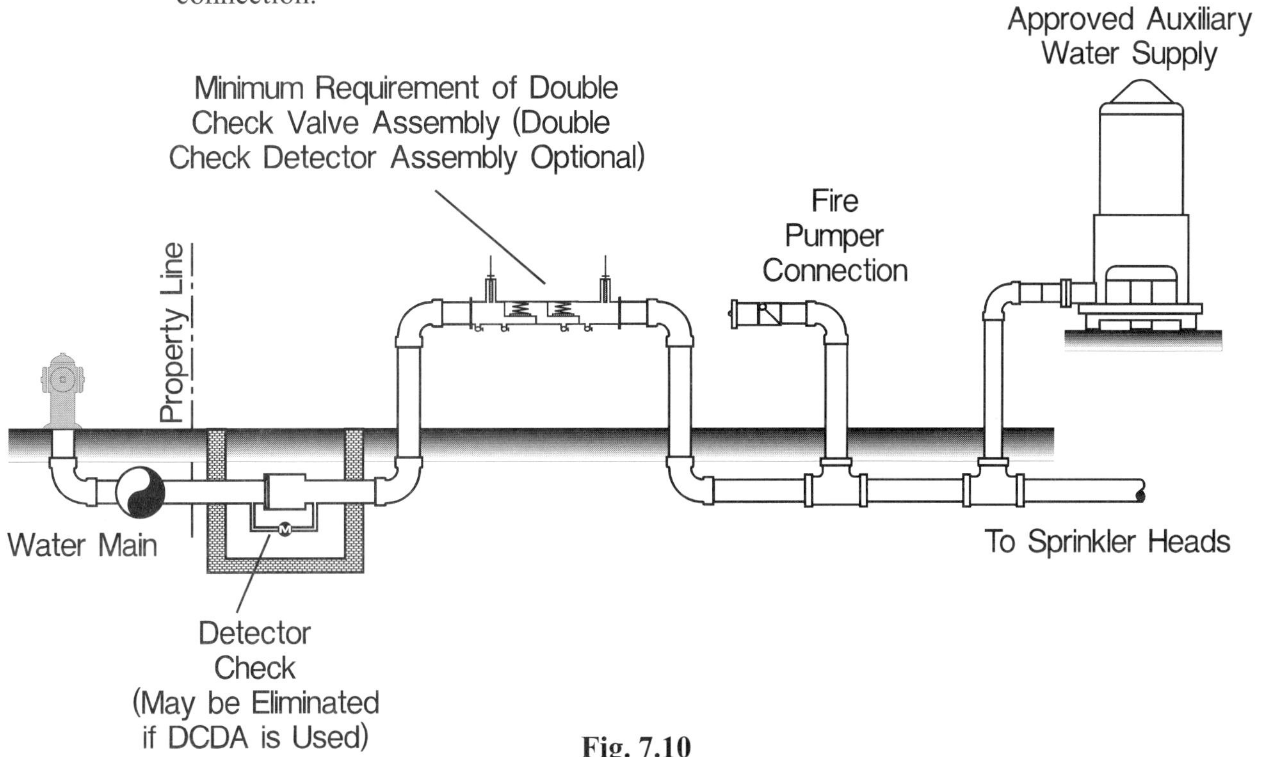

Fig. 7.10
Class 4 Fire Sprinkler System

CLASS 5 — directly supplied from public mains, and interconnected with auxiliary supplies, such as: pumps taking suction from reservoirs exposed to contamination, or rivers and ponds; driven wells; mills or other industrial water systems; or where antifreeze or other additives are used.

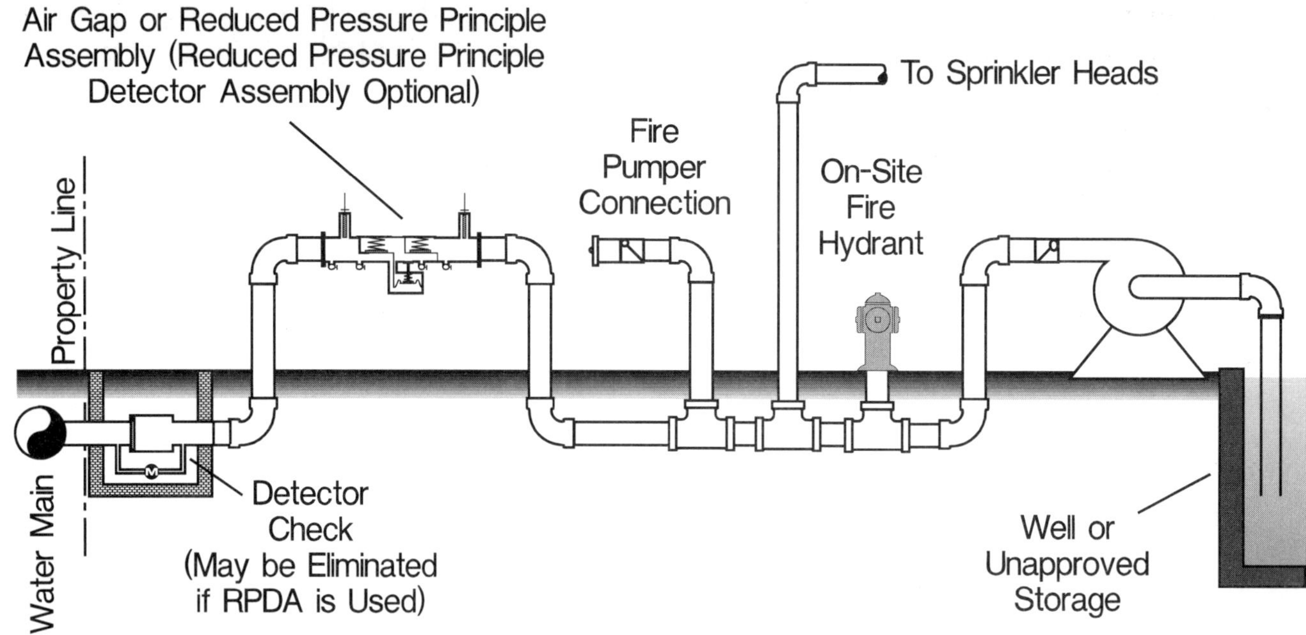

Fig. 7.11
Class 5 Fire Sprinkler System

CLASS 6 — combined industrial and fire protection systems supplied from the public water mains only, with or without gravity storage or pump suction tanks.

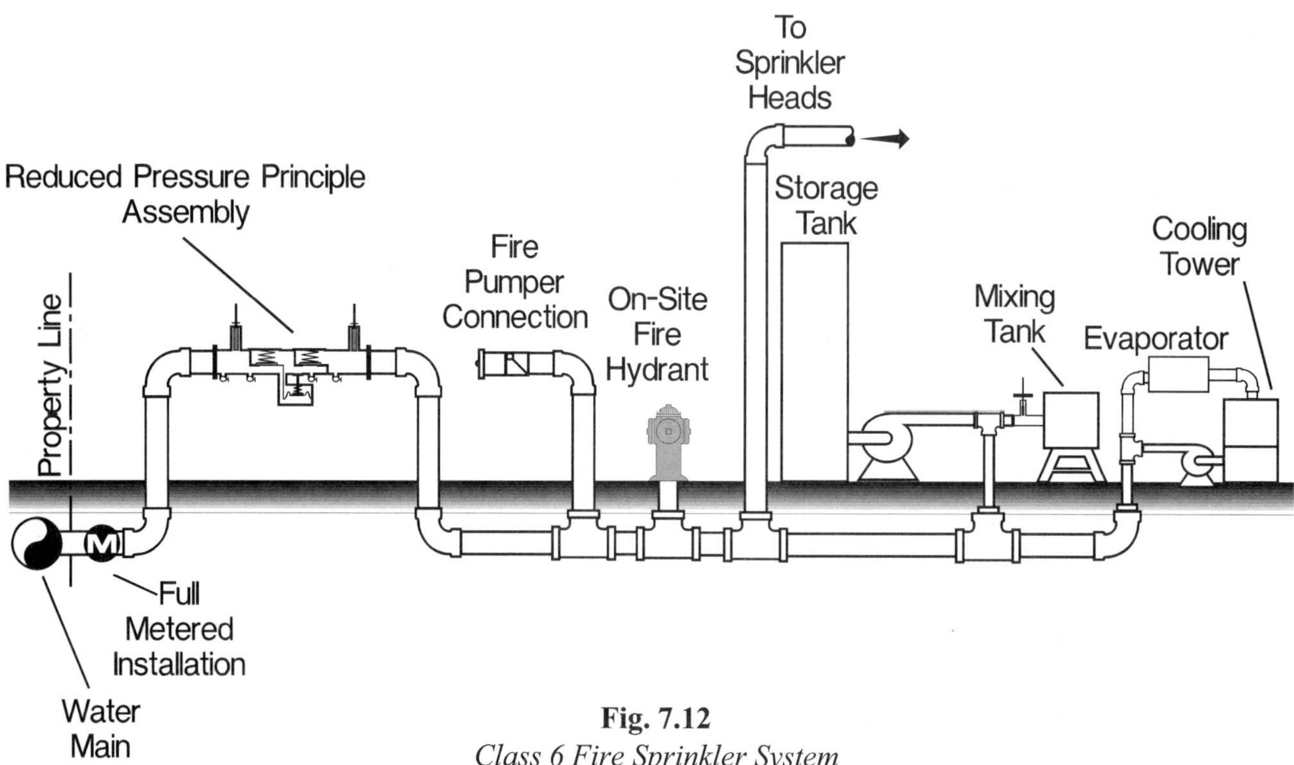

Fig. 7.12
Class 6 Fire Sprinkler System

Class 1 and 2 fire protection systems are those systems which generally and ordinarily would not require an approved backflow protection assembly at the fire system user connection to protect the public water system. However, it is recognized that special conditions (such as Multiple Services — see Section 7.2.3.21) may exist on the site of a class 1 or 2 fire sprinkler system such that an actual or potential contamination hazard is presented to the domestic water supply and under these special conditions an approved backflow prevention assembly, at the user connection for the fire sprinkler system, is warranted. Additionally, the water within fire systems which are not flushed on a regular basis may become polluted or contaminated. The water agency may want to require a water sample be taken from the end of the sprinkler system and analyzed for conformance with EPA maximum contaminant levels (MCL) for a public water system. If any MCL is exceeded, the water agency should request the health agency recommend an appropriate backflow prevention assembly.

Class 3 systems will generally require minimum protection (double check valve assembly) to prevent stagnant waters from backflowing into the public potable water system.

Class 4 systems will normally require backflow protection at the service connection. The type (air gap, reduced pressure principle assembly, or double check valve assembly) will depend on the quality of the auxiliary supply.

Class 4 and 5 systems normally would need maximum protection (air gap or reduced pressure principle assembly) to protect the public potable water system.

Class 6 system protection would depend on the requirements of both industry and fire protection, and could only be determined by a survey of the premises.

A meter (compound, detector check) should not normally be permitted as part of a backflow prevention assembly. An exception may be made, however, if the meter and backflow prevention assembly are specifically designed for that purpose.

Protection Recommended:

(a) An air gap or an approved reduced pressure principle backflow prevention assembly is recommended at the service connection of any fire fighting system where there is available an auxiliary water supply suitable for fire fighting purposes which may or may not be contaminated. This will include any fire fighting system which has a pumper connection with a usable contaminated water supply within 1700 feet of the pumper connection; it will also include any fire fighting system subject to contamination resulting from the introduction of a foaming substance, antifreeze solutions or biological or chemical additives;

(b) An approved double check valve assembly should be installed at the service connection of any fire fighting system where there is available to the premises a stored or auxiliary water supply which is not subject to contamination but may be polluted;

(c) Where a fire fighting system is internally looped and is supplied by more than one service connection it is recommended that each such service connection be protected by at least an approved double check valve backflow prevention assembly;

(d) Where the water purveyor requires the installation of a detector check on the supply to a fire fighting system, the service protection may be accomplished by the installation of an approved double check detector assembly (i.e., DCDA — see Section 4.20) in lieu of a double check valve assembly.

(e) Where the water purveyor requires the installation of a detector check on the supply to a fire fighting system, the service protection may be accomplished by the installation of an approved reduced pressure principle detector assembly (i.e., RPDA — see Section 4.39) in lieu of a reduced pressure principle backflow prevention assembly.

At any time where the fire sprinkler system piping is not an acceptable potable water system material there shall be a backflow prevention assembly isolating the fire sprinkler system from the potable water system. There are also chemicals such as Foamite (a liquid foam concentrate), used for fighting certain types of fires, which are toxic and therefore require maximum protection. Before evaluating a fire sprinkler system a review should be made of "Auxiliary Water Systems" (Section 7.2.3.3) and "Chemically Contaminated Water Systems" (Section 7.2.3.10). Also, see the definitions for "Contamination" (Section 4.15) and "Pollution" (Section 4.34).

(NOTE: Compound meters and detector checks with unapproved check valves in the bypass shall not be used as a component part of an approved double check valve assembly. The check valve in the bypass is ineffective as a backflow preventer and when it is spring or weight loaded, as required in the specifications, the effectiveness of the meter is impaired. Also, from the standpoint of installation, testing and maintenance, it is undesirable for the component parts of a double check valve assembly to be under separate ownership.)

(NOTE: Where backflow protection is required on an industrial-domestic service that is located on the same premises, backflow protection should be provided on the fire service connection. The industrial-domestic system and fire systems should have adequate protection for the highest degree of hazard affecting either system.)

7.2.3.15 Hospitals, Medical Buildings, Sanitariums, Morgues, Mortuaries, Autopsy Facilities, Nursing and Convalescent Homes and Clinics

An approved backflow prevention assembly shall be installed on the service connection to any hospital, medical building, nursing or convalescent home or clinic.

The hazards normally found in a facility of this type include cross-connections between the consumer's water system and: ***contaminated or sewer connected equipment*** such as bedpan washers, flush valve toilets and urinals, autoclaves, specimen tanks, sterilizers, pipet tube washers, cuspidors, aspirators, autopsy and mortuary equipment, etc. (NOTE: It has been found that in this type of facility little or no attention is given to the maintenance of air gaps or atmospheric vacuum breakers. It is customary to bridge an air gap by means of a hose section. Also, in multi-storied buildings the supply line to the toilets, urinals, lavatories, laboratory sinks, etc., on the lower floors may be taken off the suction side of the house pump and, as a result, sewage or other contaminated substances may be drawn into the house supply line.); ***sewer lines*** for the purpose of disposing of filter or softener backwash water from cooling systems or of providing for a quick drain for the building water lines or of flushing or blowing out obstructions, etc. (NOTE: administrative authorities will require backflow protection at the service connection to any premise on which there is located a sewage ejector or pumping station, even though there are no cross-connections); ***water cooled equipment which may be sewer-connected*** such as compressors, heat exchangers, air conditioning equipment, etc.; ***reservoirs, cooling towers and circulating systems*** which may be heavily contaminated with bird droppings, vermin, algae, bacterial slimes or with toxic water treatments compounds such as pentachlorophenol, copper sulfate, chromates, metallic glucosides, compounds of mercury, quaternary ammonium compounds, etc.; ***steam generating facilities and lines*** which may be contaminated with boiler compounds such as those chemicals listed above. (NOTE: A particular hazard is the possibility of steam getting back into the domestic system, causing either a system or health hazard.)

Protection Recommended:

An air gap or a reduced pressure principle backflow prevention assembly on the service connection to any hospital (see Section 4.27), mortuary, morgue, autopsy facility or to any multi-storied medical building or clinic.

7.2.3.16 Irrigation Systems — Premises Having Separate Systems — (i.e., Parks, Playgrounds, Cemeteries, Golf Courses, Estates, Ranches, etc.)

An approved backflow prevention assembly shall be installed on each service connection to premises on which there is an irrigation system.

The hazards normally found on premises of this type include cross-connections between the consumer's water system and: ***irrigation systems which may be equipped with pumps, injectors, pressurized tanks or vessels, or other facilities for injecting or aspirating into the irrigation system agricultural chemicals*** such as fungicides, pesticides, soil conditioners and other similar noxious, toxic or objectionable substances; ***irrigation systems subject to contamination*** from submerged inlets (sprinkler heads), auxiliary water supplies, ponds, reservoirs, swimming pools and other sources of stagnant, polluted or contaminated waters (agricultural chemicals which are broadcast on ground may contaminate water collecting around sprinkler heads); ***a system supplied from an auxiliary water supply.*** (NOTE: The administrative authority will normally require a backflow prevention assembly at the service connection to premises where there is auxiliary supply, even though there are no existing cross-connections. For more specific information regarding an auxiliary water supply and recommended protection, see Section 7.2.3.3 "Auxiliary Water Systems".) See Figs. 7.13, 8.7 and 8.8 for on-site protection methods of irrigation systems.

Protection Recommended:

(a) An air gap or a reduced pressure principle backflow prevention assembly where there is an actual or potential health hazard caused by the installation of facilities for injecting under pressure fertilizers, fungicides, pesticides, soil conditioners and other noxious, toxic or objectionable substances through the irrigation system (i.e., chemigation).

(b) An air gap or a reduced pressure principle backflow prevention assembly where there is an actual or potential health hazard caused by fertilizers, fungicides, pesticides, soil conditioners and other noxious, toxic or objectionable substances with backpressure conditions. No chemigation present.

(c) A pressure vacuum breaker (PVB or SVB) or an atmospheric vacuum breaker backsiphonage prevention assembly may be used for on site protection where there is an actual or potential health hazard caused by fertilizers, fungicides, pesticides, soil conditioners and other noxious, toxic or objectionable substances with no source of backpressure (i.e., backsiphonage potential only). See Figs. 7.13, 8.7, and 8.8

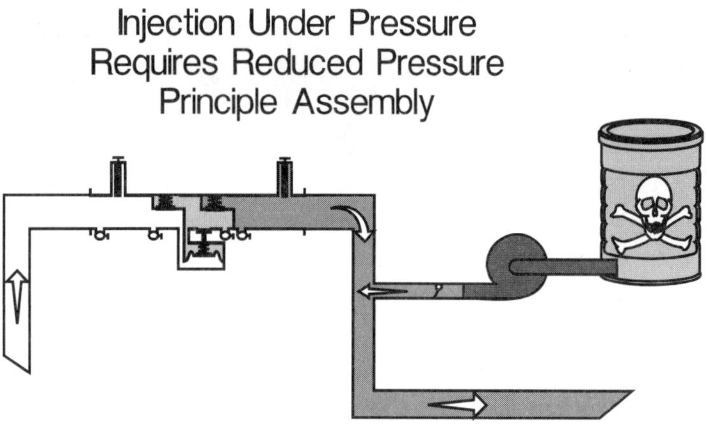

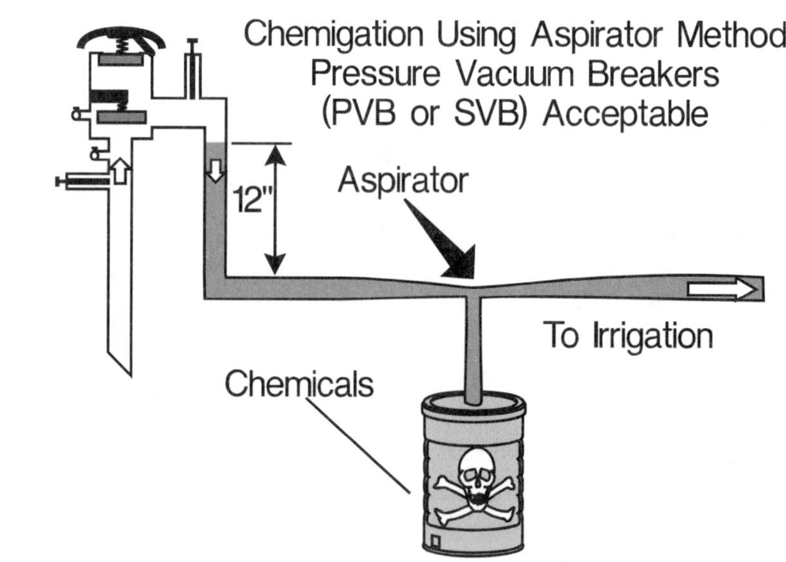

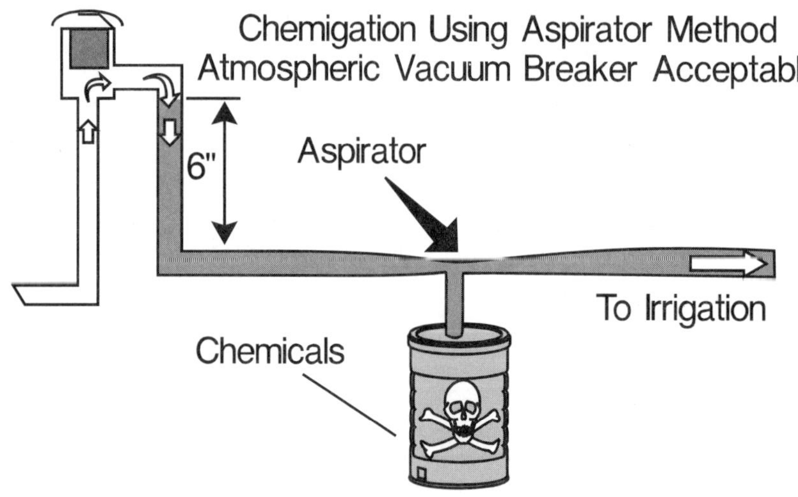

Fig. 7.13
Typical Chemigation Methods

7.2.3.17 Laundries and Dye Works

An approved backflow prevention assembly shall be installed on each service connection to any premise where a laundry or dyeing plant is operated or maintained. A laundry as used herein does not include the self-service laundry or laundromats, except where such laundry equipment constitutes a cross-connection.

The hazards normally found in a plant of this type include cross-connections between the consumer's water system and: ***laundry machines*** having under-rim or bottom inlets; ***dye vats*** in which are used toxic chemicals and dyes; ***water storage tanks*** equipped with pumps and recirculating systems; ***shrinking, bluing and dyeing machines*** with direct connections to circulating systems; ***retention and mixing tanks*** (NOTE: some of these machines or pieces of equipment have pumps which can pump contaminated fluids through cross-connections into the public water supply); ***sewage pumps*** for priming, cleaning, flushing or unclogging purposes; ***water-operated sewage sump ejectors*** for operational purposes; sewer lines for the purpose of disposing of filter or softener backwash water or water from cooling systems or of providing for a quick drain for the building lines or of flushing or blowing out obstructions, etc. (NOTE: administrative authorities require backflow protection at the service connection to any premise on which there is located a sewage ejector or pumping station, even though there are no cross-connections); ***reservoirs, cooling towers and circulating systems*** which may be heavily contaminated with bird droppings, vermin, algae, bacterial slimes or with toxic water treatment compounds such as pentachlorophenol, copper sulfate, chromates, metallic glucosides, compounds of mercury, quaternary ammonium compounds, etc.; ***steam generating facilities and lines*** which may be contaminated with boiler compounds such as those listed above.

Protection Recommended:

An air gap or a reduced pressure principle backflow prevention assembly.

7.2.3.18 Metal Manufacturing, Cleaning, Processing and Fabricating Plants

An approved backflow prevention assembly shall be installed at the service connection to any premise where metals are manufactured, cleaned, processed or fabricated and the process involves used waters and/or industrial fluids. This type of fluid may be transported through separate industrial systems, such as recirculating systems in large facilities.

The hazards normally found in a plant of this type include cross-connections between the consumer's water system and: ***reservoirs, cooling towers and circulating systems*** which may be heavily contaminated with bird droppings, vermin, algae, bacterial slimes or with toxic water treatment compounds such as pentachlorophenol, copper sulfate, chromates, metallic glucosides, compounds of mercury, quaternary ammonium compounds, etc.; ***steam generating facilities and lines*** which may be contaminated with boiler compounds such as those listed above (NOTE: A particular hazard is the possibility of steam getting back into the domestic system, causing either a system or a health hazard); ***industrial fluid systems and lines*** containing cutting and hydraulic fluids, coolants, hydrocarbon products, glycerine, paraffin, caustic and acid solutions, etc.; ***plating facilities*** involving the use of highly toxic cyanides, heavy metals in solution (such as copper, cadmium, chrome, nickel, etc.), acids and caustic solutions; ***plating solution filtering equipment*** with pumps and circulating lines; ***tanks, vats, or other vessels*** used in painting, descaling, anodizing, cleaning, stripping, oxidizing, etching, passivating, pickling, dipping or rinsing operations, or other lines or facilities needed in preparation or finishing of the products; ***water-cooled equipment which may be sewer-connected*** such as compressors, heat exchangers, air conditioning equipment, etc.; ***tanks, can and bottle washing machines and lines*** where caustics, acids, detergents and other compounds are used in cleaning, sterilizing and flushing; ***hydraulically-operated equipment*** where the public potable water pressure is used directly and may be subject to backpressure; ***equipment under hydraulic tests*** such as tanks, lines, valves, fittings and also pumps, pressure cylinders or other hydraulic facilities which may be used to provide pressures for testing purposes. (NOTE: In such cases, air, gas or hydraulic fluids may be forced back to the public system.)

Protection Recommended:

(a) An air gap or a reduced pressure principle backflow prevention assembly where there is an existing or potential health or system hazard;

(b) A double check valve assembly where there is an existing or potential pollutional hazard.

7.2.3.19 Motion Picture Studios

An approved backflow prevention assembly shall be installed at the service connection to any premise where a motion picture production studio is operated or maintained. A motion picture studio as used herein is a facility in which water is or may be used for scene tanks, lagoons, film processing laboratories, special effects, etc.

The hazards normally found in a facility of this type include cross-connections between the consumer's water system and: ***open reservoirs, lagoons, tanks or similar facilities***, used as props in the making of motion pictures (NOTE: These facilities may be heavily contaminated with body wastes, dyes, biological or chemical contaminants used in the prevention of algae and slime growths, and to color the waters for color picture purposes); ***automatic film processing machines, tanks, vats and other facilities*** used in processing film (NOTE: Toxic chemicals such as acetic acid, potassium ferricyanide and different types of aromatic or organic chemicals may be used in these facilities); ***special effects equipment*** in which chemicals and other materials may be injected into the water supply for special effects; ***sewage pumps*** for priming, cleaning, flushing or unclogging purposes; ***water-operated sewage sump ejectors*** for operational purposes; ***sewer lines*** for the purpose of disposing of filter or softener backwash water or water from cooling systems or of providing for a quick drain for the building lines or of flushing or blowing out obstructions, etc. (NOTE: Administrative authorities require backflow protection at the service connection to any premise on which there is located a sewage ejector or pumping station, even though there are no cross-connections); ***reservoirs, cooling towers and circulating systems*** which may be heavily contaminated with bird droppings, vermin, algae, bacterial slimes or with toxic water treatment compounds such as pentachlorophenol, copper sulfate, chromates, metallic glucosides, compounds of mercury, quaternary ammonium compounds, etc.; ***steam generating facilities and lines*** which may be contaminated with boiler compounds such as those listed above (NOTE: a particular hazard is the possibility of steam getting back into the domestic system, causing either a system or a health hazard); ***fire fighting systems***, including storage reservoirs which may be treated for scale formation, corrosion, algae, slime growths, etc., or fire systems which may be subject to contamination with antifreeze solutions, Foamite or other chemicals or chemical compounds used in fire fighting, or fire systems which are subject to contamination with auxiliary or used water supplies or industrial fluids; ***water-cooled equipment which may be sewer-connected*** such as compressors, heat exchangers, air conditioning equipment, etc.; ***irrigation systems which may be equipped with pumps, injectors, pressurized tanks or vessels, or other facilities for injecting into the irrigation system agricultural chemicals*** such as fungicides, pesticides, soil conditioners and other similar noxious, toxic or objectionable substances; ***irrigation systems subject to contamination*** from submerged inlets, auxiliary water supplies, ponds, reservoirs, swimming pools and other sources of stagnant, polluted or contaminated waters; ***a system supplied from an auxiliary water supply*** (NOTE: The administrative authority normally will require a backflow prevention assembly at the service connection to premises where there is an auxiliary water supply, even though there are no existing cross-connections. For more specific information regarding auxiliary water supply and recommended protection, see Section 7.2.3.3 "Auxiliary Water Systems".)

Protection Recommended:

An air gap or a reduced pressure principle backflow prevention assembly.

7.2.3.20 Multi-storied Buildings

Multi-storied buildings may be broadly grouped into three categories in terms of their internal potable water systems:

(a) Using only the service pressure to distribute the potable water throughout the structure; and, with no internal potable water reservoir;

(b) Using a booster pump to provide potable water directly to the upper floors; and,

(c) Using a booster pump to fill a covered roof reservoir from which there is a down-feed system for the upper floors.

In each of these systems it is probable that there is one or more takeoffs for industrial water within the building. And, any loss of distribution main pressure will cause backflow from these building systems unless approved backflow prevention assemblies are properly installed. Additional details are given in Section 7.2.3.6 "Buildings — Hotels, Apartment Houses, Public and Private Buildings, or any other Structure Having Unprotected Cross-Connections:" Section 7.2.3.15 "Hospitals, Medical Buildings, Sanitariums, Morgues, Mortuaries, Autopsy Facilities, Nursing and Convalescent Homes and Clinics;" and Section 7.2.3.30 "Schools and Colleges." See Fig. 7.14.

Protection Recommended:

For the categories listed above:

(a) An approved double check valve assembly where there is only a pollutional hazard; or, an approved reduced pressure principle backflow assembly where there is the potential of a contamination hazard;

(b) An air gap or an approved reduced pressure principle backflow prevention assembly where there are sewage pumping facilities or other potential health hazard conditions within the premises;

(c) An air gap or a reduced pressure principle backflow prevention assembly if there exists takeoffs on the suction side of the booster pump(s) for any sanitary facilities on the lower floors. Otherwise, an approved double check valve assembly may be used.

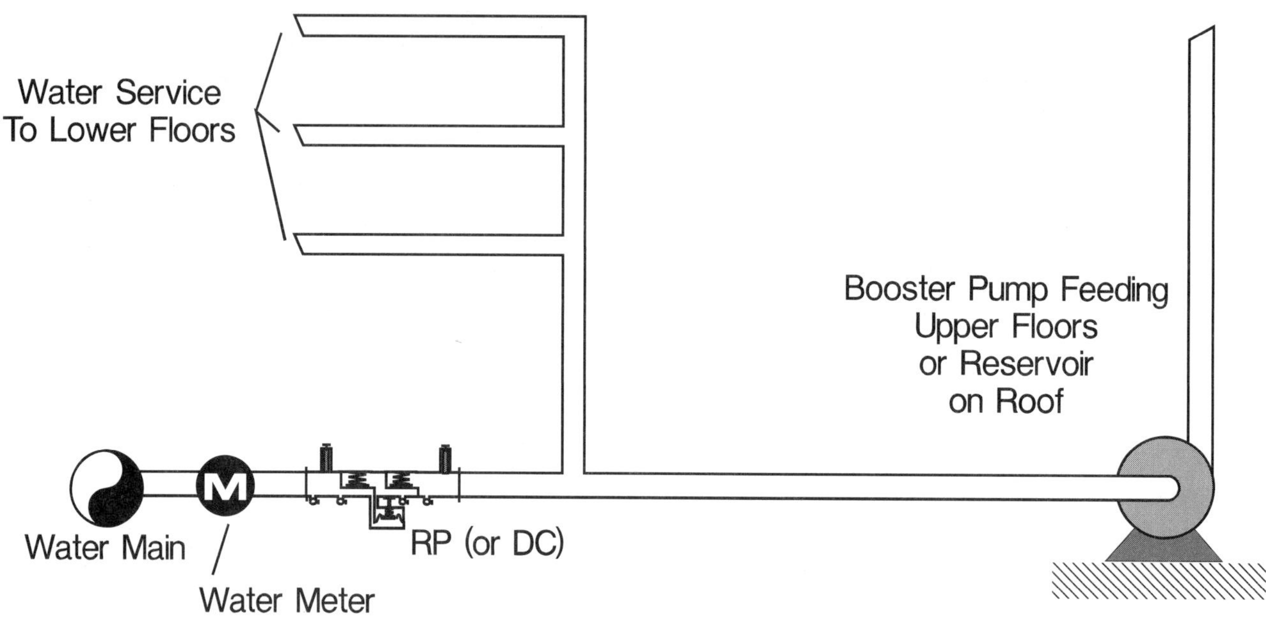

Fig. 7.14
Multi-Storied Buildings

7.2.3.21 Multiple Services — Interconnected

The internal looping of two or more services from a single water agency to a single consumer complex provides a flow-through condition due to the unequal service main pressures. The hazards found within this type of complex range from pollutional to contaminants and must be individually isolated. In the case that two or more water agencies are involved, the multiple service connections constitute an auxiliary source of water on the property in the eyes of each water agency. Hence, even without known or potential hazards within the property each service must be protected by a minimum of a double check valve assembly. See Figs. 7.15, 7.16, and 7.17.

Protection Recommended:

(a) An air gap or an approved reduced pressure principle backflow prevention assembly where contamination hazards are found or potentially exist; or,

(b) An approved double check valve backflow prevention assembly where there exists the potential of a pollutional hazard or even where all internal uses are considered to be "domestic."

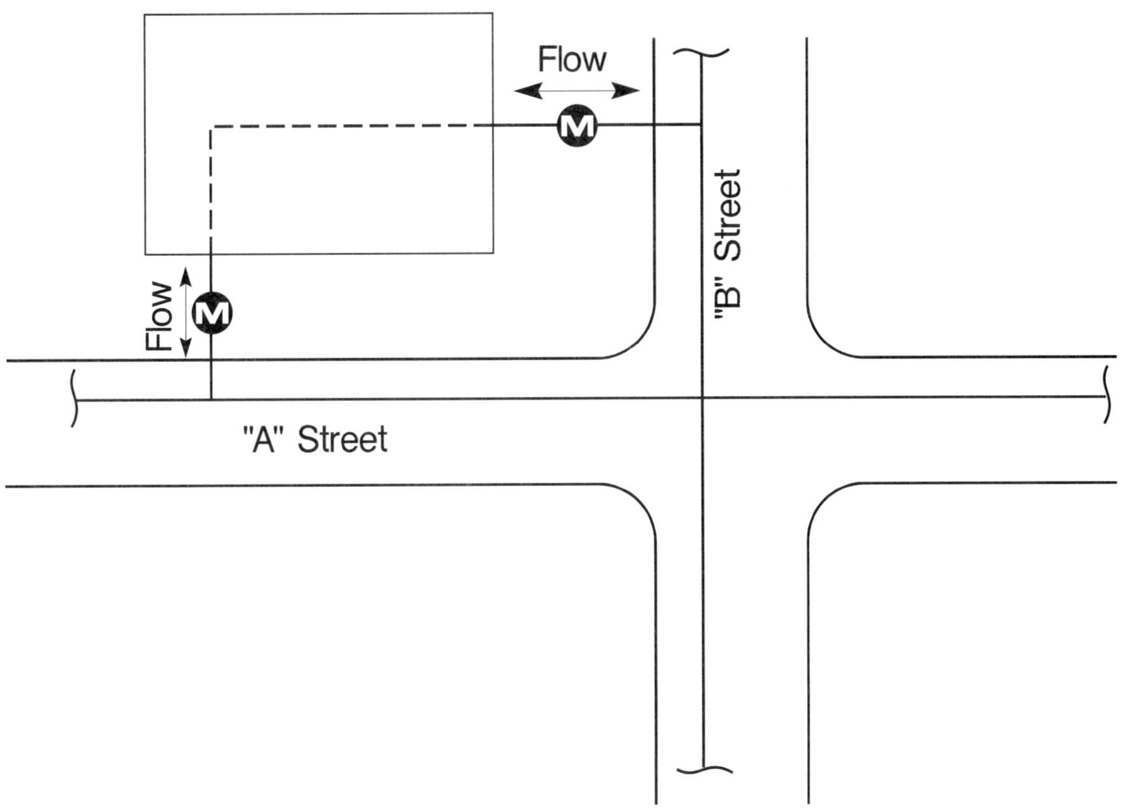

Fig. 7.15
Water Meters on Different Mains and Streets

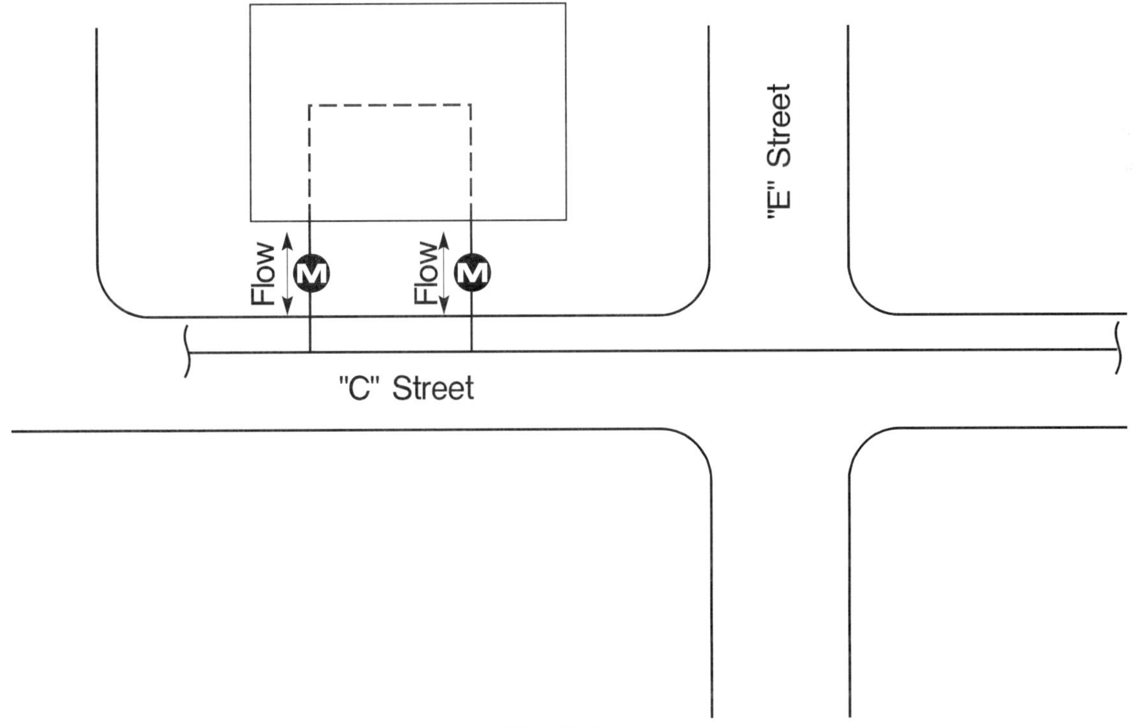

Fig. 7.16
Water Meters on Same Street a Given Distance Apart

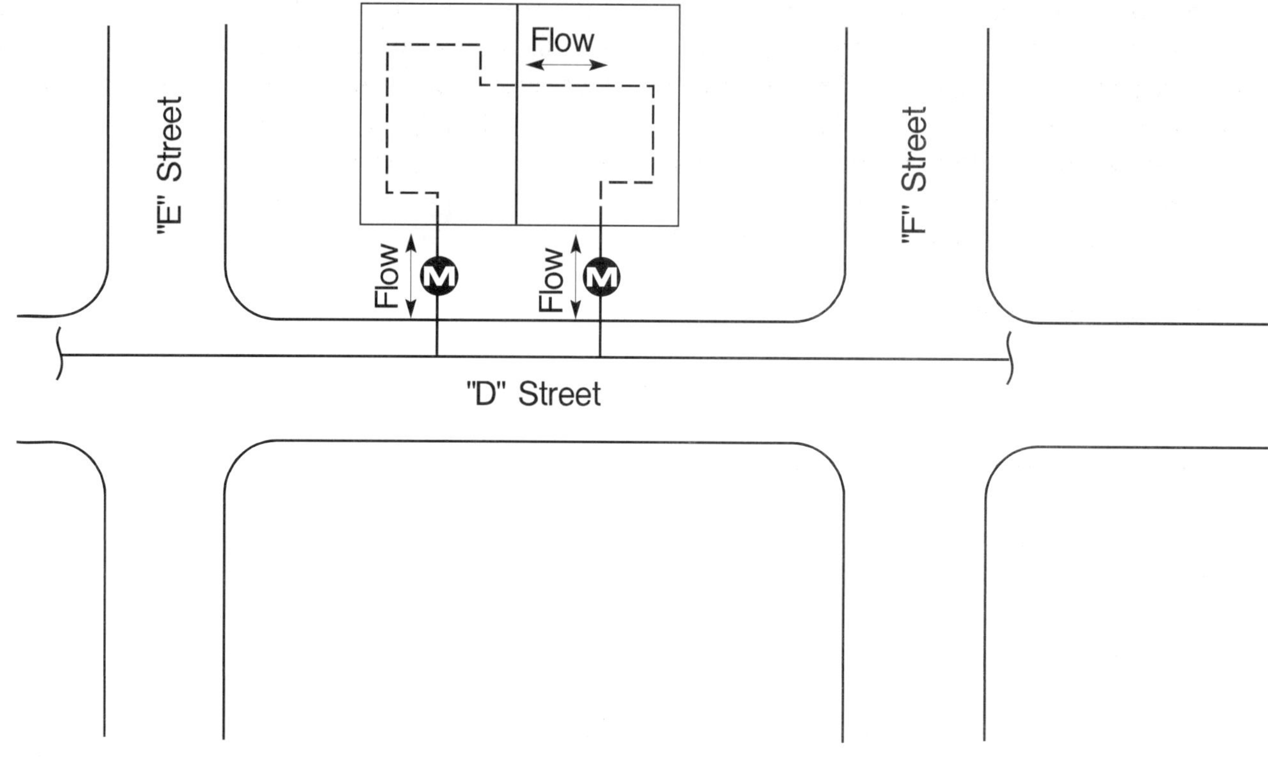

Fig. 7.17
Water Meters Adjacent or Installed Within "Short" Distance

7.2.3.22 Oil and Gas Production, Storage or Transmission Properties

An approved backflow prevention assembly shall be installed at the service connection to any premise where animal, vegetable or mineral oils and gases are produced, developed, processed, blended, stored, refined or transmitted in a pipeline, or where oil or gas tanks are maintained; or where an oxygen, acetylene, petroleum or other natural or manufactured gas production or bottling plant is operated or maintained. Such premises should also include locations where oil or gas tanks, bottling or other storage or pressure vessels are repaired, tested or maintained; or premises having dehydration or refinery facilities; or where the water service is used for slugging oil or gases through transmission lines; or where the water service is used for testing or purging oil and gas tanks or oil and gas pipelines and other like uses.

The hazards normally found in a plant of this type include cross-connections between the consumer's water system and: ***steam boilers and lines; mud pumps and mud tanks; hydraulically-operated Tretolite tanks; oil well casings*** (for dampening gas pressures); ***dehydration tanks and outlet lines from storage and dehydration tanks*** (for purging purposes); ***oil and gas lines*** (for testing, evacuating and slugging purposes); ***reservoirs cooling towers and circulating systems*** which may be heavily contaminated with bird droppings, vermin, algae, bacterial slimes or with toxic water treatment compounds such as pentachlorophenol, copper sulfate, chromates, metallic glucosides, compounds of mercury, quaternary ammonium compounds, etc.; ***steam generating facilities and lines*** which may be contaminated with boiler compounds such as those listed above (NOTE: a particular hazard is the possibility of steam getting back into the domestic system, causing either a system or a health hazard); ***industrial fluid systems and lines*** containing cutting and hydraulic fluids, coolants, hydrocarbon products, glycerine, paraffin, caustic and acid solutions, etc.; ***fire fighting systems***, including storage reservoirs which may be treated for scale formation, corrosion, algae, slime growths, etc., or fire systems which may be subject to contamination with antifreeze solutions, Foamite or other chemicals or chemical compounds used in fire fighting, or fire systems which are subject to contamination with auxiliary or used water supplies or industrial fluids; ***water-cooled equipment which may be sewer-connected*** such as compressors, heat exchangers, air conditioning equipment, etc.; ***hydraulically-operated equipment*** where the public potable water pressure is used directly and may be subject to backpressure; ***equipment under hydraulic tests*** such as tanks, lines, valves, fittings and also pumps, pressure cylinders or other hydraulic facilities which may be used to provide pressures for testing purposes. (NOTE: In such cases, air, gas or hydraulic fluids may be forced back to the public system.)

Protection Recommended:

An air gap or a reduced pressure principle backflow prevention assembly.

7.2.3.23 Paper and Paper Products Plants

An approved backflow prevention assembly shall be installed at the service connection to any premise where a paper or paper products plant (wet process) is operated or maintained. Paper or paper products plants as used herein means those plants where used waters and industrial fluids and chemicals are used in the manufacturing process.

The hazards normally found in a plant of this type include cross-connections between the consumer's water system and: ***pulp, bleaching, dyeing and processing facilities*** which may be contaminated with toxic chemicals; ***reservoirs, cooling towers and circulating systems*** which may be heavily contaminated with bird droppings, vermin, algae, bacterial slimes or with toxic water treatment compounds such as pentachlorophenol, copper sulfate, chromates, metallic glucosides, compounds of mercury, quaternary ammonium compounds, etc.; ***steam generating facilities and lines*** which may be contaminated with boiler compounds such as those listed above (NOTE: a particular hazard is the possibility of steam getting back into the domestic system, causing either a system or a health hazard); ***industrial fluid systems and lines*** containing cutting and hydraulic fluids, coolants, hydrocarbon products, glycerine, paraffin, caustic and acid solutions, etc.; auxiliary or used water supplies or industrial fluids; ***a system supplied from an auxiliary water supply*** (NOTE: The administrative authority normally will require a backflow prevention assembly at the service connection to premises where there is an auxiliary water supply, even though there are no existing cross-connections. For more specific information regarding auxiliary water supply and recommended protection, see Section 7.2.3.3 "Auxiliary Water Systems"); ***water-cooled equipment which may be sewer-connected*** such as compressors, heat exchangers, air conditioning equipment, etc.; ***fire fighting systems***, including storage reservoirs which may be treated for scale formation, corrosion, algae, slime growths, etc., or fire systems which may be subject to contamination with antifreeze solutions, Foamite or other chemicals or chemical compounds used in fire fighting, or fire systems which are subject to contamination with auxiliary or used water supplies or industrial fluids.

Protection Recommended:

An air gap or a reduced pressure principle backflow prevention assembly.

7.2.3.24 Plating Plants

An approved backflow prevention assembly shall be installed at the service connection to any premise where there is a mechanical, chemical or electrochemical plating or processing plant. This plating plant or facility may be operated or maintained either as a separate function or in conjunction with a manufacturing plant or other facility, such as an aircraft or automotive manufacturing plant. Plating as used herein includes such operations as chromium, cadmium, or other plating, galvanizing, anodizing, cleaning, stripping, oxidizing, etching, passivating or pickling, etc.

The hazards normally found in a plant of this type include cross-connections between the consumer's water system and: ***plating facilities*** involving the use of highly toxic cyanides, heavy metals in solution (such as copper, cadmium, chrome, nickel, etc.), acids and caustic solutions; ***plating solution filtering equipment*** with pumps and circulating lines; ***tanks, vats, or other vessels*** used in painting, descaling, anodizing, cleaning, stripping, oxidizing, etching, passivating, pickling, dipping or rinsing operations, or other lines or facilities needed in preparation or finishing of the products; ***steam generating facilities and lines*** which may be contaminated with boiler compounds such as those listed above (NOTE: a particular hazard is the possibility of steam getting back into the domestic system, causing either a system or a health hazard); ***water-cooled equipment which may be sewer-connected*** such as compressors, heat exchangers, air conditioning equipment, etc.

Protection Recommended:

An air gap or a reduced pressure principle backflow prevention assembly.

7.2.3.25 Power Plants

An approved backflow prevention assembly shall be installed at the service connection to any premise where a heating, ventilating, refrigerating or commercial power plant is maintained and the used water is or may be subject to chemical or bacterial deterioration or contamination. Power plants, as used herein, does not include the small heating or pressing units or low pressure closed heating systems but does include the large and complex heating, refrigerating and power plants used in large buildings and commercial or industrial plants.

The hazards normally found in a plant of this type include cross-connections between the consumer's water system and: ***sewage pumps*** for priming, cleaning, flushing or unclogging purposes; ***water-operated sewage sump ejectors*** for operational purposes; ***sewer lines*** for the purpose of disposing of filter or softener backwash water or water from cooling systems or of providing for a quick drain for the building lines or of flushing or blowing out obstructions, etc. (NOTE: administrative authorities require backflow protection at the service connection to any premise on which there is located a sewage ejector or pumping station, even though there are no cross-connections); ***reservoirs, cooling towers and circulating systems*** which may be heavily contaminated with bird droppings, vermin, algae, bacterial slimes or with toxic water treatment compounds such as pentachlorophenol, copper sulfate, chromates, metallic glucosides, compounds of mercury, quaternary ammonium compounds, etc.; ***steam generating facilities and lines*** which may be contaminated with boiler compounds such as those listed above (NOTE: a particular hazard is the possibility of steam getting back into the domestic system, causing either a system or a health hazard); ***fire fighting systems***, including storage reservoirs which may be treated for scale formation, corrosion, algae, slime growths, etc., or fire systems which may be subject to contamination with antifreeze solutions, Foamite or other chemicals or chemical compounds used in fire fighting, or fire systems which are subject to contamination with auxiliary or used water supplies or ***industrial fluids; industrial fluid systems and lines*** containing cutting and hydraulic fluids, coolants, hydrocarbon products, glycerine, paraffin, caustic and acid solutions, etc.; ***water-cooled equipment which may be sewer-connected*** such as compressors, heat exchangers, air conditioning equipment, etc.; ***hydraulically-operated equipment*** where the public potable water pressure is used directly and may be subject to backpressure; ***equipment under hydraulic tests*** such as tanks, lines, valves, fittings and also pumps, pressure cylinders or other hydraulic facilities which may be used to provide pressures for testing purposes (NOTE: in such cases, air, gas or hydraulic fluids may be forced back to the public system); ***a system supplied from an auxiliary water supply.*** (NOTE: The administrative authority normally will require a backflow prevention assembly at the service connection to the premises where there is an auxiliary water supply, even though there are no existing cross-connections. For more specific information regarding auxiliary water supply and recommended protection, see Section 7.2.3.3 "Auxiliary Water Systems".)

Protection Recommended:

An air gap or a reduced pressure principle backflow prevention assembly.

7.2.3.26 Radioactive Materials or Substances — Plants or Facilities Handling

An approved backflow prevention assembly shall be installed at the service connection to any premise where radioactive materials or substances are processed in a laboratory or plant where they may be handled in such a manner as to create a potential hazard to the water system, or where there is a reactor plant.

Protection Recommended:

An air gap or a reduced pressure principle backflow prevention assembly.

7.2.3.27 Restricted, Classified or Other Closed Facilities

An approved backflow prevention assembly shall be installed on the service connection to any facility that is not readily accessible for inspection by the water purveyor because of military secrecy requirements or other prohibitions or restrictions.

Protection Recommended:

An air gap or a reduced pressure principle backflow prevention assembly.

7.2.3.28 Rubber Plants - Natural or Synthetic

An approved backflow prevention assembly shall be installed at the service connection to any premise where natural or synthetic rubber or rubber goods or rubber tires are manufactured. This class does not include the small retreading plants, normally found in connection with garages, or small molding plants.

The hazards normally found in a plant of this type include cross-connections between the consumer's water system and: ***reservoirs, cooling towers and circulating systems*** which may be heavily contaminated with bird droppings, vermin, algae, bacterial slimes or with toxic water treatment compounds such as pentachlorophenol, copper sulfate, chromates, metallic glucosides, compounds of mercury, quaternary ammonium compounds, etc.; ***steam generating facilities and lines*** which may be contaminated with boiler compounds such as those listed above (NOTE: a particular hazard is the possibility of steam getting back into the domestic system, causing either a system or a health hazard); ***industrial fluid systems and lines*** containing cutting and hydraulic fluids, coolants, hydrocarbon products, glycerine, paraffin, caustic and acid solutions, etc.; ***fire fighting systems,*** including storage reservoirs which may be treated for scale formation, corrosion, algae, slime growths, etc. or fire systems which may be subject to contamination with antifreeze solutions, Foamite or other chemicals or chemical compounds used in fire fighting, or fire systems which are subject to contamination with auxiliary or used water supplies or industrial fluids; ***water-cooled equipment which may be sewer-connected*** such as compressors, heat exchangers, air conditioning equipment, etc.; ***hydraulically-operated equipment*** where the public potable water pressure is used directly and may be subject to backpressure; ***equipment under hydraulic tests*** such as tanks, lines, valves, fittings and pumps, pressure cylinders or other hydraulic facilities which may be used to provide pressure for testing purposes (NOTE: in such cases, air, gas or hydraulic fluids may be forced back in to the public system); ***a system supplied from an auxiliary water supply.*** (NOTE: The administrative authority normally will require a backflow prevention assembly at the service connection to the premises where there is an auxiliary water supply, even though there are no existing cross-connections. For more specific information regarding auxiliary water supply and recommended protection, see section 7.2.3.3 "Auxiliary Water Systems".)

Protection recommended:

An air gap or a reduced pressure principle backflow prevention assembly.

7.2.3.29 Sand and Gravel Plants

An approved backflow prevention assembly shall be installed at the service connection to any premise where a sand or gravel pit is operated or maintained, or where sand and gravel are classified and processed and the water is used industrially, i.e., premises where water conservation is practiced, or where there is an auxiliary water supply which is or may be used as an alternate supply to the public water system.

The hazards normally found in a plant of this type include cross-connections between the consumer's water system and : ***sand and gravel washing equipment*** supplied with a private with a private well or water pumped from sumps or retention basins on a recirculating basis. These cross-connections may be under pump pressure, in many cases considerably higher than the pressures in the public main; ***a system supplied from an auxiliary water supply*** (NOTE: The administrative authority normally will require a backflow prevention assembly at the service connection to the premises where there is an auxiliary water supply, even though there are no existing cross-connections. Form more specific information regarding auxiliary water supply and recommended protection, see Section 7.2.3.3 "Auxiliary Water Systems".)

Protection Recommended:

An air gap or a reduced pressure principle backflow prevention assembly.

7.2.3.30 Schools and Colleges

An approved backflow prevention assembly shall be installed at the service connection to any premise on which there is a high school, trade school, college or university where the water is used to supply chemical, bacteriological and biological laboratories; or where the water is used to supply separate irrigation systems; or where there are unprotected sewer cross-connections.

The hazards normally found in a plant of this type include cross-connections between the consumer's water system and: ***contaminated and/or sewer connected facilities*** such as inadequately protected flush valve toilets, urinals, aspirators, retorts, pipet tube washers, ***sterilizers, autoclaves, specimen tanks, autopsy and morgue equipment; sewer connected plumbing fixtures*** such as flush valve toilets and urinals without atmospheric vacuum breakers or with improperly maintained atmospheric vacuum breakers (NOTE: this hazard is critical because little or no attention is given to the maintenance of atmospheric vacuum breakers and frequently they are removed from the line); ***water which may be sewer-connected*** such as compressors, heat exchangers, air conditioning equipment, etc.; ***irrigation systems which may be equipped***

with pumps, injectors, pressurized tanks or vessels, or other facilities for injecting into the irrigation system agricultural chemicals such as fungicides, pesticides, soil conditioners and other similar noxious, toxic or objectionable substance; ***irrigation systems subject to contamination*** from submerged inlets, auxiliary water supplies, ponds, reservoirs, swimming pools and other sources of stagnant, polluted or contaminated waters; ***tanks, automatic film processing machines or facilities*** used in processing films, which may be contaminated with chemicals such as acetic acid, potassium ferricyanide and/or one of the many different types of the aromatic series of organic chemicals; ***a system supplied from an auxiliary water supply*** (NOTE: The administrative authority normally will require a backflow prevention assembly at the service connection to the premises where there is an auxiliary water supply, even though there are no existing cross-connections. For more specific information regarding auxiliary water supply and recommended protection, see section 7.2.3.3 "Auxiliary Water Systems".)

Protection Recommended:

An air gap or a reduced pressure principle backflow prevention assembly.

7.2.3.31 Sewage and Storm Drain Facilities, Reclaimed Water Systems

An approved backflow prevention assembly shall be installed at the service connection to any premise where the is sewer or storm drain water treatment, processing, ejector or pumping plant, gaging station or other sewage or storm water handling facility. (NOTE: This will include services to privately-owned sewage pumping or sewage ejector facilities in public or private buildings, as well as all types of water connections such as irrigation, commercial industrial and domestic temporary or regular water services, and temporary services from fire hydrants, standpipes, etc.)

The hazards normally found in a plant of this type include cross-connections between the consumer's water system and: ***sewage pumps*** for priming, cleaning, flushing or unclogging purposes; ***water-operated sewage sump ejectors*** for operation purposes; ***sewer lines*** for the purpose of disposing of filter or softener backwash water or water from cooling systems or of providing for a quick drain for the building lines or of flushing or blowing out obstructions, etc. (NOTE Administrative authorities require backflow protection at the service connection to any premise on which there is located a sewage ejector or pumping station, even though there are no cross-connections.)

A premise which has a reclaimed water system is to be treated as a premise which has a sewage supply.

Protection Recommended:

a. An air gap or a reduced pressure principle backflow prevention assembly.

b. An air gap installation must be made at the user connection where there is reclaimed water being used. On premises where reclaimed water is used and there is no interconnection with the potable water system, a reduced pressure principle backflow prevention assembly may be provided in lieu of an air gap if approved. (See Fig. 7.18.)

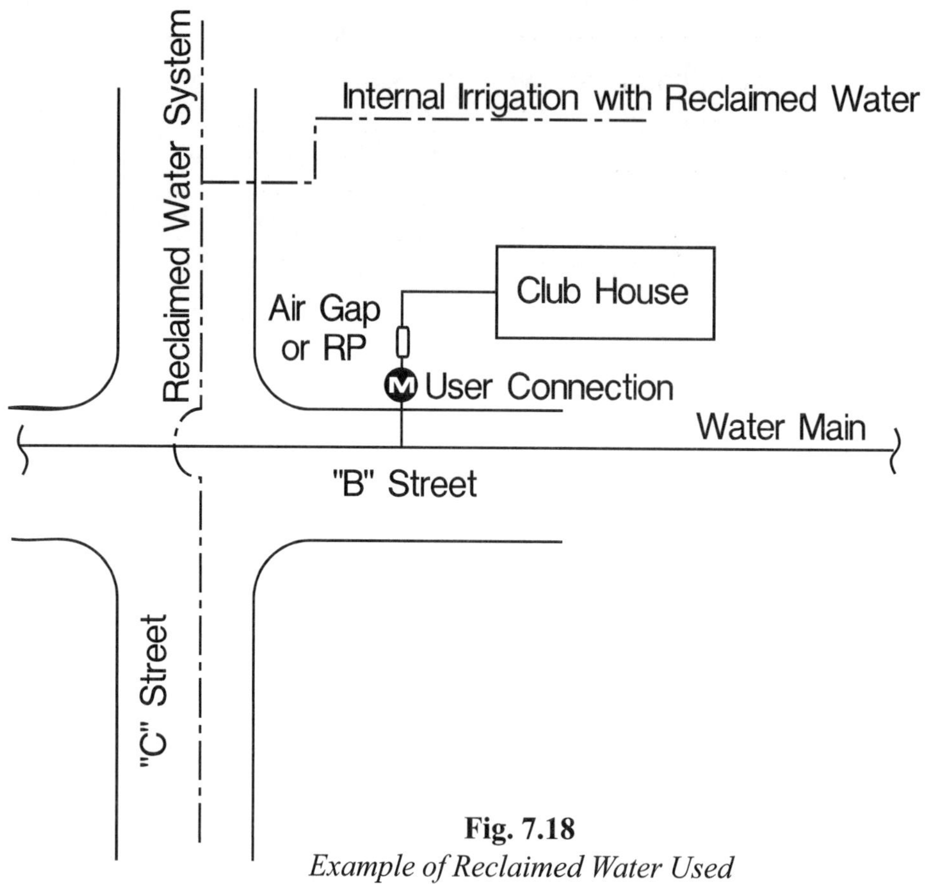

Fig. 7.18
Example of Reclaimed Water Used

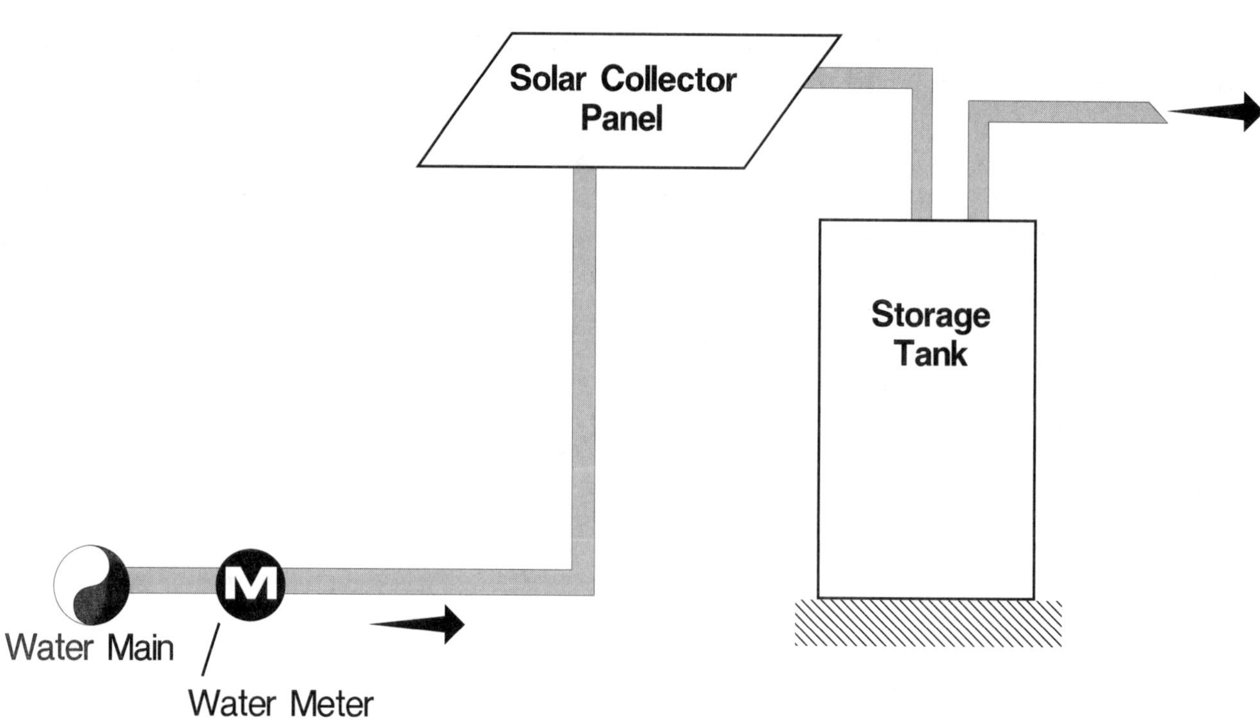

Fig. 7.19
Solar Heating System
Once Through Type

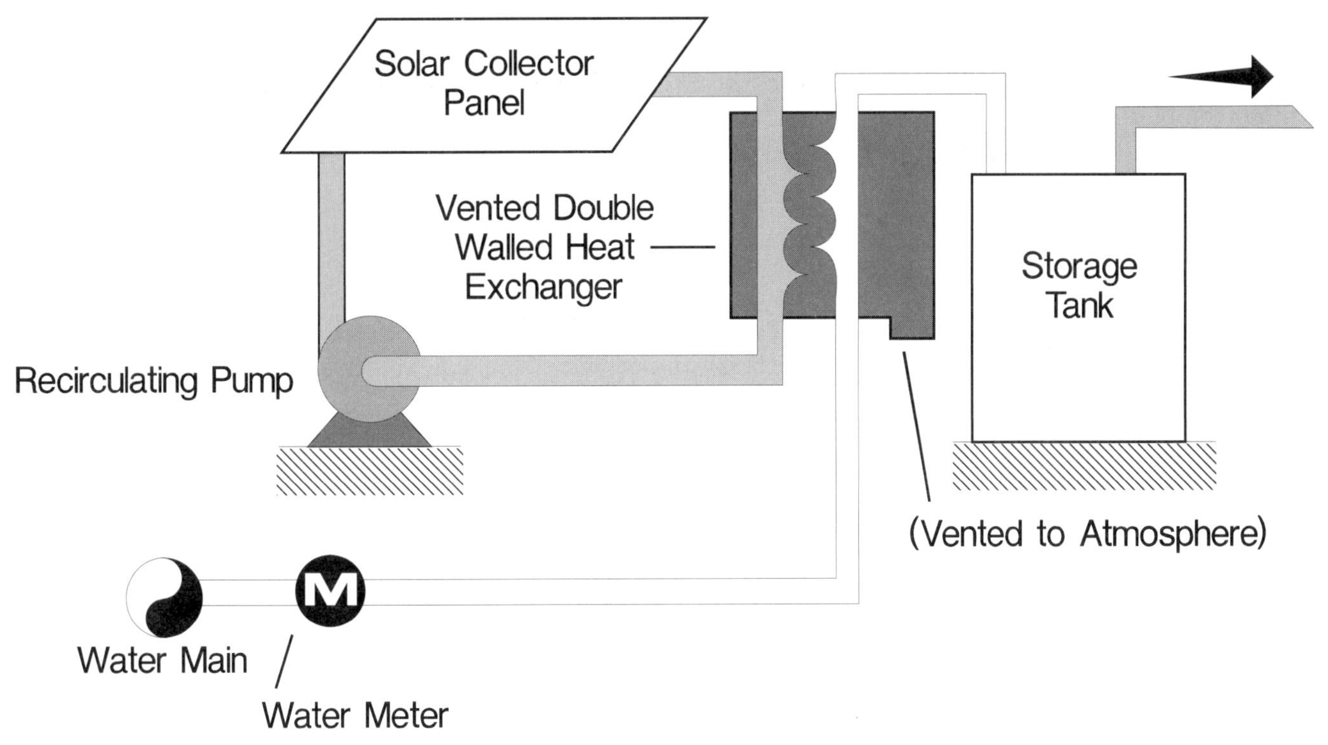

Fig. 7.20
Solar Heating System
Heat Exchanger Type

7.2.3.32 Solar Heating Systems - Direct and Auxiliary

An approved backflow prevention assembly shall be installed at the service connection to any premise having a solar heating and/or cooling system that is connected to the consumer's water system.

The hazards normally found in solar heating and/or cooling systems include cross connections between the consumer's water system and: reservoirs and/or solar collector fluids which may have antifreeze, toxic corrosion inhibitors, non-potable water, etc.; single wall heat exchangers between the consumer's water or fluids; negative pressure zones created by circulation pumps.

Some solar hot water heating systems rely on nontoxic antifreeze solution and/or nontoxic corrosion inhibitors. However, the water purveyor has no assurance that these systems will not be subsequently altered to utilize a toxic fluid. (NOTE: On swimming pool solar heating systems using the pool water as a collector fluid, a separate backflow prevention assembly is not required if an approved air gap is maintained on the pool fill line.) See Figs. 7.19 and 7.20.

Protection Recommended:

(a) For solar collector systems in which a corrosion inhibitor may be used and where there is a direct makeup connection the water service shall be protected by a reduced pressure principle backflow prevention assembly.

(b) For once through solar heating systems (i.e., domestic hot water) there is no backflow protection required.

7.2.3.33 Temporary Service - Fire Hydrants, Blow-offs, Air Valves and Other Outlets

An approved backflow prevention assembly shall be installed on the outlet where the water is or may be used to supply the consumer in a manner which would create a potential hazard to the public water system. See Fig. 7.21.

The water-using situations may be broadly grouped into the following categories:

(a) Tanks for Construction Water: Typical of such tanks are the direct filling from an outlet for use with construction water activities such as water trucks, water wagons or stationary elevated tanks. A permanent supply pipe to fill the tank with a proper air gap above the tank overflow rim can be regarded as having adequately eliminated the hazard of the tank. An automatic flow control valve utilizing a float shutoff on the piping outlet inside the tank is not a substitute for an air gap. These waters may become polluted or contaminated with bird droppings, dust, vermin, etc. or by means of chemicals which may have been introduced into the tank for dust control.

(b) Tanks for Spray Rigs: Typical of such tanks are the filling from an outlet to equipment used for pest control, hydro-mulch or agricultural spraying. A permanent external supply pipe to fill the tank with a proper air gap above the tank overflow rim can be regarded as having adequately eliminated the hazard of the tank. These waters are subject to a variety of chemical and organic additives including fertilizers, herbicides, pesticides, organic matter, etc.

(c) Mechanical Equipment: Typical of such equipment are operations involving pavement saw cutting, sand blasting, hydraulic boring or jacking, pressure testing piping, gas chlorinating, street washing, etc.

Equipment having a built-in air gap can generally be regarded as acceptable without additional backflow protection. There is a concern with the hose or piping to the equipment, such is the case with most saw cutting and sand blasting operations.

In the case of boring and jacking equipment, the operation is normally performed in a pit that may submerge the equipment in muddy water. Pressurizing with pumps and gas chlorination equipment connected to pipelines or vessels of questionable sanitary condition may create pressures in excess of the water source.

Protection Recommended:

(a) An air gap or a reduced pressure principle backflow assembly where there is an existing or potential health hazard involving a tank.

(b) A reduced pressure principle backflow prevention assembly where there is an existing or potential health hazard involving mechanical equipment or an excessive length of piping or hose.

(c) A double check valve assembly where there is an existing or potential pollutional hazard involving excessive length of piping or hose.

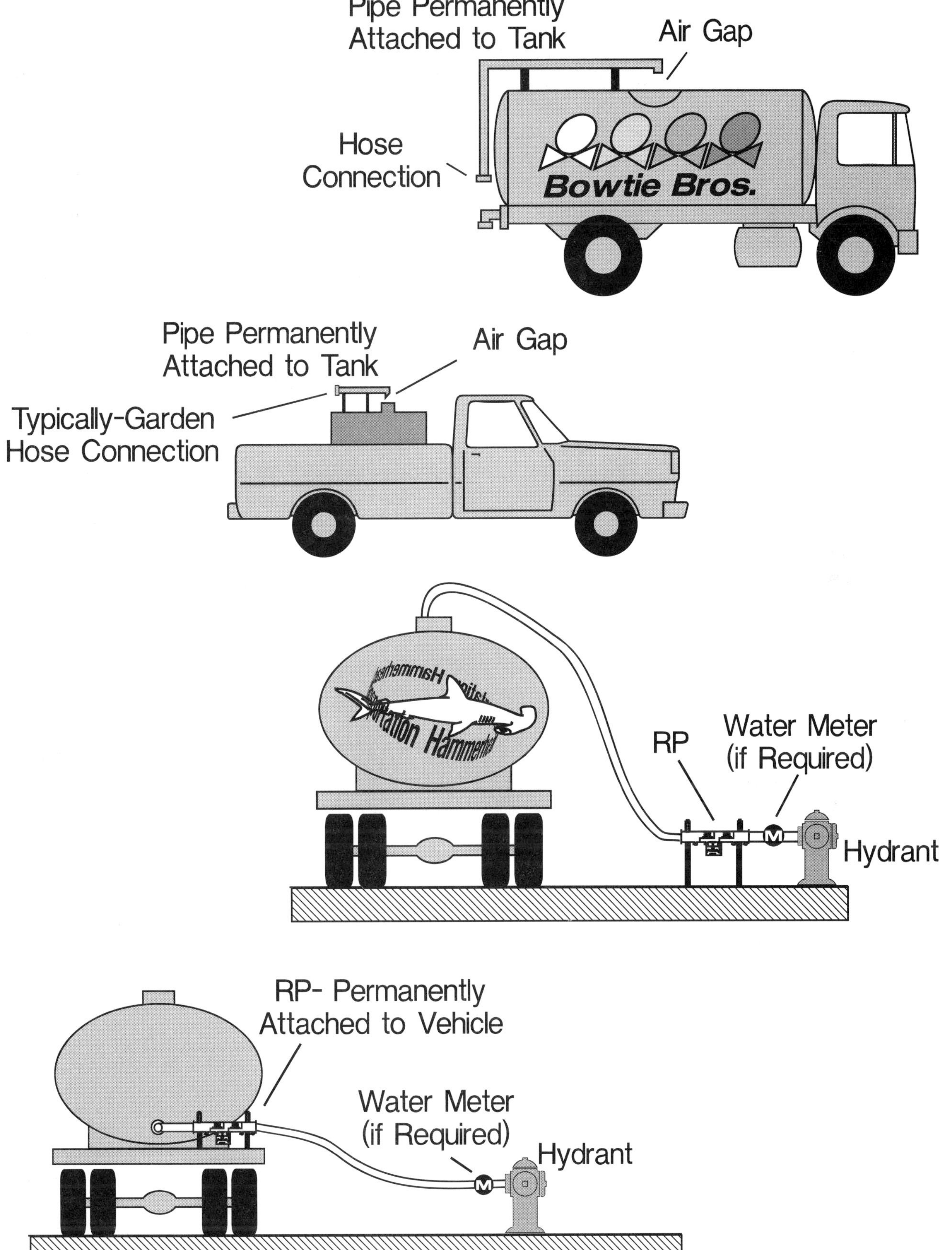

Fig. 7.21
Proper Methods of Filling Portable Spray and Cleaning Equipment

7.2.3.34 Used Water

Used water is that which has passed beyond the water purveyor's control (at the point of delivery) and may be stored, transmitted or used in such a manner as to become polluted or contaminated.

Typical of used water supplies are: waters in industrialized water systems; water in reservoirs or tanks used for fire fighting purposes; irrigation reservoirs; swimming pools; fish ponds; mirror pools; memorial and decorative fountains and cascades; cooling towers; baptismal, quenching, washing, rinsing and dipping tanks. All of these supplies, including a public potable water over which the water purveyor does not exercise sanitary control, become a potential hazard to the public water system. These waters may become polluted or contaminated because of industrial processes; contact with human body, dust, vermin, birds, etc.; or by means of chemicals and/or organic compounds which may have been introduced into the tanks, lines or systems for scale, corrosion, algae, bacterial or odor control, or for similar treatment.

In the event that backflow occurs because of inadequate protection at the service connection, polluted or contaminated used waters may be discharged into the public water system; therefore, it is necessary for the water purveyor to evaluate the potential hazard and to take steps necessary to protect the public water system in accordance with the degree of hazard found. In making such an evaluation it is not deemed necessary that the used water sources be developed and interconnected with the potable water system through cross-connections. It is deemed necessary only to determine that the water or fluids are available to the premises and of quantity sufficient so that it would be desirable and feasible for the consumer to develop and use the supply.

In addition, consumer use of water from the distribution system for heating, cooling or other purposes within the consumer's system and later returned to the distribution system is not acceptable.

Protection Recommended:

An approved backflow prevention assembly shall be installed at the service connection to any premise on which there is a used water supply or a system containing industrial fluids. This will include premises where there are reservoirs, cooling towers, recirculating systems and other used waters or industrial fluid systems. This may be accomplished in the following manner:

(a) An air gap or a reduced pressure principle backflow prevention assembly is recommended where there is a health hazard;

(b) A double check valve assembly should be used where there is only a pollution hazard.

7.2.3.35 Water Front Facilities and Industries, Marina

An approved backflow prevention assembly shall be installed at the service connection to any premise where there are piers, docks, shipyards, marinas or other water front facilities or industries, or where water from the ocean, harbor, bay, river, stream, irrigation ditch or canal, lake, etc., is available to or used on the premises, except that backflow protection may not be required on services to cafes, private residences, concessions, administration buildings, comfort stations and other similar facilities which do not have docking facilities, or which do not have outlets available for supplying water to docks or docking facilities, or which have no use for such auxiliary water supplies for irrigation, fire protection, air conditioning, cooling, swimming pool supply or other such purposes. (NOTE: The foregoing exemption will not be valid if backflow protection is required because of some other potential or existing hazard.) See Fig. 7.22.

The hazards normally found in a plant of this type include cross-connections between the consumer's water system and: ***a system supplied from an auxiliary water supply*** (NOTE: The administrative authority normally will require a backflow prevention assembly at the service connection to the premises where there is an auxiliary water supply, even though there are no existing cross-connections. For more specific information regarding auxiliary water supply and recommended protection, see Section 7.2.3.3 "Auxiliary Water Systems"); ***steam boilers and lines; mud pumps and mud tanks; hydraulically-operated Tretolite tanks; oil well casings*** (for dampening pressures); ***dehydration tanks and outlet lines from storage and dehydration tanks*** (for purging purposes); ***oil and gas tanks*** (to create hydraulic pressures and to hydraulically raise the oil and gas levels); ***oil and gas lines*** (for testing, evacuating and slugging purposes); ***reservoirs cooling towers and circulating systems*** which may be heavily contaminated with bird droppings, vermin, algae, bacterial slimes or with toxic water treatment compounds such as pentachlorophenol, copper sulfate, chromates, metallic glucosides, compounds of mercury, quaternary ammonium compounds, etc.; ***steam generating facilities and lines*** which may be contaminated with boiler compounds such as those listed above (NOTE: A particular hazard is the possibility of steam getting back into the domestic system, causing either a system or a health hazard); ***industrial fluid systems and lines*** containing cutting and hydraulic fluids, coolants, hydrocarbon products, glycerine, paraffin, caustic and acid solutions, etc.; ***fire fighting systems***, including storage reservoirs which may be treated for scale formation, corrosion, algae, slime growths, etc., or fire systems which may be subject to contamination with antifreeze solutions, Foamite or other chemicals or chemical compounds used in fire fighting, or fire systems which are subject to contamination with auxiliary or used water supplies or industrial fluids; ***water-cooled equipment which may be sewer-connected*** such as compressors, heat exchangers, air conditioning equipment, etc.; ***tanks, can and bottle washing machines and lines*** where caustics, acids, detergents and other compounds are used in cleaning, sterilizing and flushing; ***hydraulically-operated equipment*** where the public potable water pressure is used directly and may be subject to backpressure; ***equipment under hydraulic tests*** such as tanks, lines, valves, fittings and also pumps, pressure cylinders or other hydraulic facilities which may be used to provide pressures for testing purposes (NOTE: in such cases, air, gas or hydraulic fluids may be forced back to the public system), ***pulp bleaching, dyeing and processing plants*** where water systems may be contaminated with toxic chemicals; ***steam-connected facilities*** such as pressure cookers, autoclaves, retorts, etc.; ***washers, cookers, tanks, lines, flumes, and other equipment*** used for storing, washing, cleaning, blanching, cooking, flushing, fluming, or transmission of foods, fertilizers or wastes.

Protection Recommended:

(a) An air gap or a reduced pressure principle backflow prevention assembly where the auxiliary water supply available to the premises is or may be contaminated with sewage, industrial waste of a toxic nature or other contaminant which would cause a health or system hazard;

(b) A reduced pressure principle backflow prevention assembly at the point of service to a pier-head or marina with the minimum individual pier-head outlets protected by a reduced pressure principle backflow prevention assembly.

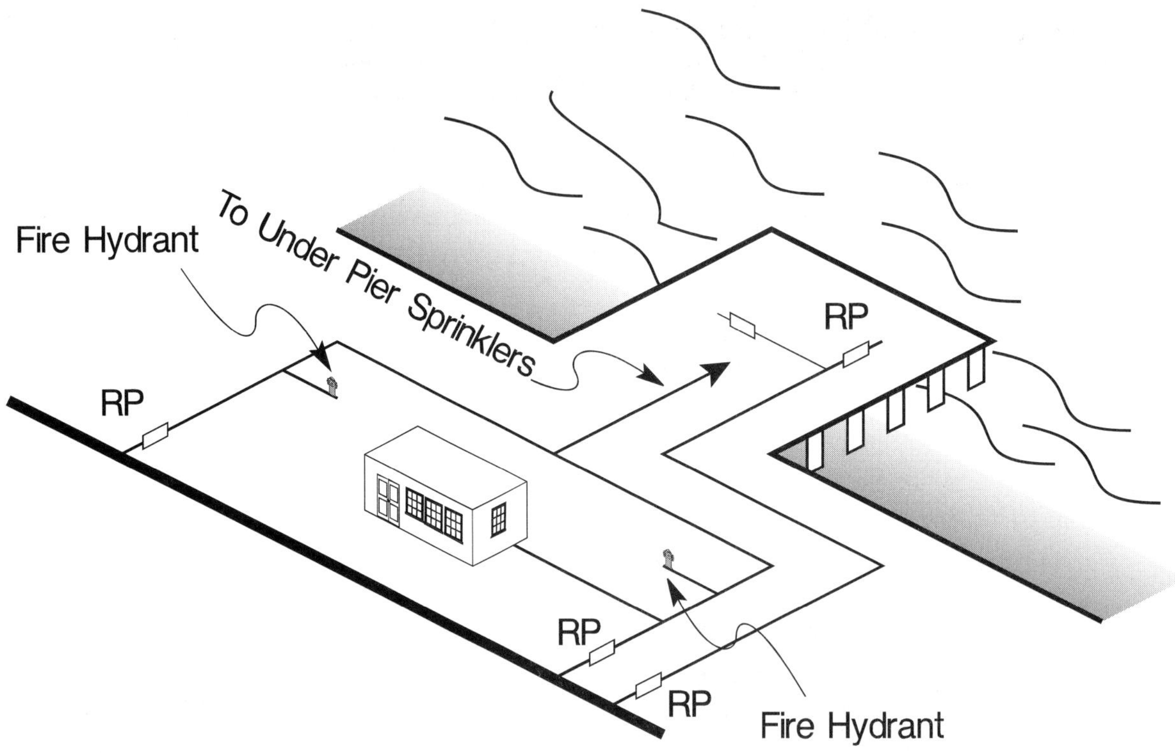

Fig. 7.22
Water Front Facilities

SECTION 8

Sample Letters, Forms, Installation Guidelines, Model Ordinance, etc.

8.1 Sample Letter: Approved Reduced Pressure Principle Backflow Prevention Assemblies

Utility or Agency Letterhead

(date)

The Consumer/Owner

____

____

You are herewith informed that you must install on *(certain designated)* water service(s) to your premises either an air gap or an approved reduced pressure principle backflow prevention assembly. This action is taken in accordance with the Federal Safe Drinking Water Act Amendments of 1986 and with the State of *(state)* and *(local)* Cross-Connection Control *(rules or regulations).* Under these *(rules or regulations)* the *(water purveyor or administrative authority)* has the primary responsibility of protecting the public potable water from backflow of dangerous substances which would endanger the public health or physically damage the public water system.

On *(date),* as part of our program to see that the *(rules or regulations)* are complied with, *(name of program specialist)* conducted a survey of your plumbing system. This survey revealed potential/actual cross-connections of the following conditions:

(List Conditions)

The above conditions present backflow hazards to the *(on premise system)* and *(to the public supply).* To correct these conditions the *(utility or agency)* requires the following:

(When service protection only, add following paragraphs)

Install an approved reduced pressure principle backflow prevention assembly downstream of the water meter a minimum of 12 inches above grade and accessible for testing and maintenance.

This letter does not address internal protection requirements. We suggest you contact the agency having jurisdiction to ensure your water system complies with plumbing codes.

(When internal and/or service protection, add following paragraph)

This letter addresses protection of certain cross-connections detected in our survey. We do not however, accept responsibility to guarantee that all cross-connections will be protected or for cross-connections that may be created in the future, due to repair or alterations made in your water system.

(Use the following paragraph only when appropriate)

A water supervisor shall be appointed who shall be responsible for the installation and use of pipelines and equipment in a manner which avoids and eliminates cross-connections in accordance with applicable laws and regulations. The (*water purveyor or administrative authority*) shall be kept informed of the identity of the person appointed as water supervisor.

It is necessary to shut off the flow of water through a backflow prevention assembly during the time it is being tested and/or repaired. If the complete interruption of water through a given service is critical to your operation, we recommend you install backflow prevention assemblies in parallel. This will allow one assembly to continue serving water while the other is being tested or repaired. A check should be made with your engineer or plumber to be sure that assemblies are properly sized for desired flows.

Note that installation of a backflow prevention assembly will prevent release of on-site pressure to the utility water mains. Therefore, it is important that a temperature/pressure relief valve and/or thermal expansion tank be properly installed to relieve any excessive increase in on-site pressure due to hot water heating systems or other activities.

Attached is a list of reduced pressure principle backflow prevention assemblies that have been evaluated and approved by the Foundation for Cross-Connection Control and Hydraulic Research of the University of Southern California. The assemblies listed thereon have been adopted by this *(purveyor or administrative authority)* as the only assemblies approved for use on the water lines under our jurisdiction.

You will be allowed *(typically 30 to 60 days)* days from the date of this letter to provide the corrective measures previously outlined.

For additional information regarding this matter you may either write to *(name of person)* at *(address)* or telephone *(phone number)* between the hours of *(specify times)*. Please contact *(name of person)* as soon as the work is done or if, for any reason, you cannot comply with the *(number)* day installation period or for clarification of any cross-connection control requirements discussed in this letter.

Sincerely,

(title)

8.2 Sample Letter: Approved Double Check Valve Assemblies

Utility or Agency Letterhead

(date)

The Consumer

____

____

You are herewith informed that you must install on *(certain designated)* water service(s) to your premises an approved double check valve backflow prevention assembly. This action is taken in accordance with the Federal Safe Drinking Water Act Amendments of 1986 and with the State of *(state)* and *(local)* Cross-Connection Control *(rules or regulations)*. Under these *(rules or regulations)* the *(water purveyor or administrative authority)* has the primary responsibility for protecting the public potable water from backflow of any pollution.

On *(date),* as part of our program to see that the *(rules or regulations)* are complied with, *(name of program specialist)* conducted a survey of your plumbing system. This survey revealed potential/actual cross-connections of the following conditions:

(List Conditions)

The above conditions present backflow hazards to the *(on premise system)* and *(to the public supply)*. To correct these conditions the *(utility or agency)* requires the following:

(When service protection only, add following paragraphs)

Install an approved double check valve backflow prevention assembly downstream of the water meter a minimum of 12 inches above grade and accessible for testing and maintenance.

This letter does not address internal protection requirements. We suggest you contact the agency having jurisdiction to ensure your water system complies with plumbing codes.

(When internal and/or service protection, add following paragraph)

This letter addresses protection of certain cross-connections detected in our survey. We do not however, accept responsibility to guarantee that all cross-connections will be protected or for cross-connections that may be created in the future, due to repair or alterations made in your water system.

(Use the following paragraph only when appropriate)

A water supervisor shall be appointed who shall be responsible for the installation and use of pipelines and equipment in a manner which avoids and eliminates cross-connections in accordance with applicable laws and regulations. The *(water purveyor or administrative authority)* shall be kept informed of the identity of the person appointed as water supervisor.

It is necessary to shut off the flow of water through a backflow prevention assembly during the time it is being tested and/or repaired. If the complete interruption of water through a given service is critical to your operation, we recommend you install backflow prevention assemblies in parallel. This will allow one assembly to continue serving water while the other is being tested or repaired. A check should be made with your engineer or plumber to be sure that assemblies are properly sized for desired flows.

Note that installation of a backflow prevention assembly will prevent release of on-site pressure to the utility water mains. Therefore, it is important that a temperature/pressure relief valve and/or a thermal expansion tank be properly installed to relieve any excessive increase in on-site pressure due to hot water heating systems or other activities.

Attached is a list of double check valve assemblies that have been evaluated and approved by the Foundation for Cross-Connection Control and Hydraulic Research of the University of Southern California. The assemblies listed thereon have been adopted by this *(purveyor or administrative authority)* as the only assemblies approved for use on the water lines under our jurisdiction.

You will be allowed *(typically 30 to 60 days)* days from the date of this letter to provide the corrective measures previously outlined.

For additional information regarding this matter you may either write to *(name of person)* at *(address)* or telephone *(phone number)* between the hours of *(specify times)*. Please contact *(name of person)* as soon as the work is done or if, for any reason, you cannot comply with the *(number)* day installation period or for clarification of any cross-connection control requirements discussed in this letter.

Sincerely,

(title)

8.3 Sample Letter: Installation and Maintenance Requirements

Utility or Agency Letterhead

(date)

The Consumer/Owner

____

____

You have been informed that you must install and maintain

- ☐ an air gap
- ☐ a reduced pressure principle assembly
- ☐ a double check valve assembly
- ☐ a pressure vacuum breaker

on the water service(s) *(to or within)* your premises. This action was taken in accordance with the Federal Safe Drinking Water Act Amendments of 1986 and with the State of *(state)* and *(local)* Cross-Connection Control *(rules or regulations)*. Under these *(rules or regulations)* the *(water purveyor or administrative authority)* has the primary responsibility for protecting the public potable water from backflow of any pollution or contamination.

AIR GAP RECEIVING TANK WITH PUMP OR GRAVITY DISTRIBUTION.

A receiving tank shall be installed on the property side of and adjacent to the water meter. The supply line between the meter and the tank must be permanently exposed for inspection purposes. There must be no outlet, tee, tap or connection of any kind to or from the supply pipe between the water meter and the opening from which the water is discharged into the receiving tank. The discharge inlet into the tank must be located at a distance of not less than two times the cross-sectional diameter of the inlet pipe above the top or overflow rim of the tank. Where required, a wind guard should be installed to prevent water loss or damage by wind-driven spray. The tank should be elevated (50 feet or more is recommended) to give an adequate gravity head; or, as an alternate, a pump may be installed to provide adequate head.

MECHANICAL BACKFLOW PREVENTION ASSEMBLIES

An ***approved*** backflow prevention assembly shall be installed on the property side of and adjacent to the water meter in accordance with *(refer to Figs. 8.1, 8.2, 8.3 and 8.4)* or an ***approved*** pressure vacuum breaker shall be installed adjacent to the specified point of use and in accordance with *(refer to Fig. 8.6)*. Where construction or equipment location present citing problems for the above noted assembly a deviation may be granted by *(administrative authority)* providing such request is made in writing prior to the installation of the assembly. On the service line there must be no outlet, tee, tap or connection of any sort to or from the supply pipe line between the meter and the protective assembly.

GENERAL:

In the event you elect to install an air gap, it will be necessary for you to report at the end of each year to us that this method of protection has remained operative and effective throughout the year without being bypassed. In the event you desire to install a mechanical backflow prevention assembly, it will be necessary for you to notify us immediately after installation in order that a field test may be made of the assembly to determine its operational characteristics. We will make the initial field test; however, any servicing required to make the assembly operate satisfactorily will be your responsibility. The *(agency)* personnel making the inspection and initial field test of the assembly will give you specific instructions as to our requirements for future test and servicing of the assembly, and will answer any questions you may have regarding rendition of service where backflow protection is required. Under the regulations of the *(administrative authority)* you are required to maintain this assembly in a continuous state of good repair and to test the assembly at intervals of one year, unless the condition of the assembly indicates the necessity for more frequent tests and servicing. We will provide you with test report forms 30 days in advance of your next periodic test date. It will be your responsibility to have it tested by a certified backflow prevention assembly tester. A report form showing the condition of the assembly and repairs made, if any, shall be prepared and forwarded to us within the 30-day period.

The enclosed drawings *(enclose copies of Figs. 8.1, 8.2, 8.3, 8.4, and 8.6)* provide installation requirements. The type of assembly or method of protection as well as any deviation from the forgoing requirements must be approved by us before installation. A list of approved backflow prevention assemblies will be supplied by us upon request.

For additional information regarding this matter, you may either write to *(name of person)* or call at *(telephone number)*.

Sincerely,

(title)

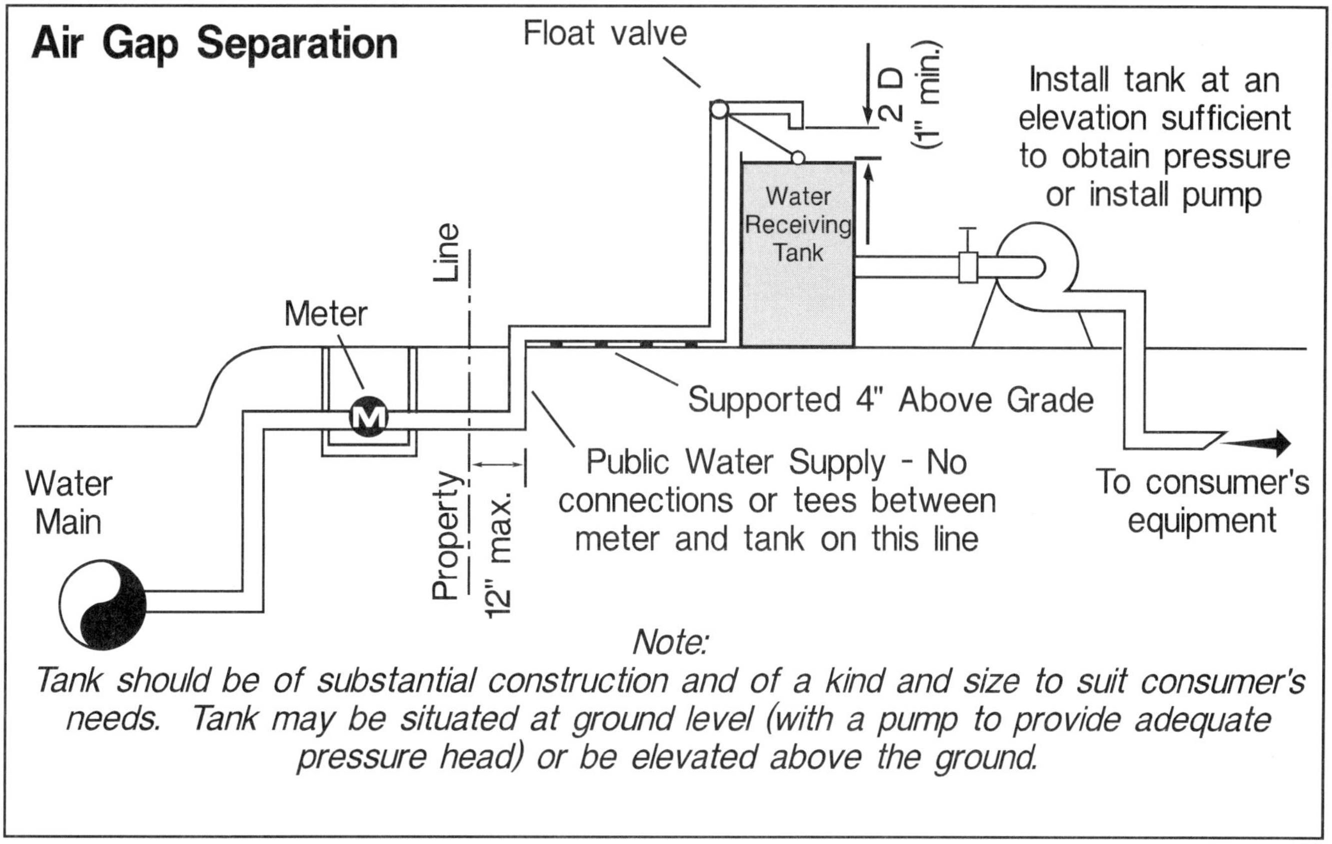

Air Gap Separation
Float valve
2 D
(1" min.)
Install tank at an elevation sufficient to obtain pressure or install pump
Water Receiving Tank
Line
Meter
Supported 4" Above Grade
Water Main
Property
12" max.
Public Water Supply - No connections or tees between meter and tank on this line
To consumer's equipment
Note:
Tank should be of substantial construction and of a kind and size to suit consumer's needs. Tank may be situated at ground level (with a pump to provide adequate pressure head) or be elevated above the ground.

Backflow Prevention Assembly
Meter
Line
Water Main
Property
Public Water Supply - No connections or tees between meter and backflow prevention assembly on this line
To consumer's equipment
Note:
Place assembly 12" or more above high water level of surrounding ground.
Note:
Resilient seated shut-off valves and testcocks are required.

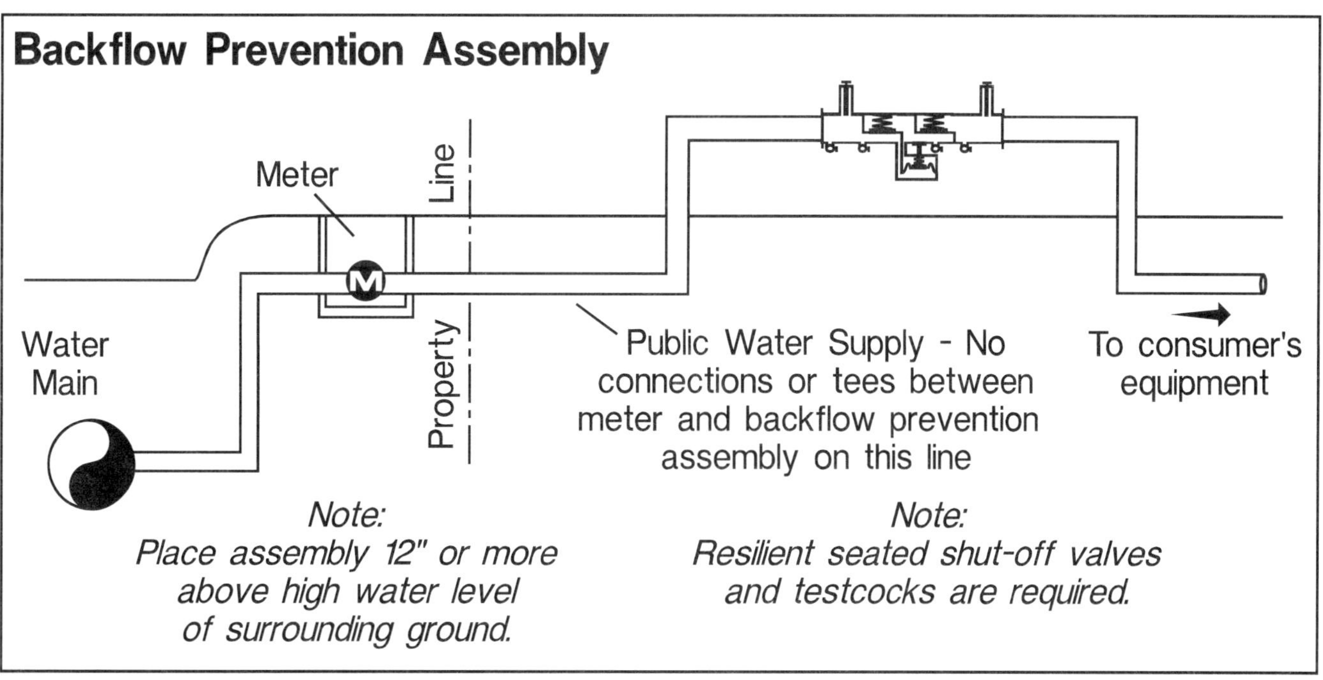

Fig. 8.1

Installation Guidelines for Air Gap and Backflow Prevention Assemblies at Water Service Connection

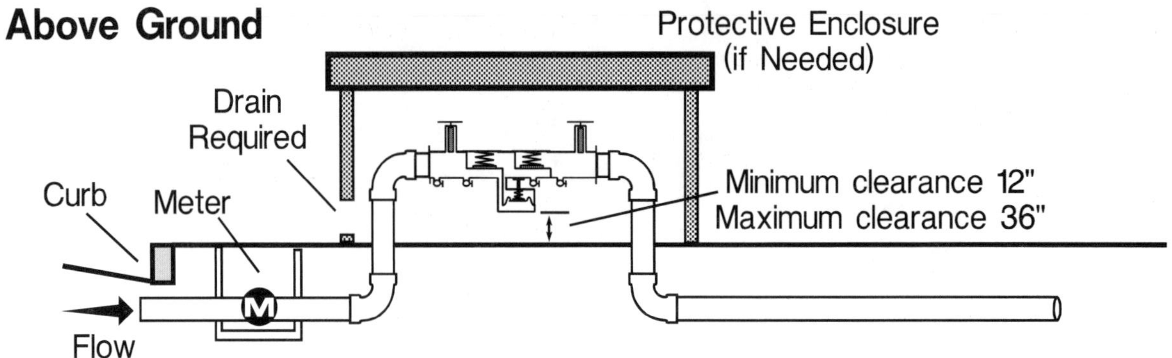

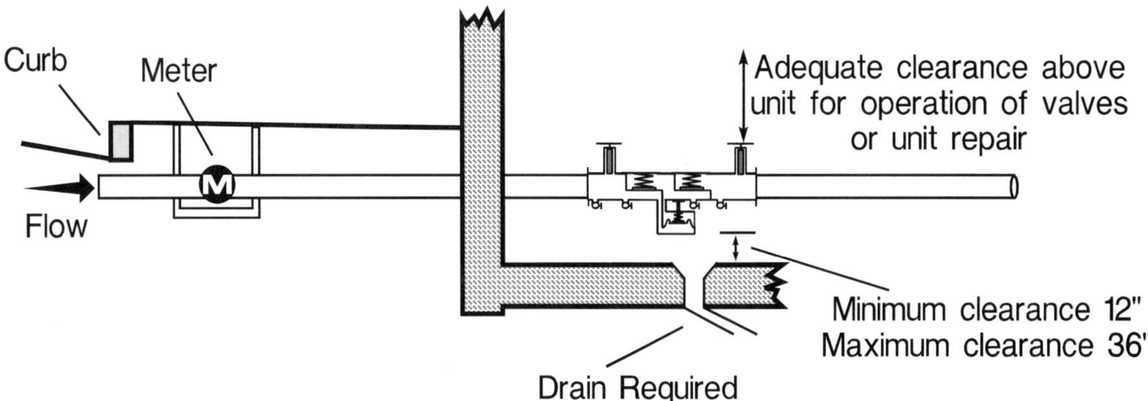

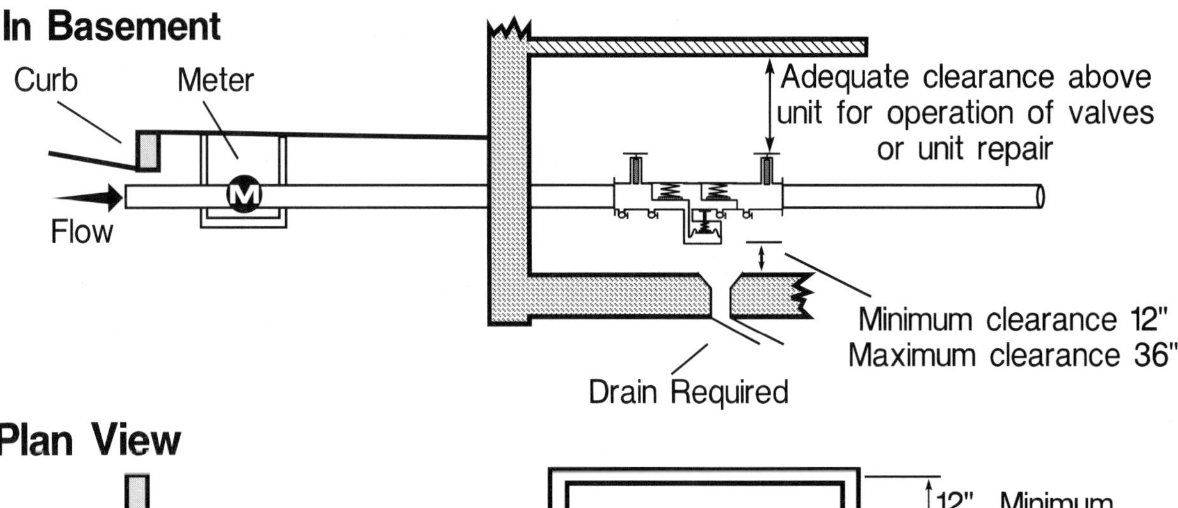

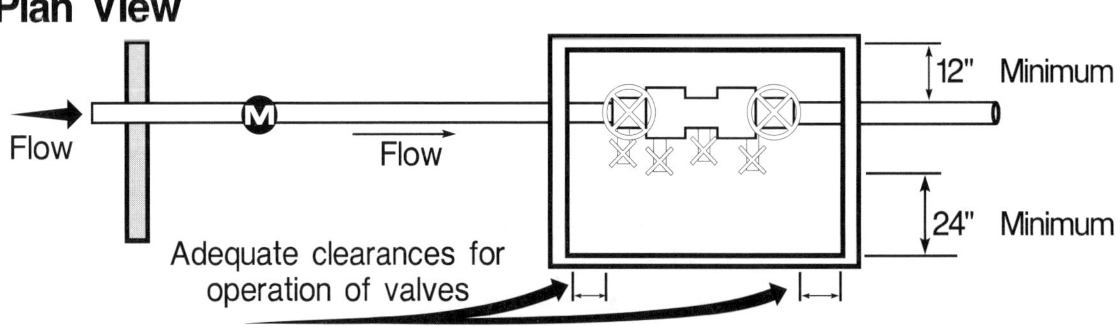

Fig. 8.2

Typical Installations with Minimum Clearances
Reduced Pressure Principle Assemblies

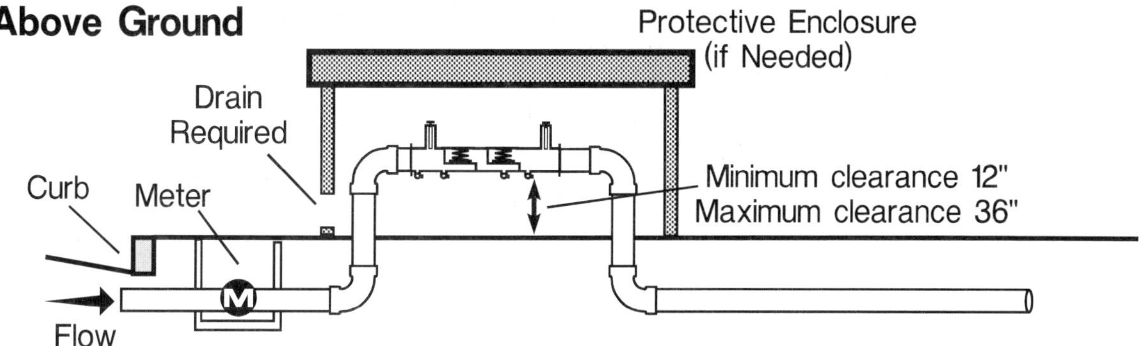

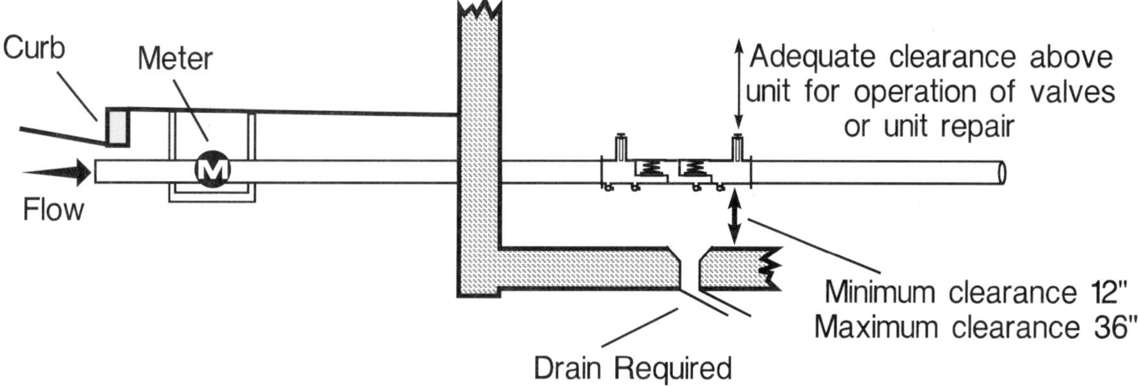

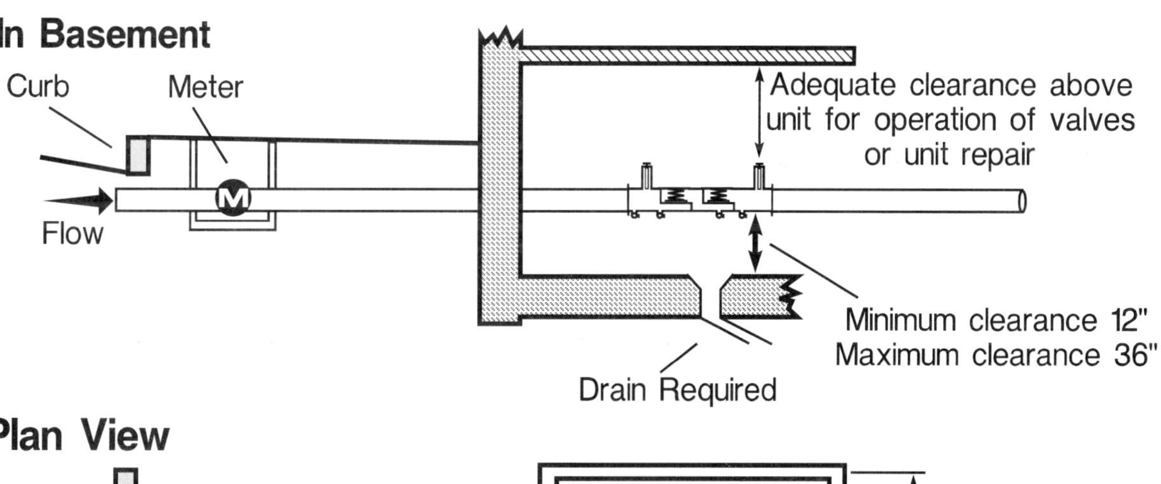

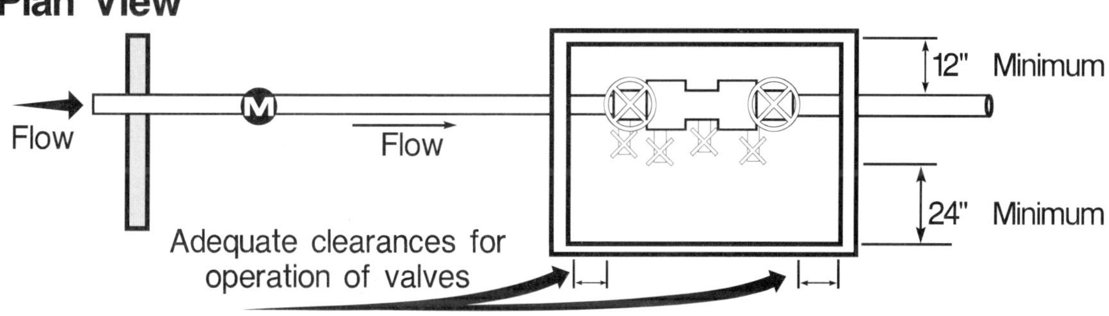

Fig. 8.3

Typical Installations with Minimum Clearances
Double Check Valve Assemblies

Service Connection Installation Guidelines

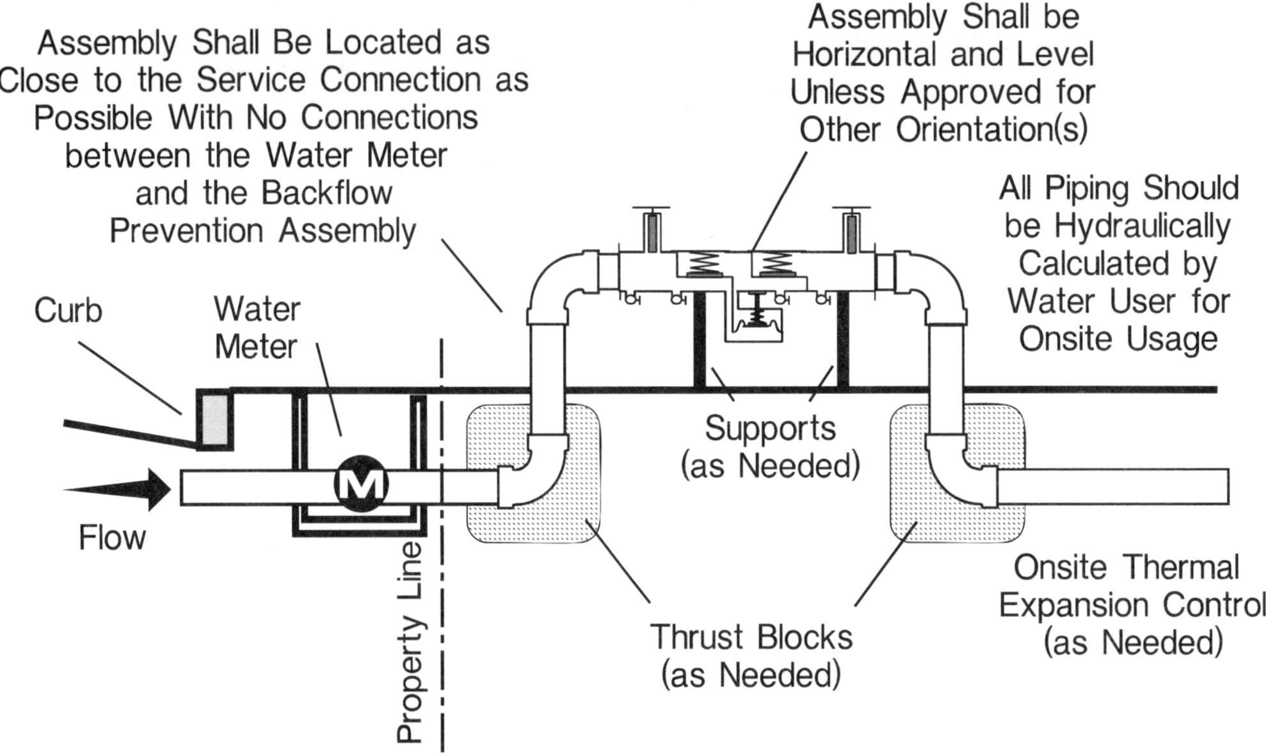

Backflow Prevention Assemblies are to be used within their rated operating conditions:

Pressure:
Backflow prevention assemblies typically have maximum working water pressures (MWWP) of 150 psi (1034 KPa) or 175 psi (1206 KPa). Assemblies are designed to operate continuously at this pressure, which is identified on the assembly.

Temperature:
Backflow prevention assemblies are designed to operate continuously at their maximum working water temperature (MWWT), which is identified on the assembly.

Rate of Flow:
Backflow prevention assemblies are designed to operate continuously up to their rated flow (i.e., gallons per minute - GPM; or liters per second - L/s) per Table 10-1.

NOTE: All installations of backflow prevention assemblies must be in compliance with state and local plumbing and building codes. Contact local administrative authority for detailed requirements.

Fig. 8.4
Service Connection Installation Guidelines

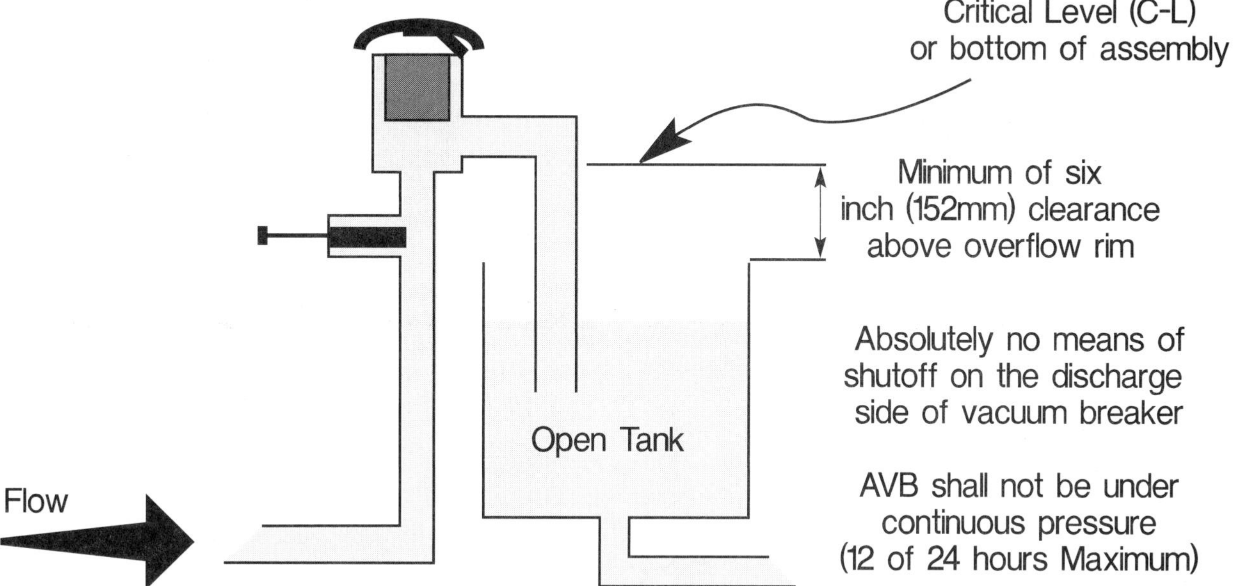

Fig. 8.5
Typical Installation
Atmospheric Vacuum Breaker (AVB)

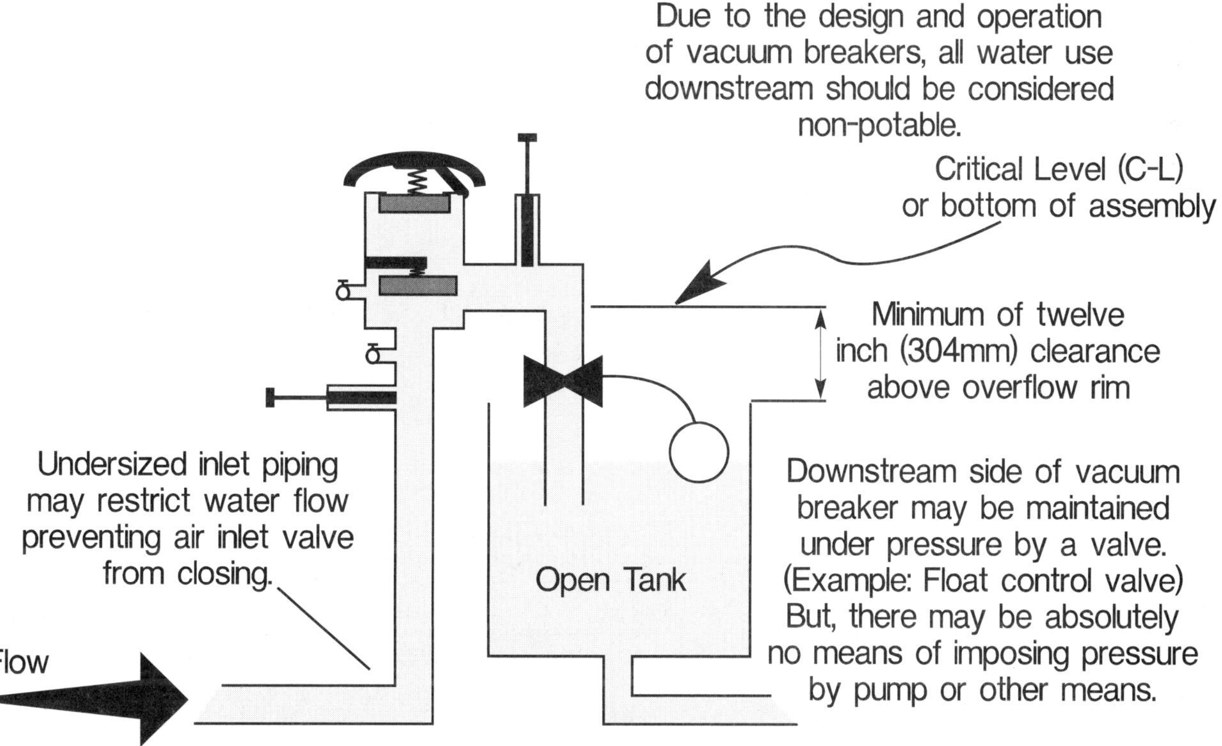

Fig. 8.6
Typical Installation
Pressure Vacuum Breaker (PVB & SVB)

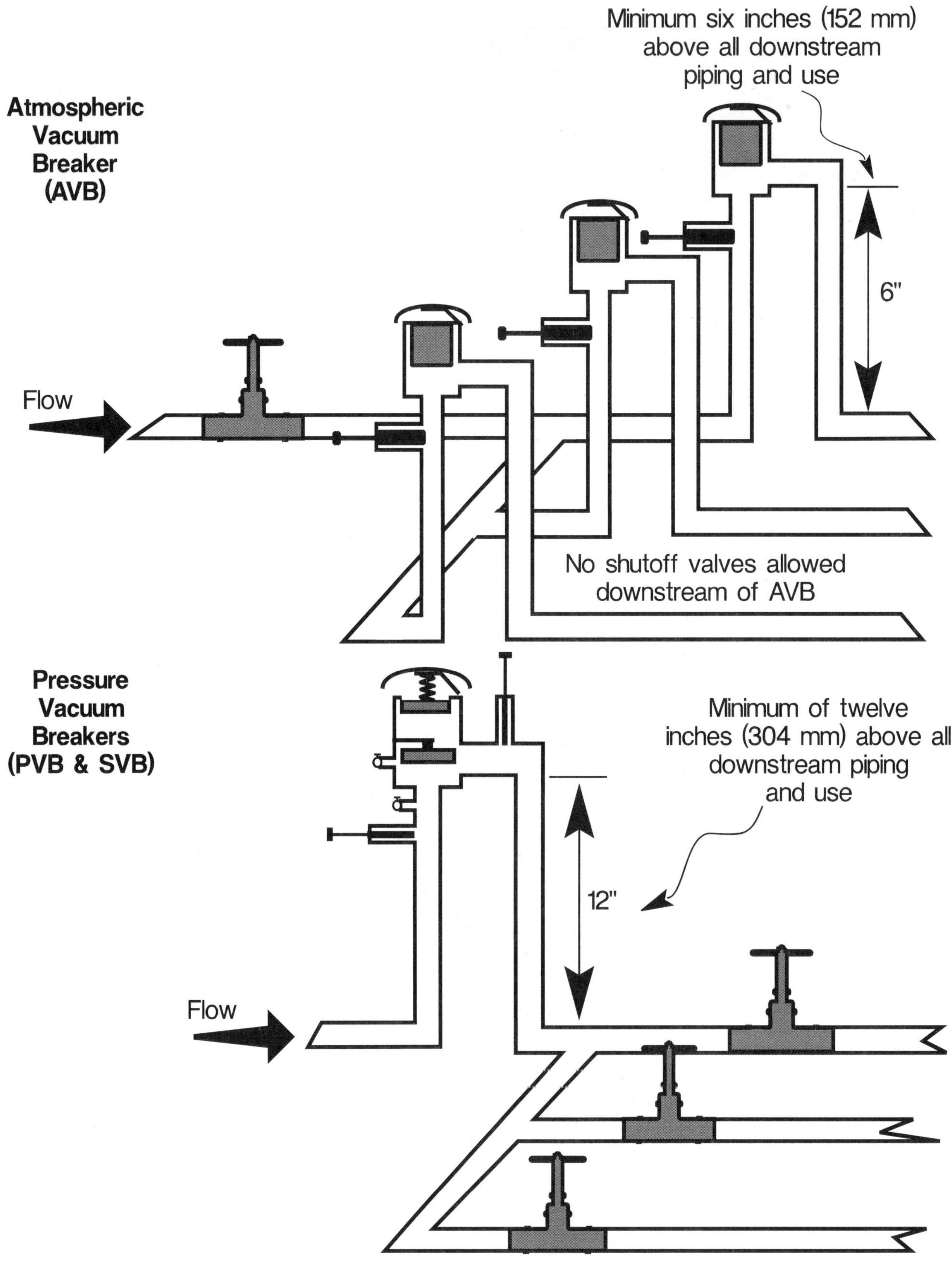

Fig. 8.7
Multi-Zone Irrigation Systems

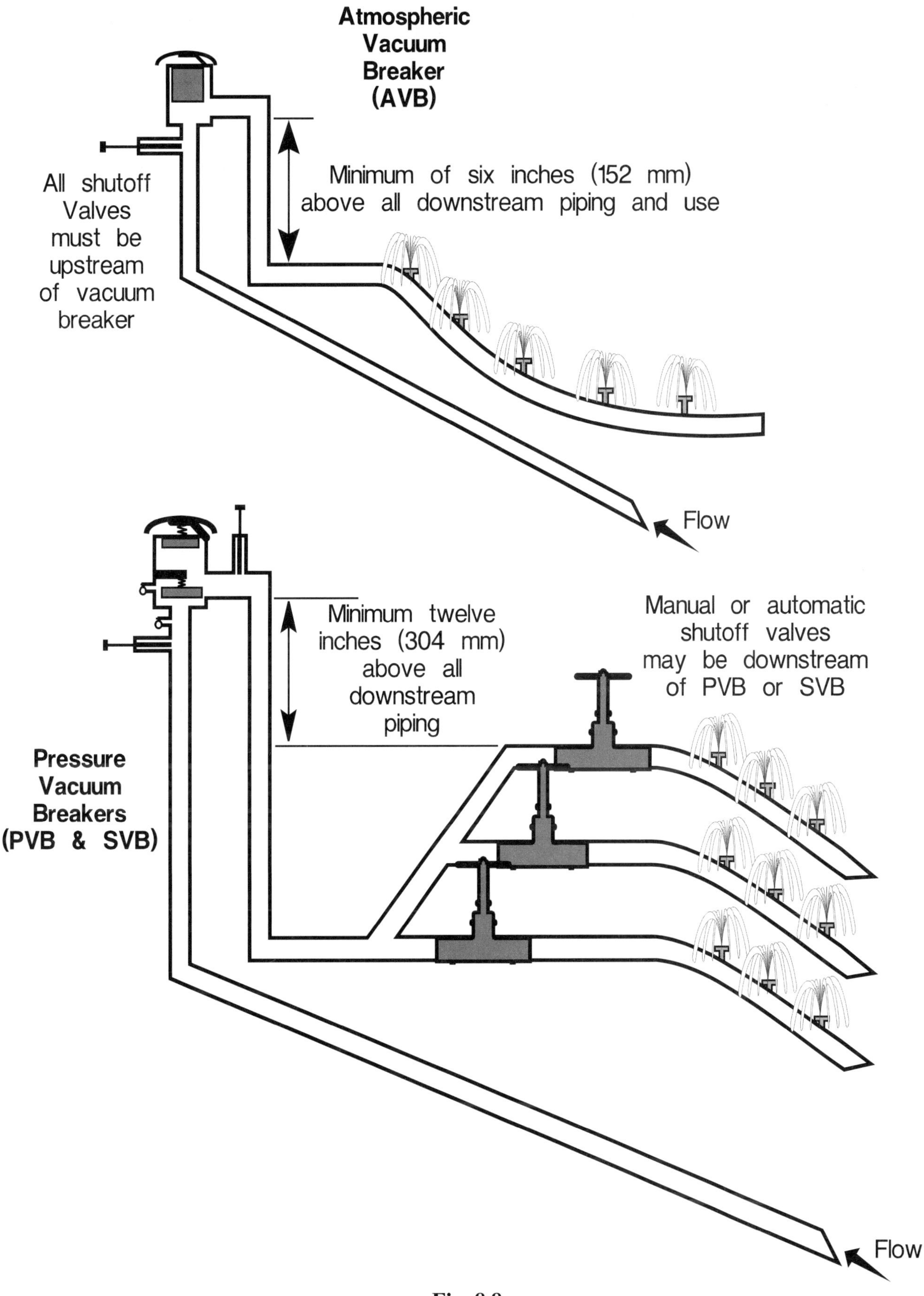

Fig. 8.8
Irrigation Systems Feeding Uphill

8.4 Guidelines for Parallel Installation of Backflow Prevention Assemblies

Two or more backflow prevention assemblies of the same type may be installed in parallel to provide uninterrupted water service to the water user, typically known as a critical service. When it is necessary to install backflow prevention assemblies in parallel, there are several hydraulic conditions which may be considered.

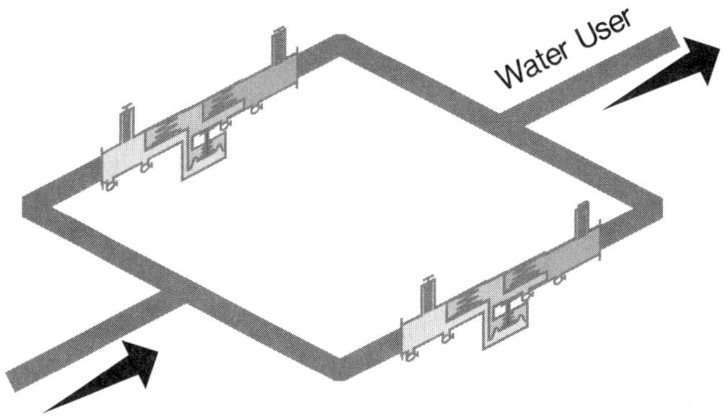

Fig. 8.9
Example of RP's Installed in Parallel

8.4.1 Full Line Size

If it is necessary to maintain maximum rated flow to the water user, then it may be necessary that each of the backflow prevention assemblies be individually sized to handle the rated flow of the inlet piping. If one of the assemblies is shut off for testing or maintenance, the other assembly(s) can maintain the full flow capacity of the water user's system.

8.4.2 Hydraulic Sizing

If the hydraulic design of the installation does not require full flow capacity through each of the assemblies, then the assemblies may be hydraulically sized so that the combined flow capacity of the assemblies in parallel will equal or exceed the flow capacity of the inlet piping.

As an example:

If two 4-inch assemblies are installed in parallel in the place of a single 6-inch unit, then the combined flow capacity would be adequate.

- ☐ Two 4-inch assemblies ----- 2 x 500 gpm = 1000 gpm
  (per Table 10-1; rated flow of 4-inch DC or RP is 500 gpm)
- ☐ Rated flow of single 6-inch = 1000 gpm

With both assemblies in operation in this installation, the water user has full flow capacity. However, during routine testing or maintenance of one of the assemblies the water user should reduce his downstream flow demand, so the rated flow of the remaining on-line assembly is not exceeded.

8.4.3 Manifold Assembly

Two or more backflow prevention assemblies may be incorporated into a manifold assembly *(see Section 10.1.2.19)* to provide uninterrupted water service to the water user. The manifold assembly is comprised of backflow prevention assemblies (DC or RP) of the same manufacturer, model, and size. Manifold adaptor fittings on both the inlet and outlet of the manifold assembly are considered integral components. The size of the manifold assembly is determined by the inlet and outlet connections of the manifold adaptor fittings. See Fig. 8.10.

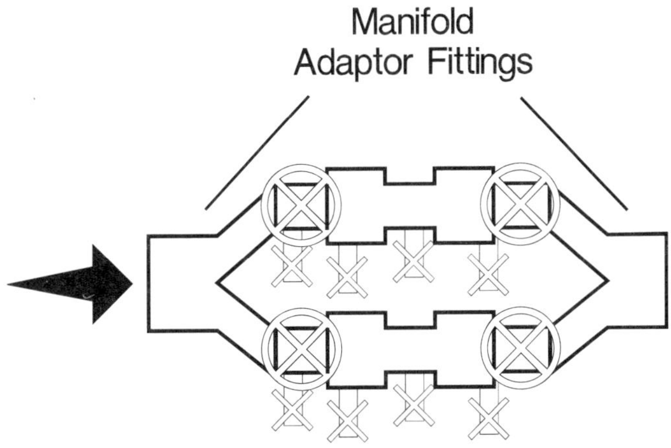

Fig. 8.10
Example of Manifold Assembly

It shall be noted that when two or more assemblies are installed in parallel, the multiple assemblies do not act in unison during normal flow conditions. One of the assemblies will act as a primary assembly, while the other(s) operates as a secondary. Only the primary assembly will open when the flow of water begins, and the secondary assembly(s) will begin to open only when the flow rate increases to a higher level. The rate of flow when the transition takes place, from one to multiple assemblies, will depend upon the loading of the check valves in the assemblies. The assembly with the lowest check valve opening points *(or cracking pressures)* will generally become the primary assembly in the manifold assembly. See Fig. 8.11.

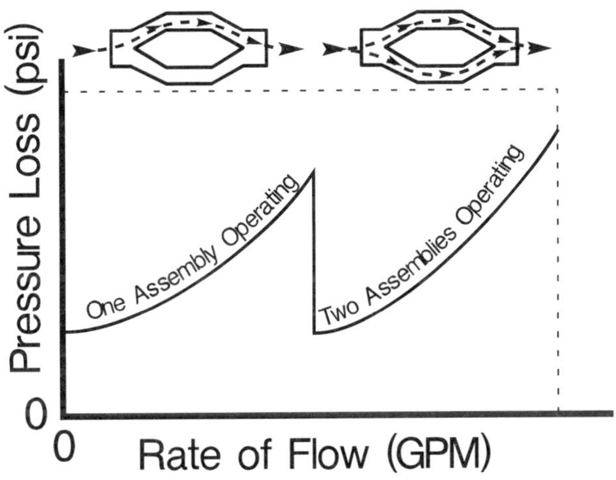

Fig. 8.11
Sample Flow Rate versus Pressure Loss Curve for Manifold Assembly

8.5 Flow Curves

Rated Flow of water through a backflow prevention assembly is the amount of water which may flow through the assembly under normal operating conditions. The rated flow varies with the type of assembly and size. The rated flows for all backflow preventers are shown on Tables 10-1 and 10-5.

Maximum Allowable Pressure Loss is the maximum amount of pressure loss permitted across an assembly at all flow rates from static up to and including rated flow. The exact limits, which vary with the size and type of assembly are shown on Tables 10-1 and 10-5. The flow curve must remain below the maximum allowable pressure loss in order for the assembly to meet the Specifications. See Fig. 8.12.

Increasing Flow and Decreasing Flow may give different amounts of pressure loss across a backflow prevention assembly. This *hysteresis* (i.e., differences in the increasing and decreasing flow curves) is caused by the slight increase in force needed to compress the springs of the backflow preventer as the check valves are opening. When making calculations for system requirements, it is necessary to use the flow curve with the greater pressure loss. See Fig. 8.13.

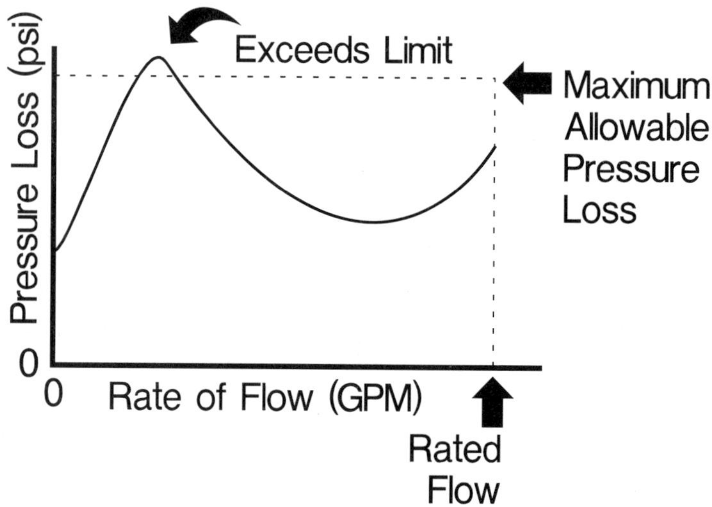

Fig. 8.12
Sample Flow Curve

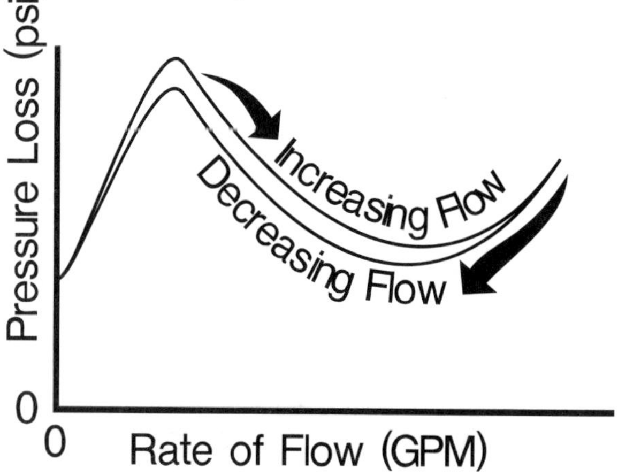

Pressure Loss (psi)
Increasing Flow
Decreasing Flow
0
0
Rate of Flow (GPM)

Fig. 8.13
Pressure Loss Due to Increasing or Decreasing Flows

8.6 Sample Letter: Periodic Test and Maintenance Report with Report Form

(NOTE: To be printed on both sides with letter on front and Test and Maintenance Report form on back.)

Utility or Agency letterhead

(Date)

The Consumer/Owner

____

____

Attn:________________________
Water Supervisor

Gentlemen:

PERIODIC TEST AND MAINTENANCE REPORT —
Backflow Prevention Assembly - Water Service
Connection No. *(number)*

The backflow prevention assembly described on the reverse side hereof is due for its periodic test, as required under our *(rule or regulation)*. Will you please have this test performed by a certified backflow prevention assembly tester possessing a valid Certification issued by the *(administrative authority)*.

If the test discloses that the assembly is not operating satisfactorily, please have the necessary repairs made and the assembly retested by a Certified tester. On completion of a test showing that the assembly is operating satisfactorily, the Certified tester shall complete the Test and Maintenance Report form on the reverse side hereof and forward it to this office no later than *(due date)*.

Additional information relative to this matter may be obtained by writing to our *(specify the office)* or by calling *(phone number)*.

Sincerely,

(title)

MANUFACTURER	MODEL	SIZE	SERIAL NUMBER

SERVICE NUMBER: LOCATION:

SERVICE NAME/ADDRESS: OWNER NAME/ADDRESS:

RP ☐
DC ☐
PVB ☐
SVB ☐
DCDA ☐
RPDA ☐

Reduced Pressure Principle Assembly

Double Check Valve Assembly

	Check Valve #1	Check Valve #2	Relief Valve	PVB/SVB
INITIAL TEST	Held at________PSID Leaked ☐	Held at_________PSID Closed Tight ☐ Leaked ☐	Opened at_________PSID Did Not Open ☐	AIR INLET Opened at _______PSID Did Not Open ☐
REPAIRS: Give details of repairs made here.	☐ Cleaned ☐ Replaced ________ ________ ________ ________ ________ ________ ________	☐ Cleaned ☐ Replaced ________ ________ ________ ________ ________ ________ ________	☐ Cleaned ☐ Replaced ________ ________ ________ ________ ________ ________ ________	CHECK VALVE Held at_________PSID Leaked ☐ ☐ Cleaned ☐ Replaced ________ ________ ________
FINAL TEST	___________PSID	___________PSID Closed Tight ☐	Opened at_________PSID	Air Inlet___________PSID Check Valve_________PSID

Comments: ________________________________

Initial Test	*Date* ________ *Time*_____ *Certified Tester No.* ________ ☐ *Passed* ☐ *Failed* *Test By (Signature)*______________ *Print Name*______________
Repair	*Date* ________ *Time*_____ *Certified Tester No.* ________ *Repaired By (Signature)*______________ *Print Name*______________
Final Test	*Date* ________ *Time*_____ *Certified Tester No.* ________ ☐ *Passed* ☐ *Failed* *Test By (Signature)*______________ *Print Name*______________

Acknowledged ______________

Fig. 8.14

Sample Test Form

8.7 Sample Letter: Follow up to Periodic Test & Maintenance Report

Utility or agency Letterhead

(date)

The Consumer/Owner

____

____

Gentlemen:

On *(date)* we mailed to *(name)* a combined form letter and Test and Maintenance Report form requesting that you have performed the necessary periodic test for the backflow prevention assembly identified on the form. This was to have been returned to us no later than *(date)*. As of today we have neither heard from you nor received your report.

Our request is in compliance with our *(rule or regulation)* number *(number)* and your failure to carry out your responsibility in this matter by *(date)* will result in discontinuance of water service to your property.

Should you desire instructions or additional information in this matter, please contact *(name)* at *(phone)*.

Sincerely,

(title)

8.8 Sample Letter: Notice of Discontinuance of Water Supply

Utility or agency Letterhead

(date)

The Consumer

____

____

Gentlemen:

You are hereby notified that in accordance with *(rule or regulation)*, the water supply to your premises, located at

(address)

will be discontinued on *(date - normally five to ten days from current date)* shall remain discontinued until you have complied with the letters of this *(name of agency and office)*. The discontinuance of your water service until you have so complied is required by *(rule or regulation)*.

Sincerely,

(title)

NOTE: This letter should be hand delivered to the consumer or mailed "return receipt requested."

8.9 Sample Letter: From Water Purveyor Regarding Periodic Test and Maintenance Report

Use Double-sided Postal Card

- - - - - One side reading as follows - - - - -

(Water Purveyor)
(Name)
(Address)

Subject: Backflow Prevention Assembly
Periodic Test & Maintenance Report

Attn:

Gentlemen:

I hereby certify that the air gap between our water system and your service connection No. *(number)* (check one ☐)

☐ Has not been bypassed, or otherwise made ineffective during the period ____________________ to ____________________.

☐ Was bypassed during the period ____________________ to ____________________ and during that time, equivalent protection, consisting of a backflow prevention assembly operating under the reduced pressure principle was installed in the line as street main protection. Your office was notified of this change on ____________________.

☐ If other than above, please explain by letter.

Firm Name____________________ Address____________________ Tel.# ____________________
Water Supervisor or
Auth. Agent____________________ Title____________________ Date____________________

1 of 2
(back side of postal card on next page)

- - - - - *One side reading as follows* - - - - -

(date)

The Consumer/Owner

____

____

Re: Backflow Prevention Assembly, Periodic Test & Maintenance Report, Water Service Connection No .*(number)*

Attn:

Gentlemen:

This will advise that the air gap described on this form is due for its periodic inspection and report as required under the *(rules or regulations).*

Will you please have this inspection performed, fill out this report form, certify as to its correctness and forward the report to this office no later than *(due date).*

Additional information relative to this matter may be obtained by writing to (*office*) or by calling *(phone).*

Sincerely,

(*title*)

2 of 2

(back side of postal card on previous page)

8.10 Sample Letter: Notice of Appointment of Water Supervisor

NOTICE OF APPOINTMENT OF

WATER SUPERVISOR

In accordance with the provisions of the *(rules or regulations)*. *(name)* has been appointed Water Supervisor. He/She will be responsible for maintaining the water system on the property located at

(address)

free from unprotected cross-connections and other sanitary defects, as required by the plumbing code, regulations and laws.

____

(Firm or company)

By:__________________________ ____________________________

(Signature) (Title)

____

(Address)

Phone: ( ) ________________________

Date:___________________________________

8.11 Sample Letter: Notice to Repair Backflow Prevention Assembly

NOTICE
TO REPAIR/MAINTAIN BACKFLOW PREVENTION ASSEMBLY

Name: ____
Address: ____
You are hereby notified on_________________________________
the ____ backflow

prevention assembly installed on property owned or controlled by you at

____

was tested and found defective as follows:

____
____
____

In accordance with the requirements set forth in *(rules or regulations)*, this assembly must be put in good operating condition on or before *(date)*.

Notify *(name)* when assembly is repaired and ready for retest.

Sincerely,

(title)

8.12 Sample:
Resolution Relative to Backflow Prevention Assembly Testers

RESOLUTION RELATIVE TO BACKFLOW PREVENTION ASSEMBLY TESTERS
ADOPTED BY *(administrative authority)*

In order to comply with requirements of the *(rules or regulations)* , under which the health agency has the overall responsibility for preventing water from unapproved sources from entering the potable water systems within water consumers' premises or the public water supply directly; and to assure that such assemblies will be adequately tested and reliable test reports made to the *{administrative authority(s)}*, the following Resolution is adopted:

No person shall be deemed to be capable of performing field tests and making reports on backflow prevention assembles, as required in the *(rules or regulations),* unless the person shall have proven their qualifications to the satisfaction of the *(administrative authority)* as hereinafter required. To determine such qualifications the *(administrative authority)* shall conduct examinations to determine the ability of any person desiring to test, troubleshoot, and make reports on backflow prevention assemblies for the purpose of complying with *(administrative authority)* requirements. The provisions of the *(administrative authority)* regulations shall not be deemed to have been complied with unless the tests have been performed and the report written by a person who has satisfactorily passed such examinations given by the *(administrative authority)* and who has received from the *(administrative authority)* a certificate or license.

The *(administrative authority)* shall make available to owners of properties on which assemblies requiring tests are maintained, a list of certified backflow prevention assembly testers qualified to make the required field tests.

APPLICATION AND QUALIFICATION OF TESTERS

APPLICATIONS

Applications shall be submitted to the *(administrative authority)* on forms prepared by the *(administrative authority)*.

Each application for certification shall be in the name of the individual desiring to take the examinations for certification.

EXAMINATIONS

The examinations given by the *(administrative authority)* shall be sufficiently comprehensive to determine that the applicant is fully capable to make accurate field tests and reports on all types of assemblies which are approved for use in the *(administrative authority's)* jurisdiction.

CLASSIFICATION AND QUALIFICATION OF TESTERS:

A. A **General Backflow Prevention Assembly Tester** shall mean any person duly certified under the *(rules and regulations)* of the *(administrative authority)* to commercially engage in the business of field testing any type of backflow prevention assembly.

QUALIFICATIONS: Shall have a valid certificate of qualifications as a journeyman plumber and shall be, or shall work under a person who has a certificate of registration as a master plumber, or who possesses a valid state contractor's license qualifying him to do plumbing contracting, or have adequate qualifications in the opinion of the *(administrative authority).*

B. A **Manufacturer's Agent** shall mean any person duly certified under the *(rules and regulations)* of the *(administrative authority)* to test any type of backflow prevention assembly made or manufactured by one company.

QUALIFICATIONS: Shall be a full-time employee or representative of the company which makes or distributes the one make of assembly for which the applicant is certified and shall be fully qualified by experience and training, as determined by the *(administrative authority)*, to test the assemblies made or distributed by the company of which the person is an employee or representative.

C. A **Limited Backflow Prevention Assembly Tester** shall mean any person duly certified under the *(rules and regulations)* of the *(administrative authority)* to test any or all types of backflow prevention assemblies at the premises owned or controlled by the individual or company, the location of which shall be specified on this certificate.

QUALIFICATIONS: Shall be a journeyman plumber, or have other qualifications which, in the opinion of the *(administrative authority),* fully qualify the applicant for the functions performed under this category.

D. **Single Assembly Tester** shall mean any person duly certified under the *(rules and regulations)* of the *(administrative authority)* to test a single backflow prevention assembly at a specific location.

QUALIFICATIONS: Shall have demonstrated ability to make the test and to perform the duties of a Water Supervisor, as required by *(rules or regulations).*

Responsibilities and Additional Requirements of Testers.

Each applicant for certification as a tester of backflow prevention assemblies shall furnish evidence to show that he has available the necessary tools and equipment to properly field test such assemblies. The tester shall be responsible for the accuracy of all tests and reports prepared by the tester.

Revocation of Certification of Tester.

The certificate issued to any tester may be revoked or suspended by the *(administrative authority)* for improper testing or reporting. Proceedings for suspension or revocation shall be in accordance with *(rules or regulations).*

I hereby certify that the above Resolution was adopted by the *(administrative authority)* at its meeting held (*date*).

(*Administrative authority*)
By ________________
Address __________________________
Date ____________________________

8.13 Sample: Backflow Prevention Assembly Tester Examination Announcement

BACKFLOW PREVENTION ASSEMBLY TESTER EXAMINATION ANNOUNCEMENT

TYPES OF CERTIFICATION:

The certificate may be issued in one of two classifications:

1. The General Backflow Prevention Assembly Tester Classification will permit the tester to perform tests on all types of backflow prevention assemblies.

2. The Limited Backflow Prevention Assembly Tester Classification will restrict the tester to a single-type backflow prevention assembly.

REQUIREMENTS:

Applicants shall have had at least two (2) years' experience in plumbing, pipefitting, or equivalent qualifications.

APPLICATIONS:

Must be received by ________________________ 19______.

EXAMINATIONS:

Candidates may be examined for a good knowledge of basic hydraulics, theory of backflow prevention, applicable laws, design of backflow prevention assemblies, and the testing and troubleshooting of backflow prevention assemblies.

The examination will be given in two parts:

1. **Written Examination:** To be taken at any one of the following;

Date	**Location**	**Time**
date1	*location1*	*time1*
date2	*location2*	*time2*
"	"	"

❖ Please indicate your choice on your application form.
❖ Time allowance, two (2) hours. No books or aids will be permitted.
❖ Successful candidates will be eligible for the performance examination below:

2. **Performance Examination:**

❖ The time, date and location will be announced later *(normally given immediately following the written examination)*.

- ❖ Candidates will be required to demonstrate field test procedures on the following operating backflow prevention assemblies:

    - ☐ Reduced pressure principle backflow prevention assembly
    - ☐ Double check valve backflow prevention assembly
    - ☐ Pressure vacuum breaker backsiphonage prevention assembly

- ❖ Field Test Procedures utilized shall be those from the *Manual of Cross-Connection Control - {latest edition}*, published by the Foundation for Cross-Connection Control and Hydraulic Research at the University of Southern California.

- ❖ Candidates will furnish all field test equipment and tools for this part of the test.

Further information may be obtained upon request from the *(administrative authority).*

8.14 Sample Form: Application for a Backflow Prevention Assembly Tester Certificate

APPLICATION FOR BACKFLOW
PREVENTION ASSEMBLY TESTER CERTIFICATION

Name:____ Date ____________

LAST FIRST MIDDLE

Address:____

____

Telephone:____________________________________

PRESENT EMPLOYMENT:

List here enough information to show fulfillment of the experience requirement. Other qualifications which you feel would fulfill the eligibility requirement should be listed on Page two under "Other Experience and Qualifications."

From:___________________________ To:_________________________________

Firm Name__________________________________ Telephone________________

Address____

____

Type of work____

____

PREVIOUS EMPLOYMENT:

From:___________________________ To:_________________________________

Firm Name__________________________________ Telephone________________

Address____

____

Type of work____

REGISTRATION:

Master Plumber Registration Number ______________________

Journeyman Certificate of Qualification Number ______________________

Plumbing Contractor's License Number ______________________

Other ______________________

OTHER EXPERIENCE OR QUALIFICATIONS:

____

____

____

____

Check in the appropriate square the certification desired:

☐ General Certificate . . . . . Any type or make of backflow prevention assembly may be tested.

☐ Limited Certificate . . . . . Restricts the tester to a single-type backflow prevention assembly.

Indicate your choice of date and location for written examination(s):

	Date	Location
1.	__________	____________________
2.	__________	____________________
3.	__________	____________________

____

Signature

8.15 Sample List: Backflow Prevention Assembly Testers

Administrative Authority Letterhead

BACKFLOW PREVENTION ASSEMBLY TESTERS CERTIFIED AS OF *(date)*

(List of Certified Testers with addresses)

NAME	CERTIFICATION NO.	ADDRESS

Note: The above Certified Backflow Prevention Assembly Testers have demonstrated their ability to field test backflow prevention assemblies to the satisfaction of the *(administrative authority)*. The certifying agency assumes no liability for the quality of work performed by individuals listed. The company or individual utilizing the backflow prevention assembly tester should determine that the Tester maintains liability insurance.

8.16 Sample Letter: Notice of Shutdown

Company Letterhead

(Date)

Memorandum To: ______________________________
From: Water Supervisor

Please be advised of the necessity for shut down of *(specify water service)* due to:

- ☐ Code compliance
- ☐ Emergency repairs
- ☐ Preventative maintenance

This shut down will affect *(specify location within plant)* on *(date)* from *(hour to hour)*.

This work is essential and must be implemented. However, if this schedule is too disruptive to your work schedule a postponement may be possible. If you have need for rescheduling please telephone *(name)* at *(extension)* before *(date)*.

Thank you for your cooperation.

Sincerely,

(title)

8.17 Sample Letter: Letter of Acknowledgment for Field Evaluation Sites

(date)

(To Whom It May Concern)

re: Field Test Site
(manufacturer)
(model and size)

It is the understanding of *(property owner, water purveyor, or health agency)* that the subject test assembly(s) will be in the field for at least twelve (12) months and during that time the assembly(s) will be tested on a nominal thirty (30) day schedule by the Foundation for Cross-Connection Control and Hydraulic Research at the University of Southern California. A normal test will usually require a shutdown of water from five to ten minutes, however during these tests the water service may be turned off for an hour or more for inspection or maintenance. During the period of the Field Evaluation, only the Foundation Staff is permitted to test, inspect, or repair the assembly(s).

At the successful conclusion of the Field Evaluation phase of the Foundation Approval program, the assembly(s) will gain Approval and become the property of *(the owner)*. Should the subject assembly(s) not complete the Field Evaluation and not gain Foundation Approval, then *(the manufacturer)* will be responsible for replacing the field test assembly(s) with Approved assembly(s).

Sincerely,

(title)

cc. Foundation for Cross-Connection Control
and Hydraulic Research
(manufacturer)
(property owner)
(water purveyor)
(health agency)

8.18 Sample Form: Field Site Application Form

Instructions: Fill in form completely and attach letters of acknowledgment. Send application to:

FCCC&HR Laboratory (or approving agency)
3022 Riverside Drive
Los Angeles, CA 90039
(213) 662-3536 or (213) 740-2032
FAX (213) 665-2055

I. Field Test Assembly

Manufacturer ____________________________________

Model ____________________________ *(circle)* RP RPDA DC DCDA PVB SVB

Size ______________________________

Serial Number *(if known)* ____________________________________

II. Field Site Information

Field site location *(i.e., name of company, park, etc .)*:__________________________________

Field site Address:____

____

On site contact person______________________________ Phone number (____) __________

Owner Name *(owner of property or manager)*:____

Owner Address: ____

____

Phone number: (____) ____________________

III. Type of Installation

Description of usage downstream of proposed field site*(i.e., park irrigation; feeds into pump that irrigates school playground; supplies domestic water for 300 apartments;etc.)*:

____

____

____

____

PLEASE COMPLETE BACK SIDE

Type of protection: ☐ Service/Meter   Meter# ________ Service /Accnt. # _________
☐ Internal

Are there any hazardous materials or chemicals in use downstream from proposed field site? *(i.e., chemical injection pump, aspirator, etc.)*____

IV. Administrative Authority (Water and/or Health Agency) having Cognizance

Name of agency: ____

Address: ____

____

Phone number: (____) ______________________FAX (____) ___________________

Contact person: ___________________________Title _________________________

V. Letters of Acknowledgment : *(See Sample Letter 8.17.)*

A. The owner or manager of the property: *(attach letter)*
B. The water and/or health agency having cognizance: *(attach letter)*

VI. Applicant Information

Manufacturer representative (print name): ____

Title _________________________

Phone number (____) _______________   FAX (____) _________________

Signature:_____________________________________   Date____________________

NOTE: This completed application must be submitted before the initial field test can be made on the subject backflow prevention assembly.

FOR OFFICE USE ONLY

ACCEPTED ☐   **EVALUATED BY:** ______________ **DATE:** ___________

REJECTED ☐

COMMENTS: ____

____

____

8.19 Sample Cross-Connection Control Program Procedure

The purpose of the following suggested procedure of cross-connection control is for the purpose of achieving greater and more meaningful involvement and participation of all administrative authorities (water purveyor, health agency, plumbing official, etc.) in the field of cross-connection control. Responsibility for specific sections may be assumed by other entities depending upon practicality and local policies.

The following program procedure is recommended:

I. Preliminary Inspection (Screening) — Water Purveyor
II. Detailed Inspection — Water Purveyor and Health Department
III. Written Report and Recommendations — Health Department
IV. Reinspection for Compliance — Water Purveyor
V. Noncompliance Evaluation — Water Purveyor and Health Department
VI. Letter for Hearing — Health Department
VII. Follow-up Contact and Hearing — Water Purveyor and Health Department
VIII. Request for Water Shut Off — Health Department

I. Preliminary Inspection (Screening)

(1) That there are no known or potential cross-connection hazards.

(2) That cross-connection hazards do exist but are adequately protected.

(3) Cross-connection hazards do exist and a detailed joint inspection is needed.

The following information should be recorded on the appropriate record during the preliminary survey:

(1) Date of inspection
(2) Name and location of premises and person contacted
(3) Types of water uses on the premises
(4) Source(s) of water
(5) Liquids under pressure
(6) All existing backflow prevention assemblies on both internal and service protection

II. Detailed Inspection

A detailed inspection of the premises by a Water Purveyor and/or the Health Department team will evaluate the hazards existing. The detailed inspection also establishes measures of protection by either requesting internal plumbing changes or the installation of backflow prevention assemblies commensurate with the degree of hazard found. The most reasonable way of achieving abatement will be arrived at by agreement between the Water Purveyor and the Health Department as to the type and location of assemblies to be installed.

III. Written Report and Recommendation

The report and recommendations will be written by the Health Department. The report will be based on the findings of the detailed joint inspection and will be sent to the consumer outlining the cross-connections found and the manner in which they are to be corrected. The report establishes a time limit for compliance. This would generally be thirty (30) days. A copy of the report will be sent to the Water Purveyor.

IV. Reinspection for Compliance

The Water Purveyor's program specialist will resurvey the premises at the end of the time allowed to verify compliance or noncompliance. Findings of this survey will be given to the Health Department.

V. Noncompliance Evaluation

If the reinspection shows noncompliance by the consumer, the Water Purveyor and Health Department will jointly meet with the consumer to discuss reasons for noncompliance.

VI. Letter of Hearing

If the noncompliance evaluation and all reasonable avenues of approach fail to bring about compliance, the Health Department will send a letter to the consumer requesting him to show cause why his water should not be shut off and why legal action should not be initiated against him.

VII. Follow-up Contact and Hearing

If, after seven (7) days from the date of the letter of Hearing, additional follow-up attempts fail to secure compliance, the consumer will be called to a joint office hearing with the Water Purveyor and Health Department.

VIII. Request for Water Shut Off

If the hearing is unsuccessful, the Health Department will request the Water Purveyor, by letter, to shut off the water service to the consumer until compliance is obtained.

8.20 Model Ordinance

AN ORDINANCE
for the
CONTROL OF BACKFLOW AND CROSS-CONNECTIONS

Amendments to the *(local or state authority)* Code of *(city or state)*

(chapter and section)

Section 1. CROSS-CONNECTION CONTROL — GENERAL POLICY

1.1 **Purpose.** The purpose of this Ordinance is:

1.1.1 To protect the public potable water supply of *(political jurisdiction)* from the possibility of contamination or pollution by isolating within the consumer's internal distribution system(s) or the consumer's private water system(s) such contaminants or pollutants which could backflow into the public water systems; and,

1.1.2 To promote the elimination or control of existing cross-connections, actual or potential, between the consumer's in-plant potable water system(s) and non-potable water system(s), plumbing fixtures and industrial piping systems; and,

1.1.3 To provide for the maintenance of a continuing Program of Cross-Connection Control which will systematically and effectively prevent the contamination or pollution of all potable water systems.

1.2 **Responsibility.** The *(Water Commissioner or State Health Official)* shall be responsible for the protection of the public potable water distribution system from contamination or pollution due to the backflow of contaminants or pollutants through the water service connection. If, in the judgement of said *(Water Commissioner or Health Official)* an approved backflow prevention assembly is required *(at the consumer's water service connection; or, within the consumer's private water system)* for the safety of the water system, the *(Water Commissioner or Health Official)* or his designated agent shall give notice in writing to said consumer to install such an approved backflow prevention assembly(s) at a specific location(s) on his premises. The consumer shall immediately install such an approved backflow prevention assembly(s) at the consumer's own expense; and, failure, refusal or inability on the part of the consumer to install, have tested and maintained said assembly(s), shall constitute grounds for discontinuing water service to the premises until such requirements have been satisfactorily met.

Section 2. DEFINITIONS

2.1 **Water Commissioner or Health Official**. The (*Commissioner or Health Official*) in charge of the (*Water Department or Health Department*) of the (*Political Jurisdiction*) is invested with the authority and responsibility for the implementation of an effective cross-connection control program and for the enforcement of the provisions of this ordinance.

2.2 **Approved**.

a. The term "approved" as herein used in reference to a water supply shall mean a water supply that has been approved by the health agency having jurisdiction.

b. The term "approved" as herein used in reference to an air gap, a double check valve assembly, a reduced pressure principle backflow prevention assembly or other backflow prevention assemblies or methods shall mean an approval by the administrative authority having jurisdiction.

2.3 **Auxiliary Water Supply**. Any water supply on or available to the premises other than the purveyor's approved public water supply will be considered as an auxiliary water supply. These auxiliary waters may include water from another purveyor's public potable water supply or any natural source(s) such as a well, spring, river, stream, harbor, etc., or used waters or industrial fluids. These waters may be contaminated or polluted or they may be objectionable and constitute an unacceptable water source over which the water purveyor does not have sanitary control.

2.4 **Backflow.** The term "backflow" shall mean the undesirable reversal of flow of water or mixtures of water and other liquids, gases or other substances into the distribution pipes of the potable supply of water from any source or sources. See terms Backsiphonage (2.6) and Backpressure (2.5).

2.5 **Backpressure.** The term "backpressure" shall mean any elevation of pressure in the downstream piping system (by pump, elevation of piping, or steam and/or air pressure) above the supply pressure at the point of consideration which would cause, or tend to cause, a reversal of the normal direction of flow.

2.6 **Backsiphonage.** The term "backsiphonage" shall mean a form of backflow due to a reduction in system pressure which causes a subatmospheric pressure to exist at a site in the water system.

2.7 **Backflow Preventer**. An assembly or means designed to prevent backflow.

2.7.1 **Air gap.** The term "air gap" shall mean a physical separation between the free flowing discharge end of a potable water supply pipeline and an open or non-pressure receiving vessel. An "approved air gap" shall be at least double the diameter of the supply pipe measured vertically above the overflow rim of the vessel — in no case less than 1 inch (2.54 cm).

2.7.2 **Reduced Pressure Principle Backflow Prevention Assembly.** The term "reduced pressure principle backflow prevention assembly" shall mean an assembly containing two independently acting approved check valves together with a hydraulically operating, mechanically independent pressure differential relief valve located between the check valves and at the same time below the first check valve. The unit shall include properly located resilient seated test cocks and tightly closing resilient seated shutoff valves at each end of the assembly. This assembly is designed to protect against a non-health (i.e., pollutant) or a health hazard (i.e., contaminant). This assembly shall not be used for backflow protection of sewage or reclaimed water.

2.7.3 Double Check Valve Backflow Prevention Assembly. The term "double check valve backflow prevention assembly" shall mean an assembly composed of two independently acting, approved check valves, including tightly closing resilient seated shutoff valves attached at each end of the assembly and fitted with properly located resilient seated test cocks. (See Specifications, Section 10 for additional details.) This assembly shall only be used to protect against a non-health hazard (i.e., pollutant).

2.8 Contamination. The term "contamination" shall mean an impairment of the quality of the water which creates an actual hazard to the public health through poisoning or through the spread of disease by sewage, industrial fluids, waste, etc.

2.9 Cross-Connection. The term "cross-connection" shall mean any unprotected actual or potential connection or structural arrangement between a public or a consumer's potable water system and any other source or system through which it is possible to introduce into any part of the potable system any used water, industrial fluid, gas, or substance other than the intended potable water with which the system is supplied. Bypass arrangements, jumper connections, removable sections, swivel or change-over devices and other temporary or permanent devices through which or because of which backflow can or may occur are considered to be cross-connections.

a. The term "direct cross-connection" shall mean a cross-connection which is subject to both backsiphonage and backpressure.

b. The term "indirect cross-connection" shall mean a cross-connection which is subject to backsiphonage only.

2.10 Cross-Connections — Controlled. A connection between a potable water system and a non-potable water system with an approved backflow prevention assembly properly installed and maintained so that it will continuously afford the protection commensurate with the degree of hazard.

2.11 Cross-Connection Control by Containment. The term "service protection" shall mean the appropriate type or method of backflow protection at the service connection, commensurate with the degree of hazard of the consumer's potable water system.

2.12 Hazard, Degree of. The term "degree of hazard" shall mean either a pollutional (non-health) or contamination (health) hazard and is derived from the evaluation of conditions within a system.

2.12.1 Hazard — Health. The term "health hazard" shall mean an actual or potential threat of contamination of a physical or toxic nature to the public potable water system or the consumer's potable water system that would be a danger to health.

2.12.2 Hazard — Plumbing. The term "plumbing hazard" shall mean an internal or plumbing type cross-connection in a consumer's potable water system that may be either a pollutional or a contamination type hazard. This includes but is not limited to cross-connections to toilets, sinks, lavatories, wash trays and lawn sprinkling systems. Plumbing type cross-connections can be located in many types of structures including homes, apartment houses, hotels and commercial or industrial establishments. Such a connection, if permitted to exist, must be properly protected by an appropriate type of backflow prevention assembly.

2.12.3 Hazard — Pollutional. The term "pollutional hazard" shall mean an actual or potential threat to the physical properties of the water system or the potability of the public or the consumer's potable water system but which would not constitute a health or system hazard, as defined. The maximum degree or intensity of pollution to which the potable water system could be degraded under this definition would cause a nuisance or be aesthetically objectionable or could cause minor damage to the system or its appurtenances.

2.12.4 Hazard — System. The term "system hazard" shall mean an actual or potential threat of severe danger to the physical properties of the public or the consumer's potable water system or of a pollution or contamination which would have a protracted effect on the quality of the potable water in the system.

2.13 Industrial Fluids. The term "industrial fluids" shall mean any fluid or solution which may be chemically, biologically or otherwise contaminated or polluted in a form or concentration which would constitute a health, system, pollutional or plumbing hazard if introduced into an approved water supply. This may include, but not be limited to: polluted or contaminated used waters; all types of process waters and "used waters" originating from the public potable water system which may deteriorate in sanitary quality; chemicals in fluid form; plating acids and alkalies; circulated cooling waters connected to an open cooling tower and/or cooling waters that are chemically or biologically treated or stabilized with toxic substances; contaminated natural waters such as from wells, springs, streams, rivers, bays, harbors, seas, irrigation canals or systems, etc.; oils, gases, glycerine, paraffins, caustic and acid solutions and other liquid and gaseous fluids used industrially, for other processes, or for fire fighting purposes.

2.14 Pollution. The term "pollution" shall mean an impairment of the quality of the water to a degree which does not create a hazard to the public health but which does adversely and unreasonably affect the aesthetic qualities of such waters for domestic use.

2.15 Water — Potable. The term "potable water" shall mean any public potable water supply which has been investigated and approved by the health agency. The system must be operating under a valid health permit. In determining what constitutes an approved water supply, the health agency has final judgment as to its safety and potability.

2.16 Water — Non-potable. The term "non-potable water" shall mean a water supply which has not been approved for human consumption by the health agency having jurisdiction.

2.17 Water — Service Connection. The term "service connection" shall mean the terminal end of a service connection from the public potable water system, (i.e., where the water purveyor may lose jurisdiction and sanitary control of the water at its point of delivery to the consumer's water system). If a water meter is installed at the end of the service connection, then the service connection shall mean the downstream end of the water meter.

2.18 Water — Used. The term "used water" shall mean any water supplied by a water purveyor from a public potable water system to a consumer's water system after it has passed through the service connection and is no longer under the control of the water purveyor. See Section 7.2.3.33.

Section 3. REQUIREMENTS

3.1 Water System

3.1.1 The water system shall be considered as made up of two parts: The Water Purveyor's System and the Consumer's System.

3.1.2 Water Purveyor's System shall consist of the source facilities and the distribution system; and shall include all those facilities of the water system under the complete control of the purveyor, up to the point where the consumer's system begins.

3.1.3 The source shall include all components of the facilities utilized in the production, treatment, storage, and delivery of water to the distribution system.

3.1.4 The distribution system shall include the network of conduits used for the delivery of water from the source to the consumer's system.

3.1.5 The consumer's system shall include those parts of the facilities beyond the termination of the water purveyor's distribution system which are utilized in conveying potable water to points of use.

3.2 Policy

3.2.1 No water service connection to any premise shall be installed or maintained by the water purveyor unless the water supply is protected as required by *(political jurisdiction)* laws and regulations and this *(name of legal document).* Service of water to any premise shall be discontinued by the water purveyor if a backflow prevention assembly required by this *(name of legal document)* is not installed, tested and maintained, or if it is found that a backflow prevention assembly has been removed, bypassed, or if an unprotected cross-connection exists on the premises. Service will not be restored until such conditions or defects are corrected.

3.2.2 The consumer's system should be open for inspection at all reasonable times to authorized representatives of the *(Water or Health agency name)* to determine whether unprotected cross-connections or other structural or sanitary hazards, including violations of these regulations, exist. When such a condition becomes known, the *(Water Commissioner or Health Officer)* shall deny or immediately discontinue service to the premises by providing for a physical break in the service line until the consumer has corrected the condition(s) in conformance with the *(political jurisdiction)* statutes relating to plumbing and water supplies and the regulations adopted pursuant thereto.

3.2.3 An approved backflow prevention assembly shall also be installed on each service line to a consumer's water system at or near the property line or immediately inside the building being served; but, in all cases, before the first branch line leading off the service line wherever the following conditions exist:

a. In the case of premises having an auxiliary water supply which is not or may not be of safe bacteriological or chemical quality and which is not acceptable as an additional source by the *(Water Commissioner or Health Officer)*, the public water system shall be protected against backflow from the premises by installing an approved backflow prevention assembly in the service line commensurate with the degree of hazard.

b . In the case of premises on which any industrial fluids or any other objectionable substance is handled in such a fashion as to create an actual or potential hazard to the public water system, the public system shall be protected against backflow from the premises by installing an approved backflow prevention assembly in the service line commensurate with the degree of hazard. This shall include the handling of process waters and waters originating from thewater purveyor's system which have been subject to deterioration in quality.

c. In the case of premises having (1) internal cross-connections that cannot be permanently corrected or protected against, or (2) intricate plumbing and piping arrangements or where entry to all portions of the premises is not readily accessible for inspection purposes, making it impracticable or impossible to ascertain whether or not dangerous cross-connections exist, the public water system shall be protected against backflow from the premises by installing an approved backflow prevention assembly in the service line.

3.2.4 The type of protective assembly required under subsections 3.2.3a, b, and c shall depend upon the degree of hazard which exists as follows:

a. In the case of any premise where there is an auxiliary water supply as stated in subsection 3.2.3.a of this section and it is not subject to any of the following rules, the public water system shall be protected by an approved air gap or an approved reduced pressure principle backflow prevention assembly.

b. In the case of any premise where there is water or substance that would be objectionable but not hazardous to health, if introduced into the public water system, the public water system shall be protected by an approved double check valve backflow prevention assembly.

c. In the case of any premise where there is any material dangerous to health which is handled in such a fashion as to create an actual or potential hazard to the public water system, the public water system shall be protected by an approved air gap or an approved reduced pressure principle backflow prevention assembly. Examples of premises where these conditions will exist include sewage treatment plants, sewage pumping stations, chemical manufacturing plants, hospitals, mortuaries and plating plants.

d. In the case of any premise where there are unprotected cross-connections, either actual or potential, the public water system shall be protected by an approved air gap or an approved reduced pressure principle backflow prevention assembly at the service connection.

e. In the case of any premise where, because of security requirements or other prohibitions or restrictions, it is impossible or impractical to make a complete in-plant cross-connection survey, the public water system shall be protected against backflow from the premises by either an approved air gap or an approve reduced pressure principle backflow prevention assembly on each service to the premise.

3.2.5 Any backflow prevention assembly required herein shall be a make, model and size approved by the *(Water Commissioner or Health Official)*. The term "Approved Backflow Prevention Assembly" shall mean an assembly that has been manufactured in full conformance with the standards established by the American Water Works Association entitled:

AWWA/ANSI C510-92[1] Standard for Double Check Valve Backflow Prevention Assemblies;

AWWA/ANSI C511-92[1] Standard for Reduced Pressure Principle Backflow Prevention Assemblies;

and, have met completely the laboratory and field performance specifications of the Foundation for Cross-Connection Control and Hydraulic Research of the University of Southern California (USC FCCCHR) established in:

Specifications of Backflow Prevention Assemblies — Section 10 of the most current edition of the *Manual of Cross-Connection Control.*

Said AWWA and USC FCCCHR standards and specifications have been adopted by the *(Water Commissioner or Health Official)*. Final approval shall be evidenced by a "Certificate of Compliance" for the said AWWA standards; or "Certificate of Approval" for the said USC FCCCHR Specifications; issued by an approved testing laboratory.

The following testing laboratory has been qualified by the *(Water Commissioner or Health Officer)* to test and approve backflow prevention assemblies:

Foundation for Cross-Connection Control and Hydraulic Research
University of Southern California
KAP-200 University Park MC-2531
Los Angeles, California 90089-2531

Testing laboratories other than the laboratory listed above will be added to an approved list as they are qualified by the *(Water Commissioner or Health Officer).*

Backflow preventers which may be subjected to backpressure or backsiphonage that have been fully tested and have been granted a Certificate of Approval by said qualified laboratory and are listed on the laboratory's current list of approved backflow prevention assemblies may be used without further test or qualification.

3.2.6 It shall be the duty of the consumer at any premise where backflow prevention assemblies are installed to have a field test performed by a certified backflow prevention assembly tester upon installation and at least once per year. In those instances where the *(Water Commissioner or Health Officer)* deems the hazard to be great enough he may require field tests at more frequent intervals. These tests shall be at the expense of the water user and shall be performed by *(Water Department)* personnel or by a certified tester approved by the *(Water Commis-*

[1]Prior to 1989 the AWWA/ANSI C506 Standard covered both the double check valve assembly and the reduced pressure principle backflow prevention assembly.

sioner or Health Officer). It shall be the duty of the *(Water Commissioner or Health Officer)* to see that these tests are made in a timely manner. The consumer shall notify the *(Water Commissioner or Health Officer)* in advance when the tests are to be undertaken so that an official representative may witness the field tests if so desired. These assemblies shall be repaired, overhauled or replaced at the expense of the consumer whenever said assemblies are found to be defective. Records of such tests, repairs and overhaul shall be kept and made available to the *(Water Commissioner or Health Officer).*

3.2.7 All presently installed backflow prevention assemblies which do not meet the requirements of this section but were approved devices for the purposes described herein at the time of installation and which have been properly maintained, shall, except for the testing and maintenance requirements under subsection 3.2.6, be excluded from the requirements of these rules so long as the *(Water Commissioner or Health Officer)* is assured that they will satisfactorily protect the water purveyor's system. Whenever the existing device is moved from the present location or requires more than minimum maintenance or when the *(Water Commissioner or Health Officer)* finds that the maintenance constitutes a hazard to health, the unit shall be replaced by an approved backflow prevention assembly meeting the requirements of this section.

3.2.8 The *(Water Commissioner or Health Officer)* is authorized to make all necessary and reasonable rules and policies with respect to the enforcement of this ordinance. All such rules and policies shall be consistent with the provisions of this ordinance and shall be effective *(number)* days after being filed with the *(clerk or secretary)* of the *(political jurisdiction).*

The foregoing ordinance was first read at the meeting of the *(name of governing body)* of the *(political jurisdiction)* of ____________________ on the ________ day of ____________________, 19_____ and adopted by the following called vote on motion of *(Official).*

Ayes:
Noes:
Abstaining: Approver____________________________
Absent: (*Official Title*)

Attest:

____
(*clerk or secretary*)

SEAL

SECTION 9
Backflow Prevention Assembly Field Test Procedures and Gage Accuracy Verification

As part of a complete cross-connection control program, proper field test procedures for the required initial and subsequent annual tests must be used. In this Section the field test procedures are detailed as follows:

Section 9.1 — Preliminary Steps

Section 9.2 — Reduced Pressure Principle Backflow Prevention Assembly (RP)

Section 9.3 — Double Check Valve Backflow Prevention Assembly (DC)

Section 9.4 — Pressure Vacuum Breaker Backsiphonage Prevention Assembly (PVB)

Section 9.5 — Spill-Resistant Pressure Vacuum Breaker Backsiphonage Prevention Assembly (SVB)

Section 9.6 — Reduced Pressure Principle - Detector Backflow Prevention Assembly (RPDA)

Section 9.7 — Double Check - Detector Backflow Prevention Assembly (DCDA)

Properly calibrated gage equipment is essential to insure accurate data acquisition. Therefore, methods to inspect the accuracy of the gages used in the above test procedures are supplied in Section 9.8. Gaging equipment should be checked for accuracy at least once a year, and re-calibrated when inaccuracy is greater than ±0.2 psid. Local administrative authorities should be consulted regarding possible local calibration requirements.

Section 9.8.1 — Differential gage accuracy check — (water column)

Section 9.8.2 — Differential gage accuracy check — (mercury manometer)

9.1 Preliminary Steps

9.1.1 The tester must observe the condition of the test gage equipment during all steps of the following field test procedures. Visually inspect the test gage equipment for obvious leakage or damage. The gage should zero out when not pressurized; needle valves and fittings must be drip tight; gage should be drained after testing to protect against freezing.

9.1.2 The tester must observe general safety procedures during all aspects of the field test and maintenance procedures. Including, but not limited to:

Personal Safety
Confined Spaces
Elevated platforms
Electrical hazards
Tools

9.1.3 As a prelude to each of the field test procedures it is essential to follow some basic steps.

9.1.3.1 Notification — Owner of the assembly and/or the onsite personnel must be notified that water service will be shut off during the test procedure. Special arrangements may have to be made so that interruption of service will not create a hardship on the user.

Note: Fire Sprinkler Systems
The tester shall request that the owner or occupant notify the authority having jurisdiction, the fire department, if required, and the alarm receiving facility before shutting down a fire sprinkler system or its water supply. The notification prior to testing shall include the purpose for the shutdown of the system, the component(s) involved, and the estimated time required.

9.1.3.2 Identify — Make sure that the proper assembly is being tested by checking identification on the assembly for make (manufacturer), model, size and serial number. Record all of this information, as well as the test data, before leaving the location.

9.1.3.3 Inspect — Inspect the assembly for the required components for the field test procedure (i.e., upstream and downstream shutoff valves, and properly located test cocks).

9.1.3.4 Observe — Carefully observe area around the assembly for telltale signs of leakage *(i.e., moss or algae growth, plant life, or soil erosion)*. This should supply the tester with additional information regarding the condition of the assembly before the test is performed. *Example:* Wet spot under relief valve port of reduced pressure principle backflow prevention assembly is an indication of relief valve activity, possibly from pressure fluctuations or fouling of the assembly. Proper testing will define the problem.

9.1.4 Field Test Reporting

9.1.4.1 Physical Identification

Physical identification of the backflow prevention assembly must be recorded. This information should include, but not be limited to:

Assembly - Manufacturer, model, serial number, size
Location - Address, physical location

9.1.4.2 Field Test Results

The field tests results from the field test procedure must be accurately recorded on the appropriate Field Test Form. (See Fig. 8.14 for example of Field Test Form.) The *Field Test Form* will typically detail the following:

a. Initial Test Results
b. Repair/Maintenance performed
c. Retest Results

9.1.4.3 Tester Information

a. Tester identification (i.e., certification No. or license No.)
b. Tester's signature and printed name, phone number
c. Date and Time of test

9.1.4.4 Distribution of Field Test Forms

a. Administrative Authority(s)
b. Owner
c. Tester

9.1.5 Maintenance and Repairs

Consult the manufacturer's repair/maintenance manuals before attempting any disassembly, large spring loads may be present in some designs.

To maintain the backflow prevention assembly in proper operating condition, the tester/repair person must follow some basic steps.

a) Use properly operating and calibrated gage equipment (see Section 9.8)

b) Use proper field test procedures (see Sections 9.2, 9.3, 9.4, 9.5, 9.6, and 9.7)

c) Consult manufacturer's repair/maintenance manuals when disassembly is required. Specialized tools (i.e., not commercially available) may be required for some repair/ maintenance procedures.

d) Use only manufacturer's replacement parts.

e) Immediately following repair, maintenance, or replacement procedures, the backflow prevention assembly must be retested and reports properly distributed.

f) When returning an assembly to normal operating condition, *slowly* repressurize the assembly and then the downstream piping and equipment to avoid waterhammer.

9.1.6 Manual of Cross-Connection Control Ninth Edition Modifications

To give the tester more detailed information about the field test procedures, a number of improvements have been incorporated in this Ninth Edition of the *Manual of Cross-Connection Control.* Please note the following:

RP (Section 9.2)

Additional detailing for each of the tests has been added.

DC (Section 9.3)

The test procedure has been changed to a *direction of flow* test utilizing the differential pressure gage. The duplex gage method has been eliminated.

PVB (Section 9.4)

Additional detailing for each of the tests has been added.

SVB (Section 9.5)

New Section

RPDA and DCDA (Sections 9.6 and 9.7)

New Sections

Illustrated Field Test Procedures

The field test procedures have been illustrated to enhance the understanding of the procedures. The Legend below will be helpful in understanding the procedures on the following pages.

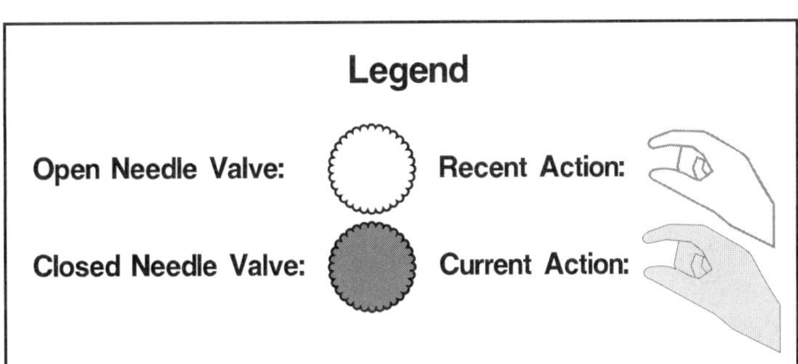

Appendix

Additional field testing information for each of the field test procedures has been included in Appendix A.

RP - Alternate Gage Configurations - Text for field test procedures utilizing gages with two (2) or three (3) needle valve configurations.

RP - An optional direction of flow test has been added for the check valve No. 2.

PVB and SVB - A means of determining backpressure from the piping downstream of the PVB and SVB is detailed.

9.2 Reduced Pressure Principle Backflow Prevention Assembly (RP)

9.2.1 Equipment Required:

a) Differential Pressure Gage—Minimum range 0 -15 PSID (0.1 or 0.2 psid graduations)

b) Three 6 ft. lengths—minimum 1/4"∅ high pressure hose

c) 1/4" Needle valves, for fine control of flows

d) Three 1/4" IPS x 45° SAE flare connectors—brass

e) Adapter fittings for each test cock size—brass 1/8" x 1/4", 1/4" x 1/2", 1/4" x 3/4"

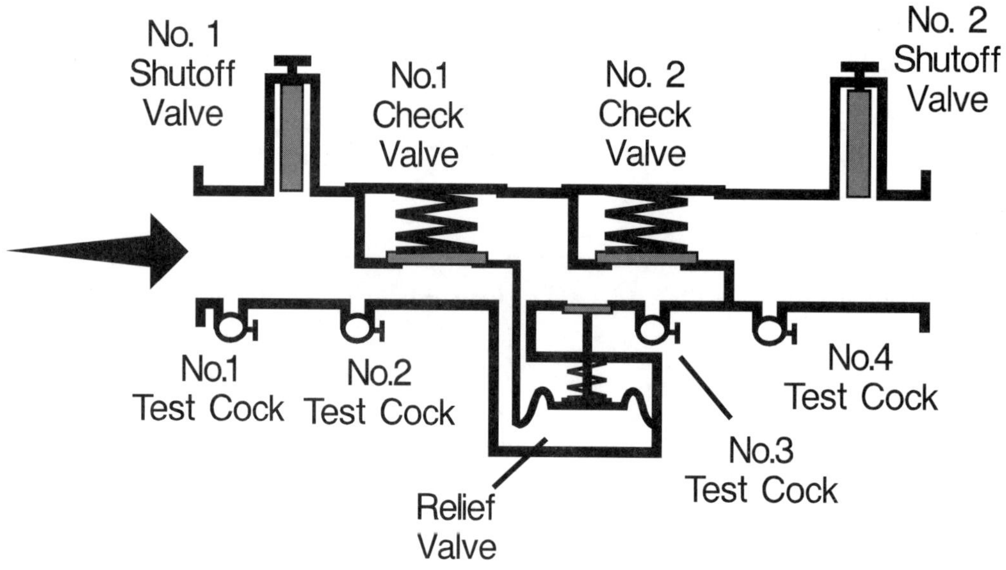

Fig. 9.1
Reduced Pressure Principle Assembly with All Components

9.2.2 Field Test Procedure

Test No. 1 - *Relief Valve Opening Point*

Purpose: To test the operation of the differential pressure relief valve.

Requirement: The differential pressure relief valve must operate to maintain the zone between the two check valves at least 2 psi less than the supply pressure.

NOTE: It is important that during this test the tester does not cause the relief valve to discharge before step ***j*** below. *(See Troubleshooting Section 9.2.3.1 - Exercising the Differential Pressure Relief Valve.)*

Steps:

Follow all preliminary steps detailed in Section 9.1

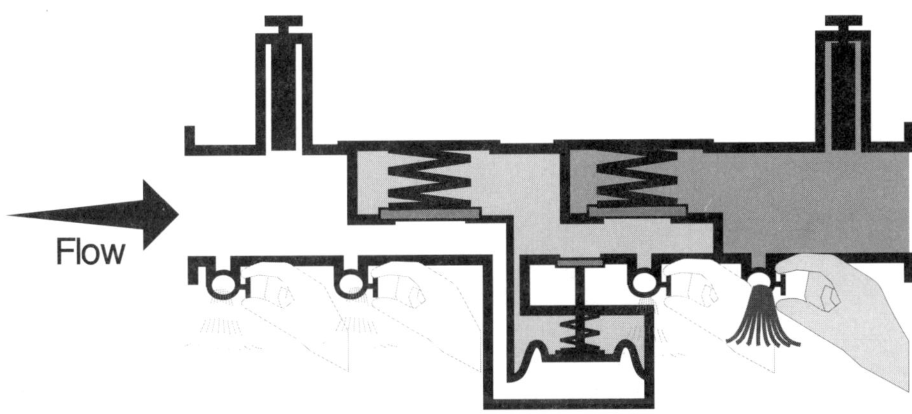

a. Open No. 4 test cock to establish flow through the unit, then flush water through test cocks No. 1, No. 2 (open No. 2 test cock slowly), and No. 3, by opening and closing each test cock one at a time, to eliminate foreign material. Be careful not to activate the relief valve during this process. Close test cock No. 4.

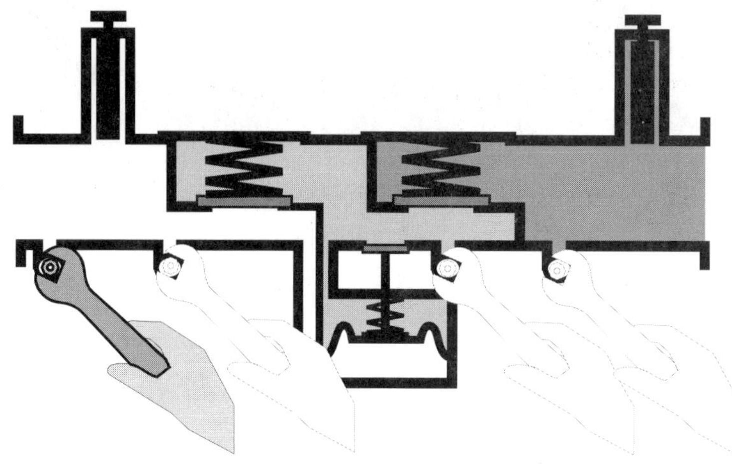

b. Install appropriate fittings to test cocks.

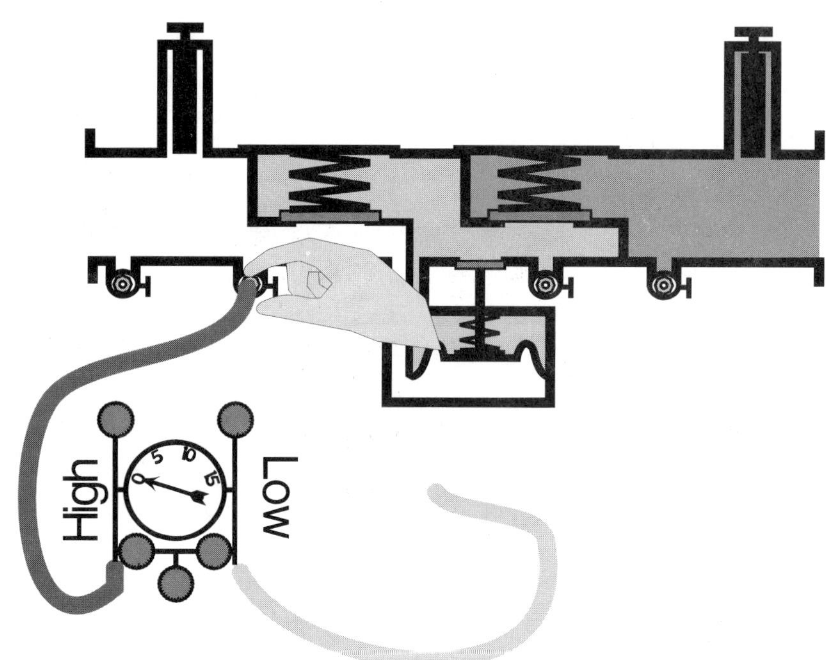

c. Attach hose from the high side of the differential pressure gage to the No. 2 test cock.

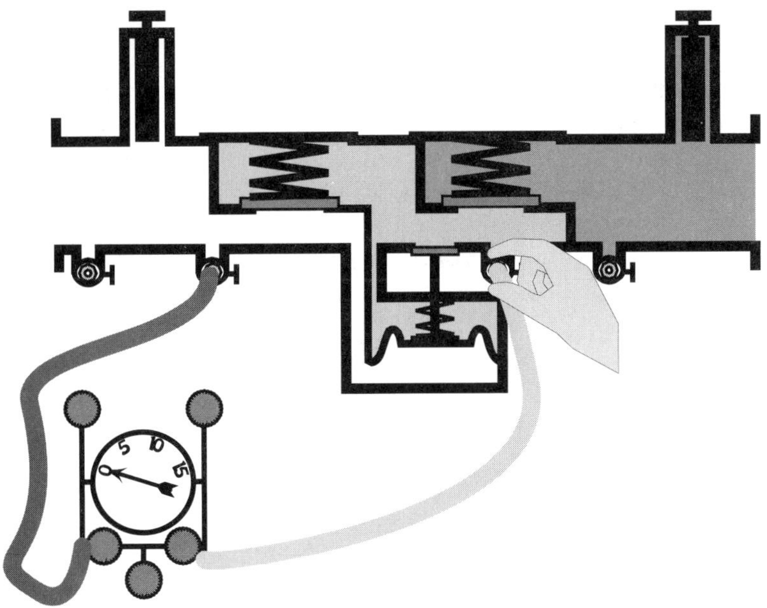

d. Attach hose from the low side of the differential pressure gage to the No. 3 test cock.

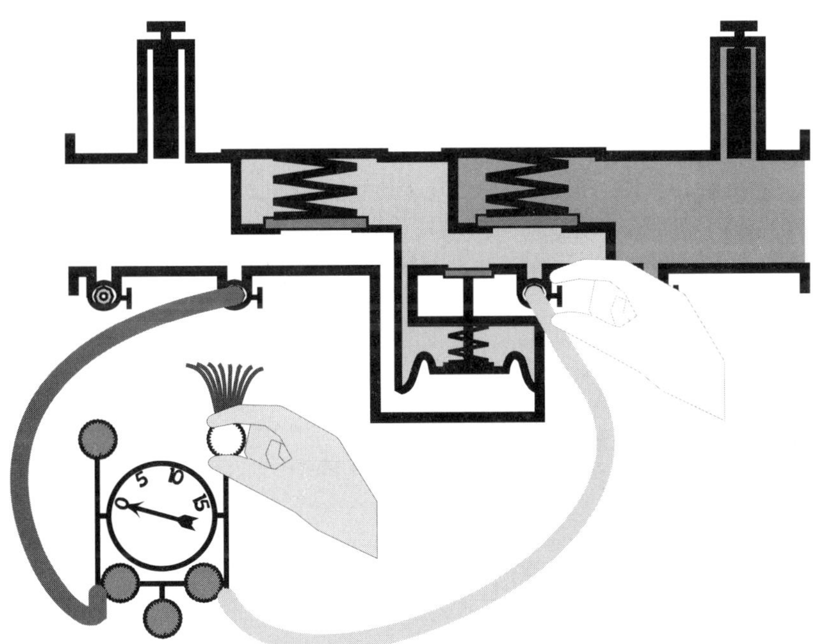

e. Open test cock No. 3 slowly and then bleed all air from the hose and gage by opening the low side bleed needle valve.

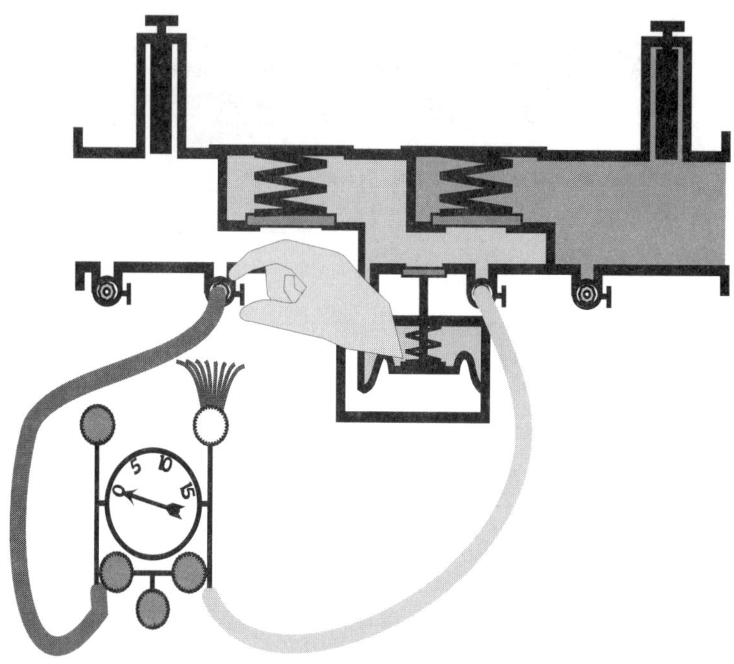

f. Maintain the low side bleed needle valve in the open position while test cock No. 2 is opened **slowly**.

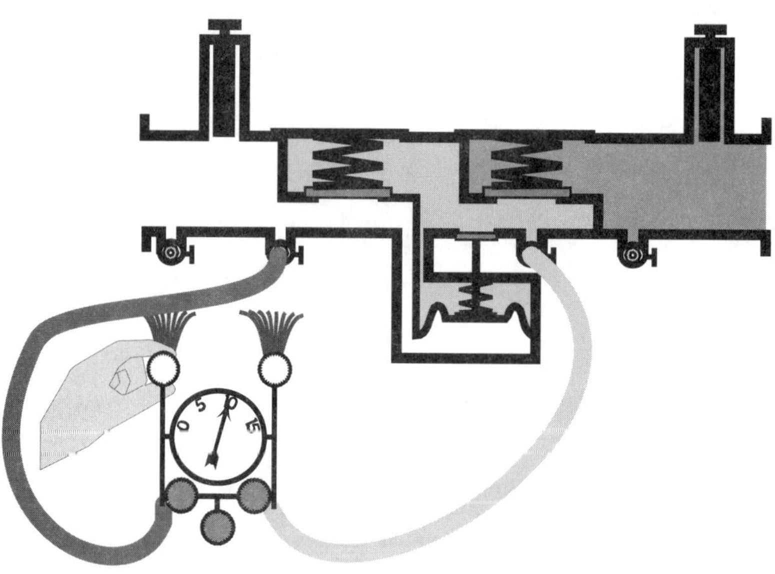

g. Open the high side bleed needle valve to bleed all air from the hose and gage.

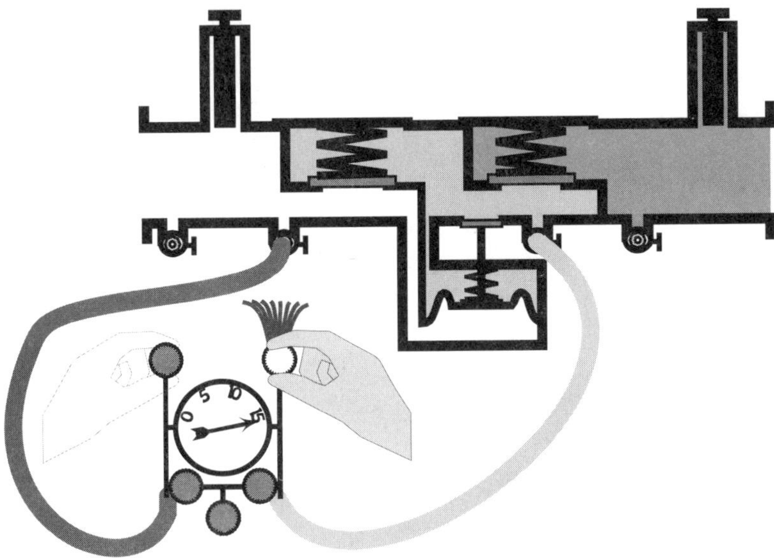

h. Close the high side bleed needle valve, then close the low side bleed needle valve after the gage reading has reached the upper end of the scale.

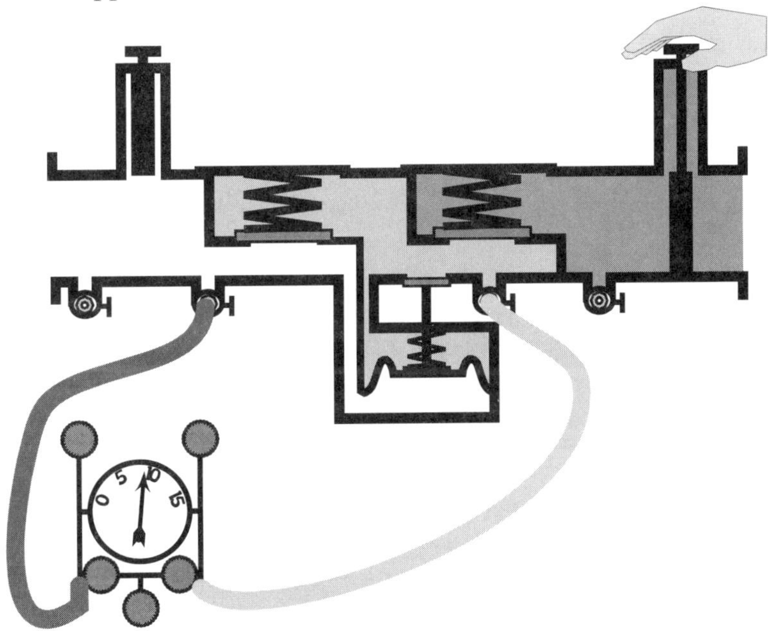

i. Close No. 2 shutoff valve. Should the gage reading drop to the low end of the gage scale and the differential pressure relief valve discharge continuously, then the No. 1 check valve is leaking. Tests No. 1, No. 2 and No. 3 may not be completed. *(See Troubleshooting Section 9.2.3.2 - Leaking No. 1 Check Valve.)* However, should the gage reading remain above the differential pressure relief valve opening point (i.e., relief valve does not discharge), then observe the gage reading. This is the apparent pressure drop across the No. 1 check valve. During Tests No. 1, No. 2, and No. 3 of this procedure the differential pressure gage is on-line showing the pressure drop across the No. 1 check valve.

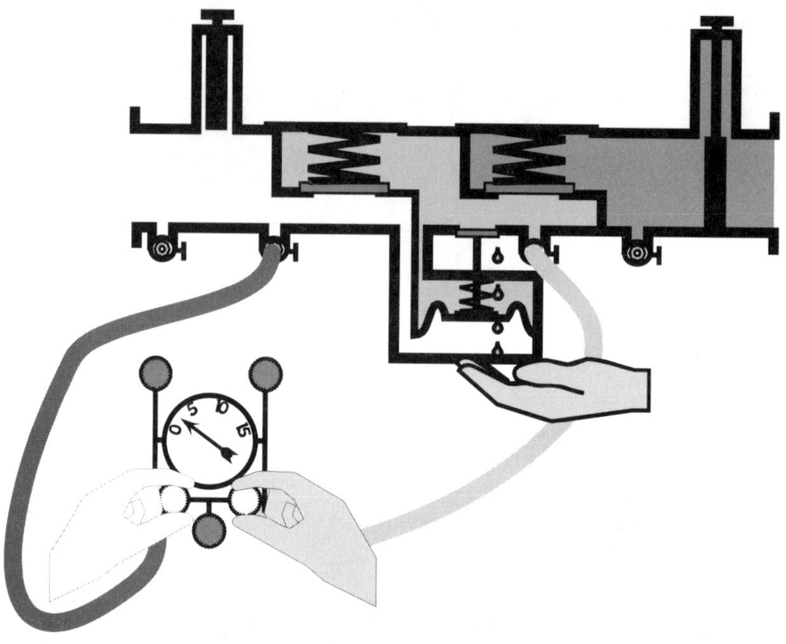

j. Open the high side control needle valve approximately one turn, and then open the low side control needle valve no more than one-quarter (1/4) turn to bypass water from the No. 2 test cock to the No. 3 test cock. If the low side control needle valve must be opened more than one-quarter (1/4) turn to lower the differential pressure reading to the relief valve opening point, then see *Troubleshooting Section 9.2.3.3 - Instructions for Leaking No. 2 Shutoff Valve.* Observe the differential pressure reading as it **slowly** drops to the relief valve opening point. Record this opening point value when the first discharge of water is detected.

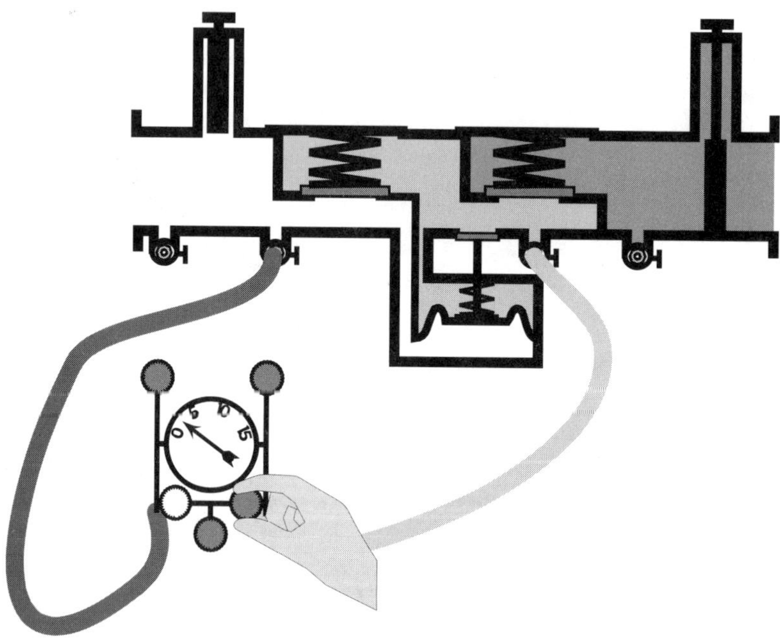

k. Close the low side control needle valve.

Test No. 2 - *Tightness of No. 2 Check Valve*

Purpose: To test the No. 2 check valve for tightness against backpressure.

Requirement: The No. 2 check valve shall be tight against backpressure.

Steps:

a. Maintain the No. 2 shutoff valve in a closed position (from test No. 1).

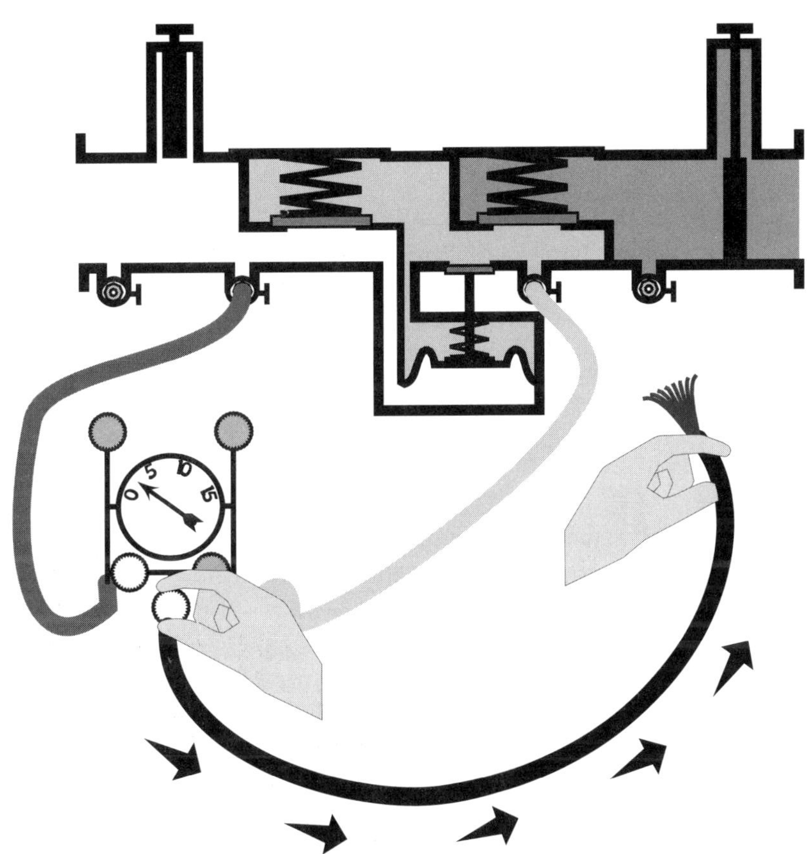

b. Vent all of the air through the bypass hose by opening the bypass needle valve. Close the bypass needle valve only (The high side control needle valve is to remain open).

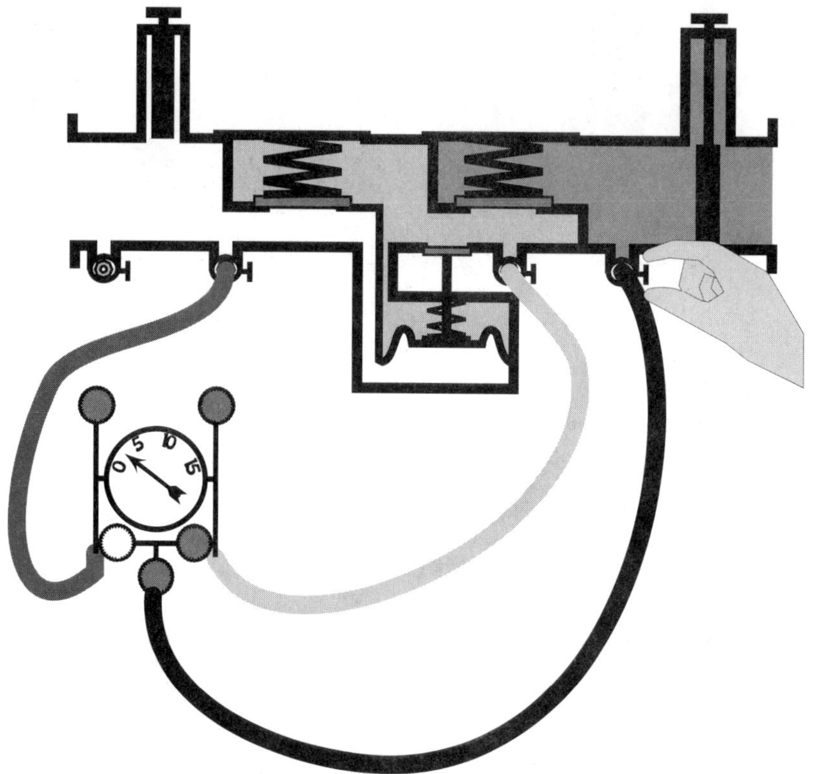

c. Attach the bypass hose from the gage to the No. 4 test cock, then open the No. 4 test cock.

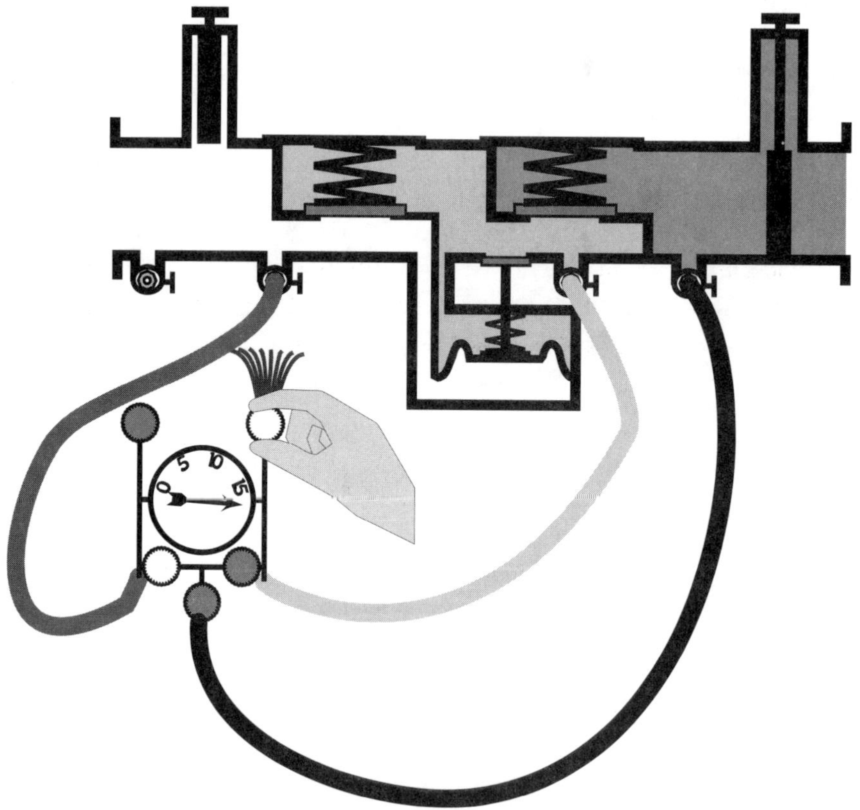

d. Bleed water from the zone by opening the low side bleed needle valve on the gage in order to re-establish the normal reduced pressure within the zone. Once the gage reading reaches a value above the apparent No. 1 check valve pressure drop, close the low side bleed needle valve.

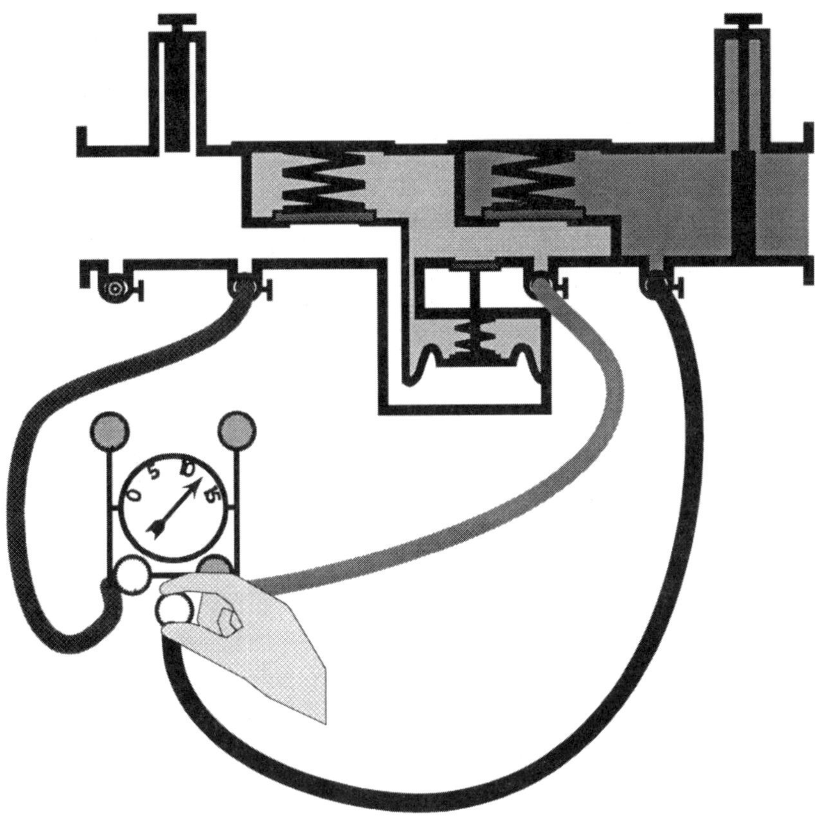

e. Open the bypass needle valve.

- If the indicated differential pressure reading remains steady then the No. 2 check valve is reported as "closed tight." Go to Test No. 3.

- If the differential pressure reading falls to the relief valve opening point, bleed water through the low side bleed needle valve until the gage reading reaches a value above the apparent No. 1 check valve pressure drop. If the gage reading settles above the relief valve opening point, record the No. 2 check valve as "closed tight," and proceed to Test No. 3. If the differential pressure reading falls to the relief valve opening point again, then the No. 2 check valve is noted as "leaking," and Test No. 3 below cannot be completed.

- If the differential pressure reading drops, but stabilizes above the relief valve opening point, the No. 2 check valve can still be reported as "closed tight." *See the Troubleshooting Section 9.2.3.4 - Disc Compression - No. 2 Check Valve.*

- If the reading on the gage increases to the high end of the scale, *see Troubleshooting Section 9.2.3.5 - Backpressure Condition.*

Test No. 3 - *Tightness of No. 1 Check Valve*

Purpose: To determine the tightness of check valve No. 1, and to record the static pressure drop across check valve No. 1.

Requirement: The static pressure drop across check valve No. 1 should be at least 3.0 psi greater than the relief valve opening point (Test No. 1). This 3.0 psi buffer will prevent the relief valve from discharging during small fluctuations in line pressure. A buffer of less than 3.0 psi does not imply a leaking check valve No. 1 (i.e., allowing backflow to occur), but rather is an indication of how well check valve No. 1 is sealing. *(See Appendix Section A.2.1 - 3.0 psi Buffer.)*

Steps:

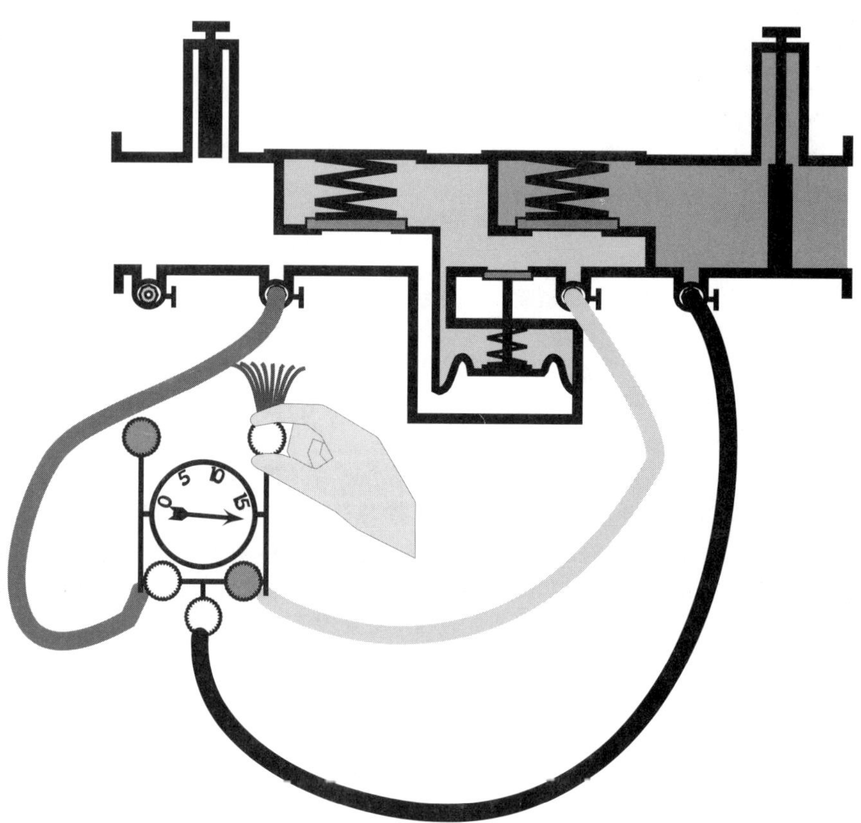

a. With the bypass hose connected to test cock No. 4 as in step ***e*** of Test No. 2 *(high side control needle valve and bypass needle valve remaining open),* bleed water from the zone through the low side bleed needle valve on the gage until the gage reading exceeds the apparent No. 1 check valve pressure drop. Close the low side bleed needle valve. After the gage reading settles, the steady state differential pressure reading indicated (reading is not falling on the gage) is the actual static (i.e., no flow) pressure drop across check valve No. 1 and is to be recorded as such.

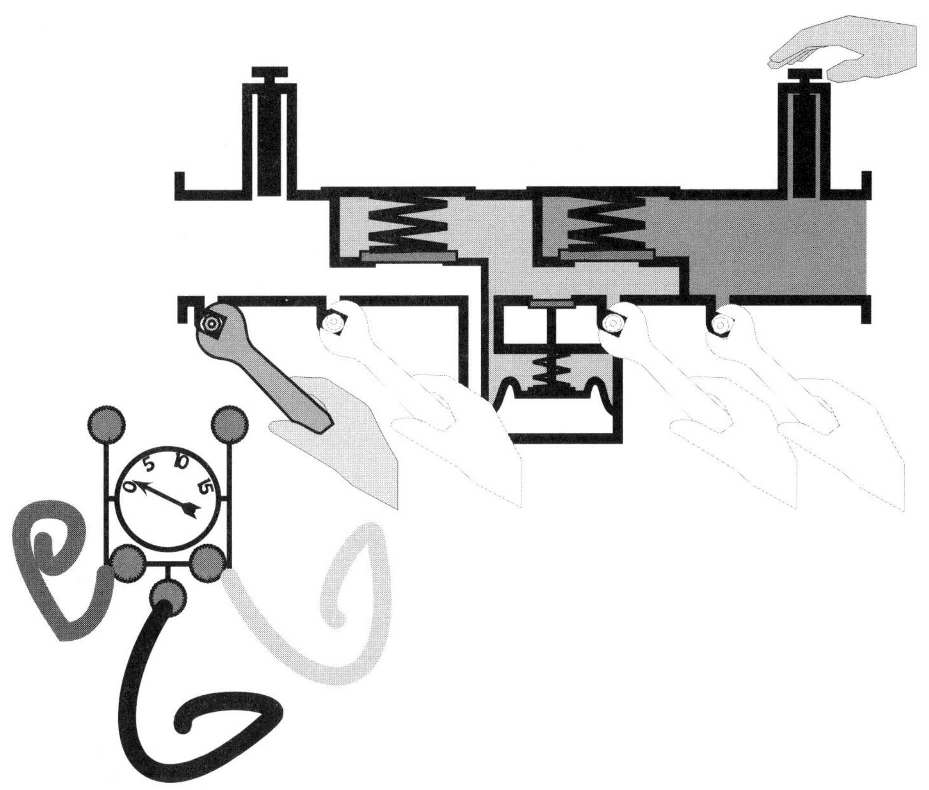

b. Close all test cocks, slowly open shutoff valve No. 2, and remove all test equipment.

Values recorded from above Tests No. 1, No. 2, and No. 3			***Acceptable Results***
Test No. 1:	Relief valve opening point —	___.___ psid	2.0 psid or greater
Test No. 2:	Check valve No. 2 —	Tight/Leak	Tight
Test No. 3:	Check valve No. 1 —	___.___ psid	A Recommended Value of at Least 3.0 psi Greater than the Relief Valve Opening Point

9.2.3 Troubleshooting

9.2.3.1 Exercising the Differential Pressure Relief Valve

It is one of the objectives of the field test procedure to determine the opening point value of the differential pressure relief valve, *the first time that it opens*. If the relief valve is activated (caused to open) before step ***j*** of Test No. 1, then a misleading value may be recorded. By causing the relief valve to open prematurely, the exercising of the moving components in the relief valve will generally produce higher relief valve opening point values. If the initial opening point would have been below the minimum 2.0 psid, but the tester activates the relief valve and then records the opening point value as greater than 2.0 psid, the tester may inadvertently pass an assembly which is not functioning properly.

In normal field operation, the differential pressure relief valve may not get exercised prior to the occurrence of a backflow condition. Therefore, the corresponding field test should evaluate the unit under the same conditions.

9.2.3.2 Leaking No. 1 Check Valve

If during step ***i*** of Test No. 1, the No. 1 check valve is found to be leaking, this will not allow the tester to proceed with the balance of the field test procedure. Since the relief valve is already discharging the relief valve opening point can not be accurately determined. However, the indicated gage reading with the relief valve discharging is an *approximate* value of the relief valve opening point.

The No. 2 check valve can not be evaluated since the gage reading is already at the relief valve opening point.

9.2.3.3 Instructions for Leaking No. 2 Shutoff Valve *(See Fig. 9-2)*

If the differential pressure reading on the gage during Test No. 1, step ***j***, does not change (i.e., lower), or the low side control needle valve must be opened more than one quarter (1/4) turn, it is likely that the No. 2 shutoff valve is leaking. The No. 2 shutoff valve should be re-opened and closed in an effort to get a better seal. This may particularly occur when testing units which do not have resilient seated shutoff valves. Make sure that hose fittings are drip-tight.

In tests No. 1, No. 2, and No. 3 above, a leaking No. 2 shutoff valve may affect the accuracy of the recorded values. A small leak in the No. 2 shutoff valve can be tolerated as long as the hose capacity is enough to satisfy the leak of the No. 2 shutoff valve. In test No. 1, step ***j***, if the low side needle valve must be opened more than one-quarter (1/4) turn then the above procedure may provide invalid data. If the volume of the leak of the No. 2 shutoff requires more than one-quarter (1/4) turn of the needle valve, then fittings should be placed in the No. 1 and No. 4 test cocks to accomodate an additional ***temporary*** bypass hose from the No. 1 test cock to the No. 4 test cock. A 1/2"∅ or 3/4"∅ hose may be needed to satisfy the leak in larger assemblies.

Attach temporary bypass hose to No. 1 test cock and bleed all air from the bypass hose by opening No. 1 test cock, then close. Attach the other end of the temporary bypass hose to No. 4 test cock. Open No. 1 test cock to pressurize hose. Slowly open No. 4 test cock, observing the reading on the differential pressure gage. If the reading increases, this may indicate a leaking No. 2 shutoff valve and a backpressure condition, *see Section 9.2.3.5 - Backpressure Condition.* Should the reading begin to drop, be prepared to record the reading at which the relief valve opens. However, should the reading stabilize above the relief valve opening point return to *Test No. 1, step j.* If the reading on the gage still will not drop to the relief valve opening point, then the leak through the No. 2 shutoff valve is greater than the capacity of the temporary bypass hose. Repair or replacement of the No. 2 shutoff valve is necessary before an accurate test can be completed.

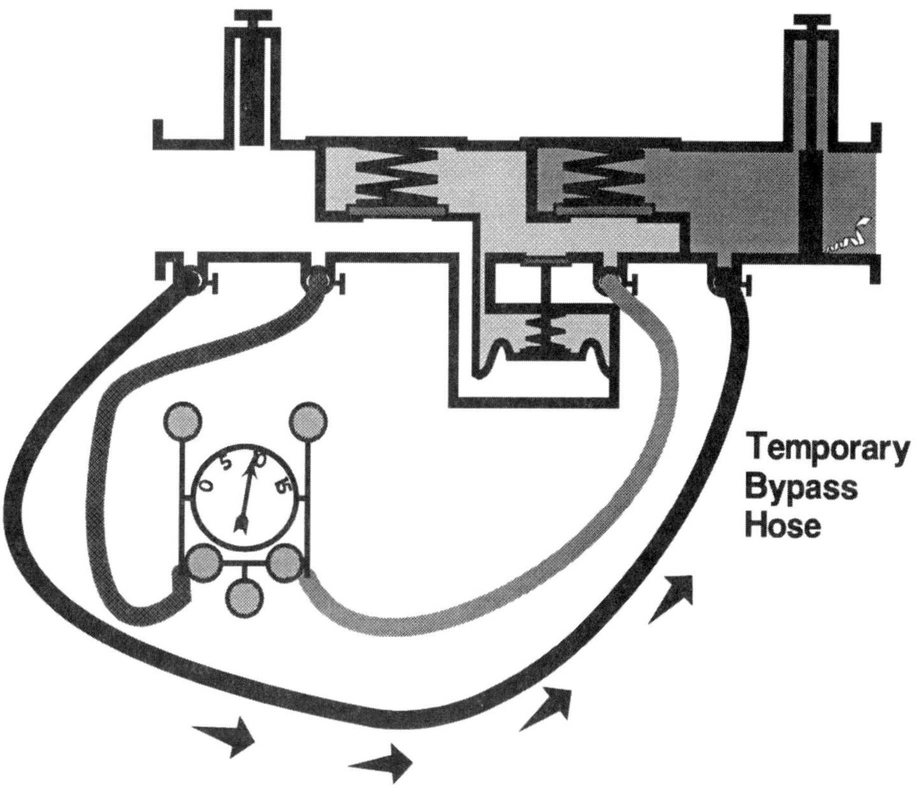

Fig. 9.2
Compensation for No. 2 Shutoff Valve Leak

9.2.3.4 Disc Compression — No. 2 Check Valve *(see Fig. 9.3)*

As high side pressure from the No. 2 test cock is being transferred to the No. 4 test cock (i.e., backside of the No. 2 check valve) during step ***e*** of Test No. 2, the differential reading on the gage may drop. This lowering of the No. 1 check valve reading can be caused by the small backpressure created behind the No. 2 check valve. This backpressure will cause the No. 2 check valve seat to imbed more deeply into the elastomer disc.

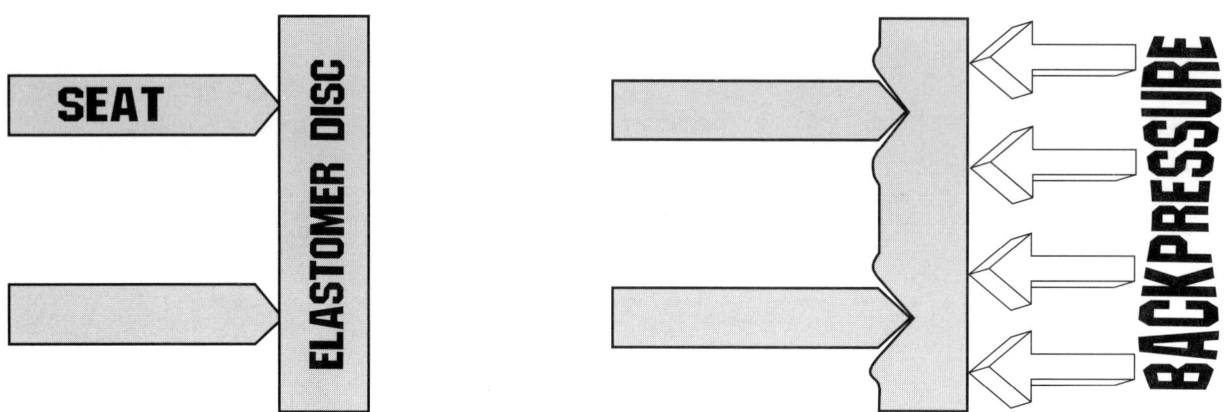

Fig. 9.3
Disc Compression - Second Check Valve

This decreases the trapped volume between the two check valves (i.e., the zone of reduced pressure) and, in turn, increases the pressure in the zone. An increase of the zone pressure will lower the differential pressure across the No. 1 check valve. To eliminate this false reading on the gage, the excess pressure built up in the zone must be bled off. This is done by opening the low side bleed needle valve (see step ***a,*** Test No. 3).

9.2.3.5 Backpressure Condition *(See Fig. 9.4.)*

As the bypass needle valve is opened during step ***e*** of Test No. 2, the differential pressure reading on the gage may increase. This may be caused by a pressure downstream of the No. 2 check valve which is greater than the line pressure at the No. 2 test cock - i.e., backpressure condition. The backpressure condition would pass through the bypass hose from the No. 4 test cock and raise the pressure on the high side of the gage, and therefore increase the reading on the gage. This backpressure will cause a backflow condition through the gage if the gage is left on line. For the backpressure condition to reach the No. 4 test cock, the No. 2 shutoff valve would have to be leaking.

To determine if there is a backpressure condition, close the No. 2 test cock, then observe the reading:

- ☐ If the gage reading drops to the low end of the scale, backpressure is not present. Return to step ***e,*** Test No. 2.

- ☐ If the gage reading increases to the high end of the scale then backpressure is present, close the No. 4 test cock. The rest of the test procedure can not be completed, however, the following information can be reported:

  - No. 2 Check Valve is holding tight against a backpressure - Evidenced by the relief valve not discharging upon arrival.

  - No. 1 Check Valve is holding tight, and the apparent differential pressure reading, Test No. 1 step ***i*** - can be reported as the actual static pressure drop across check valve No. 1.

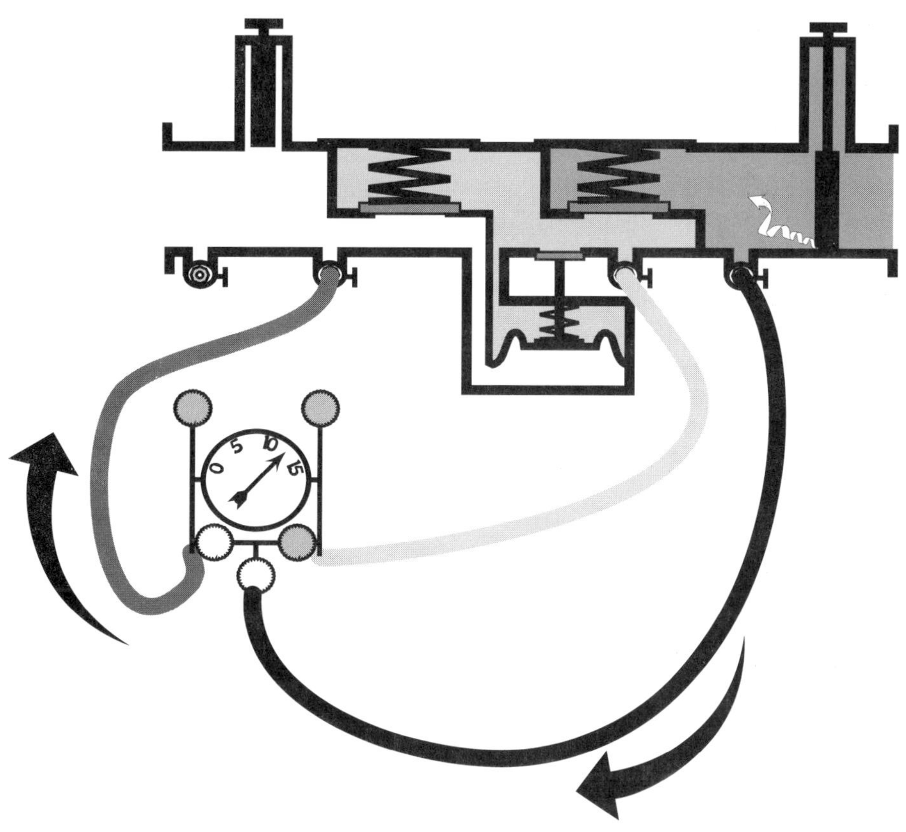

Fig. 9.4
Backflow Condition due to Backpressure

9.2.3.6 Troubleshooting Summary

NOTE: Many problems can be corrected by cleaning the internal components of the backflow prevention assemblies. Carefully observe condition of components.

Problem	May be caused by
Relief Valve discharges continuously	1. Faulty No. 1 check 2. Faulty No. 2 check with backpressure condition 3. Faulty Relief Valve 4. Plugged Relief Valve Sensing Line(s)
Relief Valve discharges intermittently	1. Properly working assembly with large pressure fluctuations 2. No. 1 check buffer is too small (i.e., less than 3.0 psi), with small line pressure fluctuation 3. Water hammer
Relief Valve discharges after No. 2 Shutoff Valve is closed (Test No. 1)	1. Normally indicates faulty No. 1 check a. dirty or damaged disc b. dirty or damaged seat 2. Leak through Relief Valve diaphragm
Relief Valve would not open, differential on the gage would not drop (Test No. 1)	1. Leaky No. 2 Shutoff Valve with flow through the assembly
Relief Valve would not open, differential drops to zero (Test No. 1)	1. Relief Valve stuck closed due to corrosion or scale 2. Relief valve sensing line(s) plugged
Relief Valve opens too high (with sufficiently high No. 1 check reading)	1. Faulty Relief Valve a. Dirty or damaged disc b. Dirty or damaged seat
No. 1 Check reading too low (less than 3.0 psi buffer) (Tests No. 1 and No. 3)	1. Dirty or damaged disc 2. Dirty or damaged seat 3. Guide members hanging up
Leaky No. 2 Check (Test No. 2)	1. Dirty or damaged disc 2. Dirty or damaged seat 3. Guide members hanging up

Repair note: Lubricants shall only be used to assist with the reassembly of components, and shall be non-toxic.

9.3 Double Check Valve Backflow Prevention Assembly (DC)

9.3.1 Equipment Required:

a) Differential Pressure Gage—Minimum range 0 -15 PSID (0.1 or 0.2 psid graduations)

b) One 6 ft. length—minimum 1/4"∅ high pressure hose

c) 1/4" Needle valves, for fine control of flows

d) One 1/4" IPS x 45° SAE flare connector—brass

e) Adapter fittings for each test cock size—brass 1/8" x 1/4", 1/4" x 1/2", 1/4" x 3/4"

f) Street ell, pipe nipple or tube (for length, see Section 9.3.3)

g) Bleed-off valve (see Appendix Section A.4.1—Bleed-Off Valve Arrangement)

NOTE: For both of the following tests the differential pressure gage must be held at the same level as the assembly being tested. Be sure that hoses not being used are also kept at this level

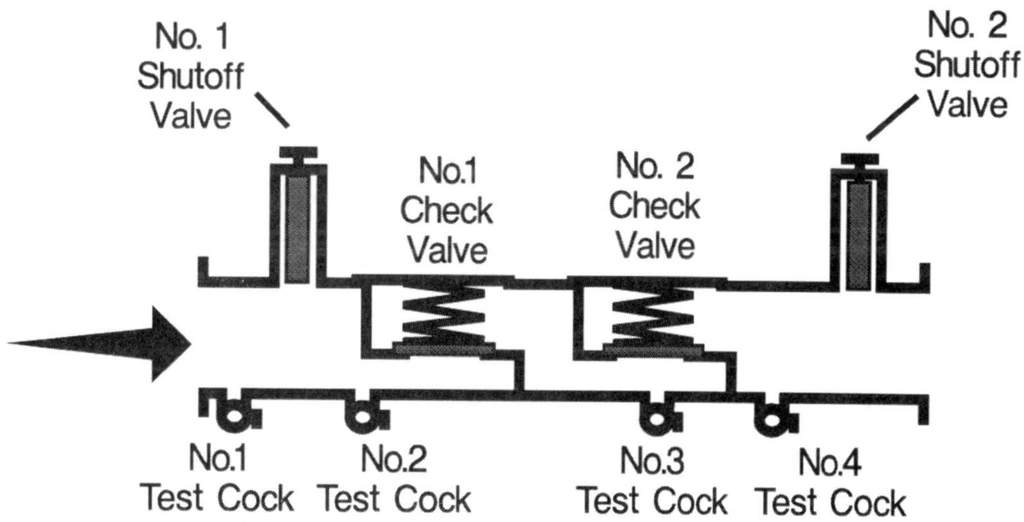

Fig. 9.5
Double Check Valve Assembly

9.3.2 Field Test Procedure

Test No. 1 - *Tightness of No. 1 Check Valve*

Purpose: To determine the static pressure drop across check valve No. 1.

Requirement: The static pressure drop across check valve No. 1 shall be at least 1.0 psid.

Steps:

Follow all preliminary steps in Section 9.1

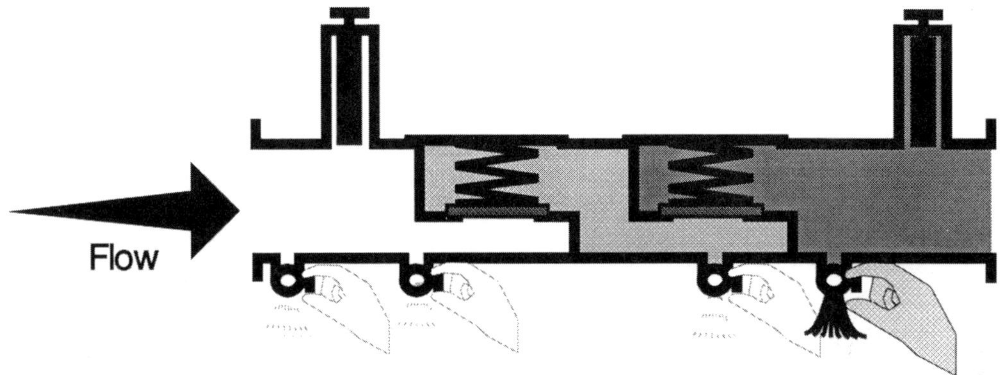

a. Bleed water through all four test cocks, one at a time, to eliminate foreign material.

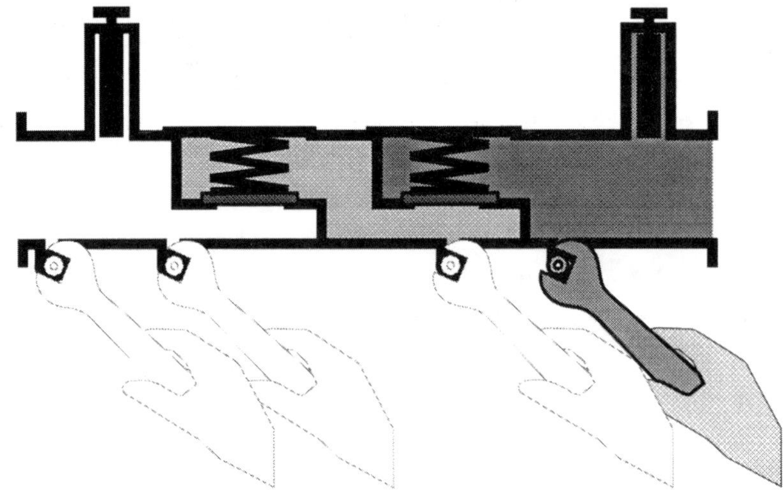

b. Install appropriate fittings.

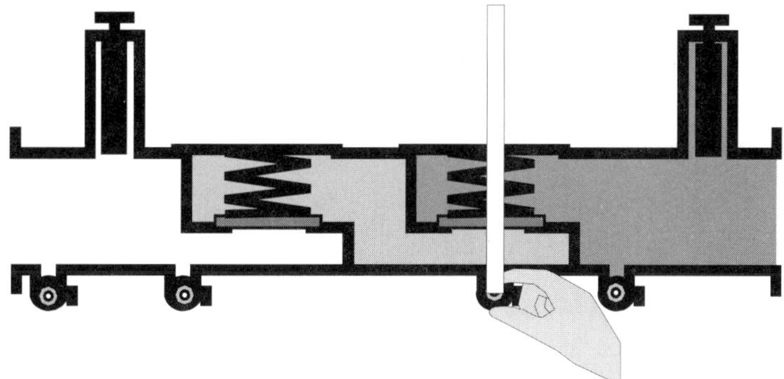

c. If test cock No. 3 is not at the highest point of the check valve body, then a vertical tube or pipe must be installed on test cock No. 3 so that it rises to the top of the check valve body. *(See Section 9.3.3.1 - Location of Gage Equipment.)*

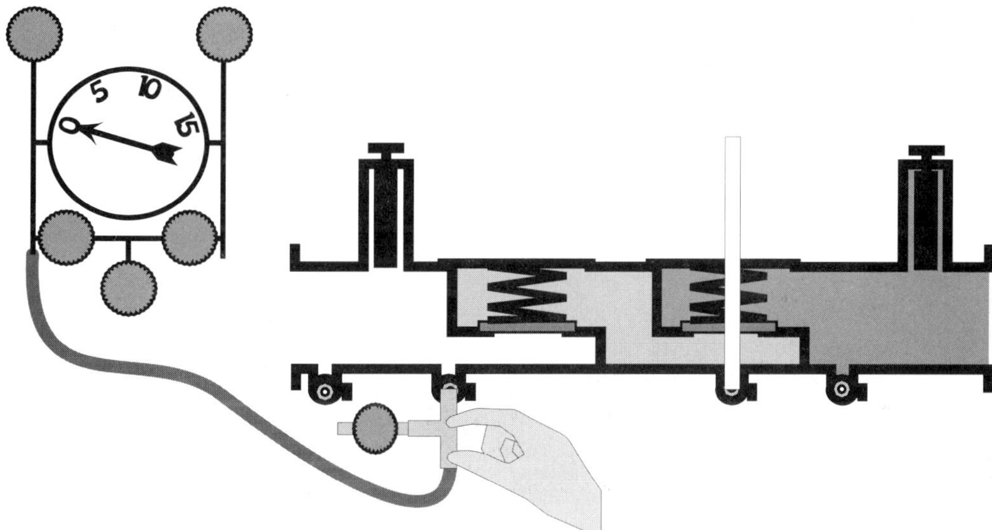

d. Attach bleed valve arrangement and hose from the high side of the differential pressure gage to the No. 2 test cock.

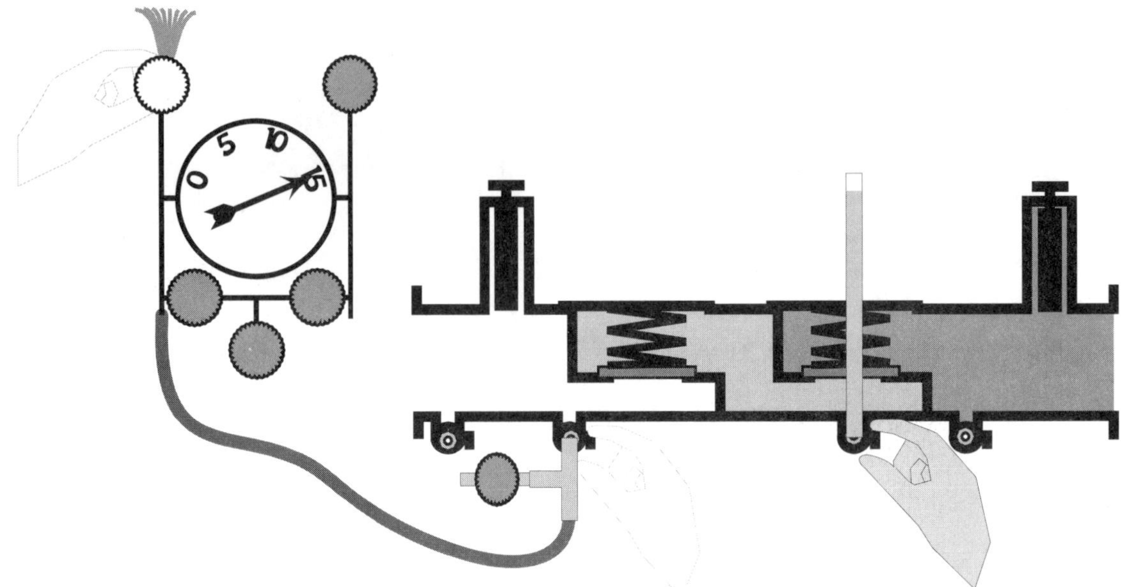

e. Open test cock No. 2 and bleed all air from the hose and gage by opening the high side bleed needle valve, then close the high side bleed needle valve. If a tube is attached to test cock No. 3, open test cock No. 3 to fill the tube, then close test cock No. 3.

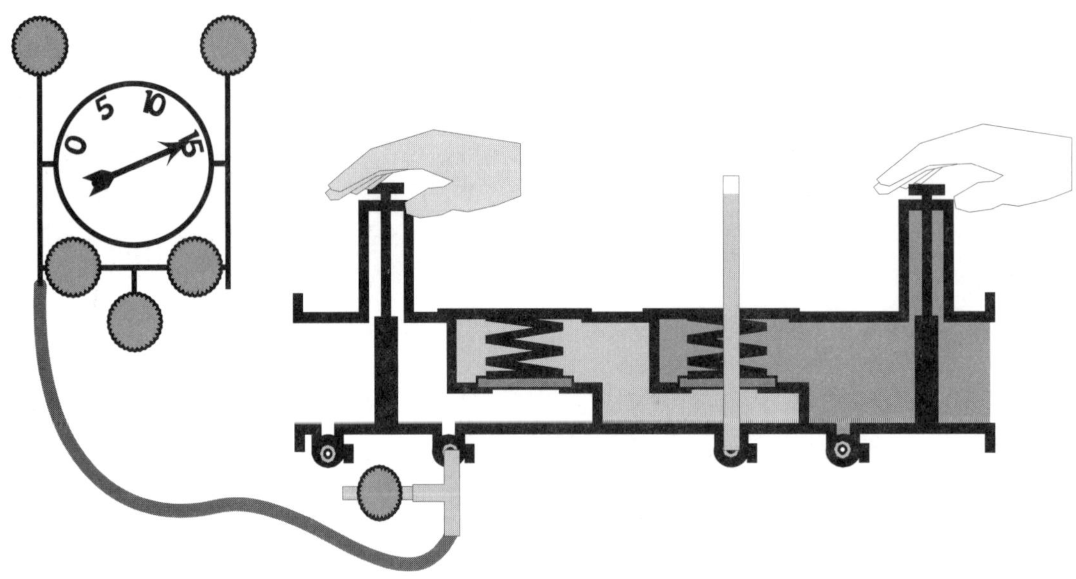

f. Close No. 2 shutoff valve, then close No. 1 shutoff valve.

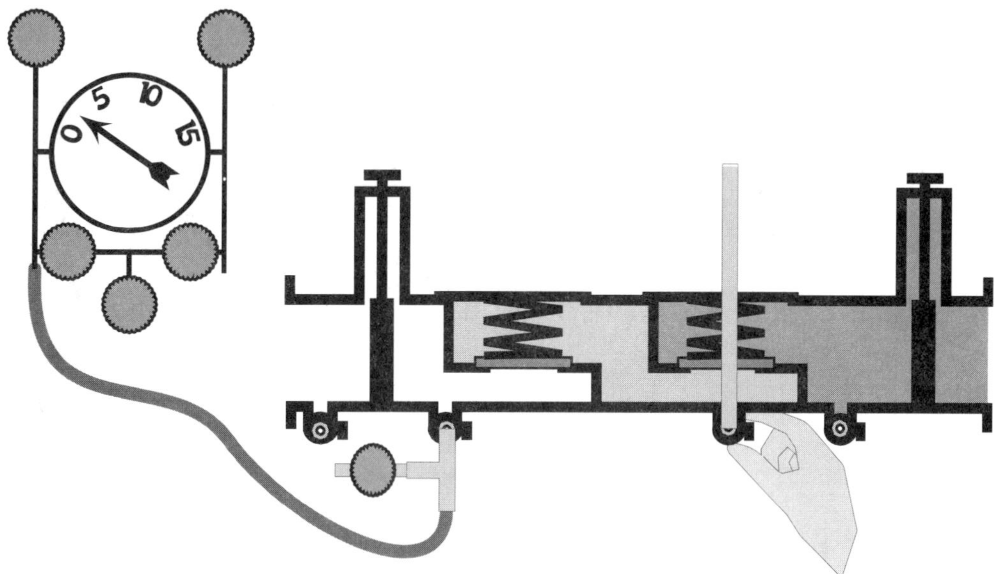

g. Slowly open test cock No. 3. After the gage reading stabilizes and water stops running out of test cock No. 3, the steady state differential pressure reading indicated on the gage is the static pressure drop across check valve No. 1 and is to be recorded as such. Proceed to step ***h.*** Should there be a continuous discharge of water from test cock No. 3, *see Troubleshooting Section 9.3.3.2 - Leaking shutoff valves - water discharges from No. 3 test cock.* Should the water level at test cock No. 3 recede, *see Troubleshooting Section 9.3.3.3 - Leaking shutoff valves, water recedes from No. 3 test cock.*

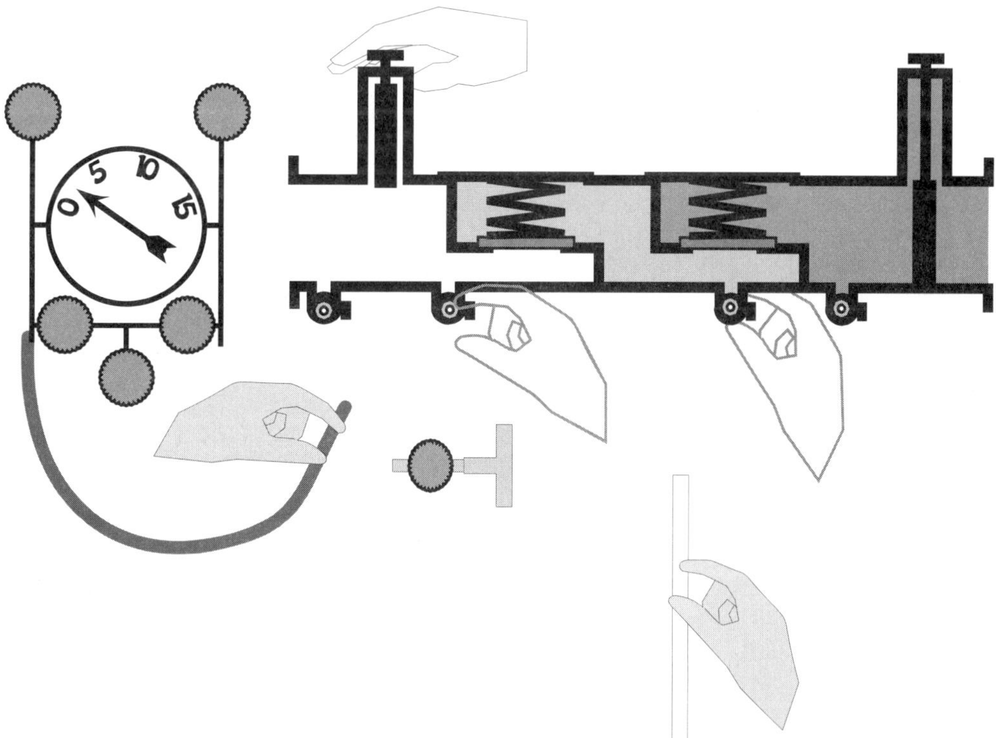

h. Close all test cocks, open shutoff valve No. 1, and remove all test equipment.

Test No. 2 - *Tightness of No. 2 Check Valve*

Purpose: To determine the static pressure drop across check valve No. 2.

Requirement: The static pressure drop across check valve No. 2 shall be at least 1.0 psid.

Steps:

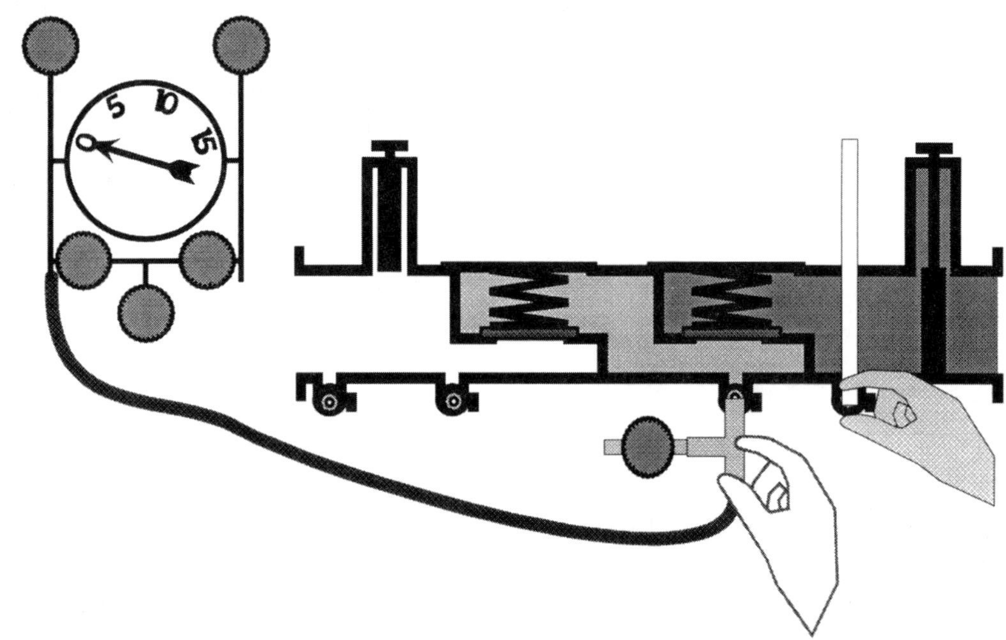

a. Attach bleed-off valve arrangement and hose from the high side of the differential pressure gage to the No. 3 test cock. If test cock No. 4 is not at the highest point of the check valve body, then a vertical tube or pipe must be installed on test cock No. 4 so that it rises to the top of the check valve body. *(See Section 9.3.3.1 -- Location of Gage Equipment.)*

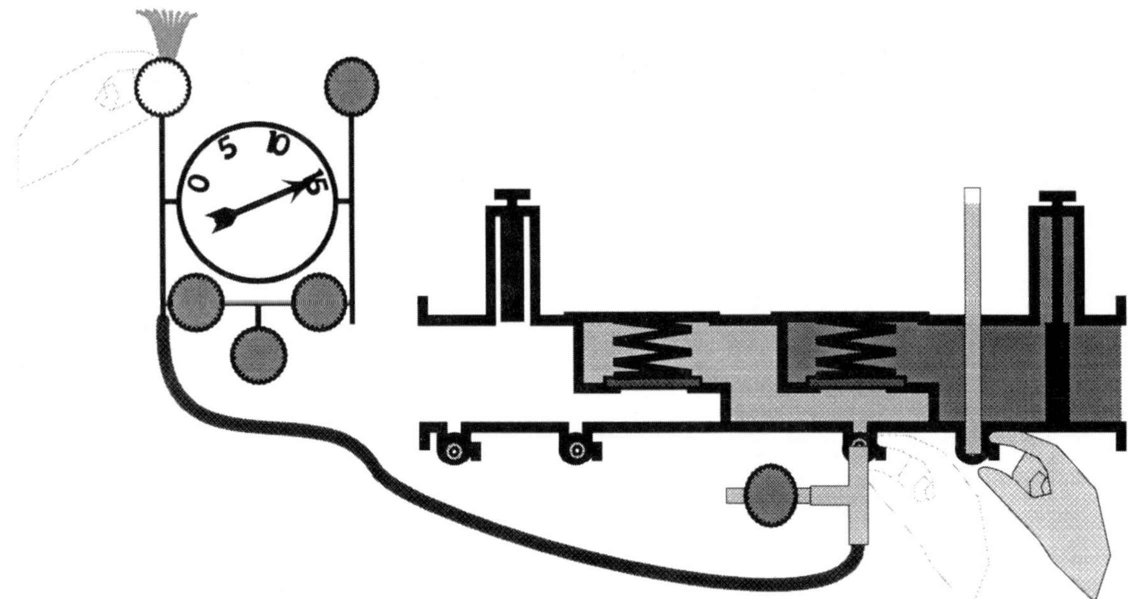

b. Open test cock No. 3, and bleed all air from the hose and gage by opening the high side bleed needle valve, and then closing the high side bleed needle valve. If a tube is attached to test cock No. 4, open test cock No. 4 to fill the tube, then close test cock No. 4.

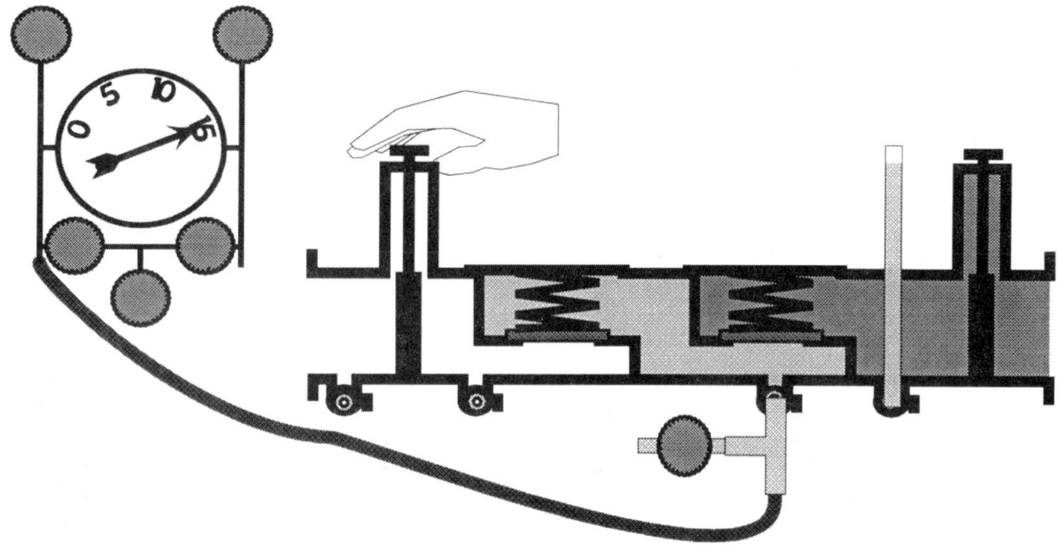

c. Close No. 1 shutoff valve.

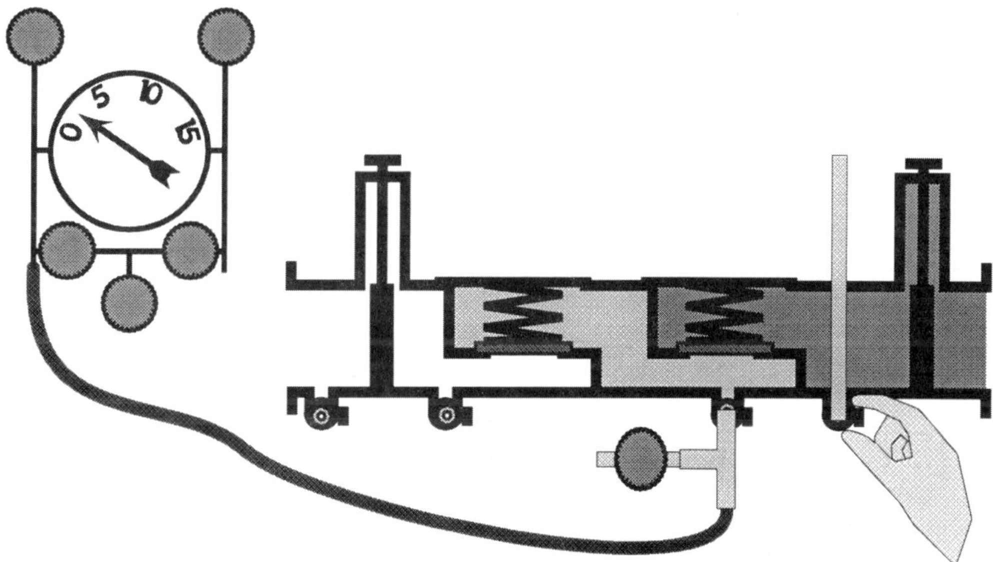

d. Slowly open test cock No. 4. After the gage reading stabilizes and water stops running out of test cock No. 4, the steady state differential pressure reading indicated is the static pressure drop across check valve No. 2 and is to be recorded as such. Go to Step ***f.*** Should there be a continuous flow of water from test cock No. 4, *see Troubleshooting Section 9.3.3.4 - No. 2 Check Test - Leaking Shutoff Valves.* If the water recedes from the No. 4 test cock proceed to step ***e.***

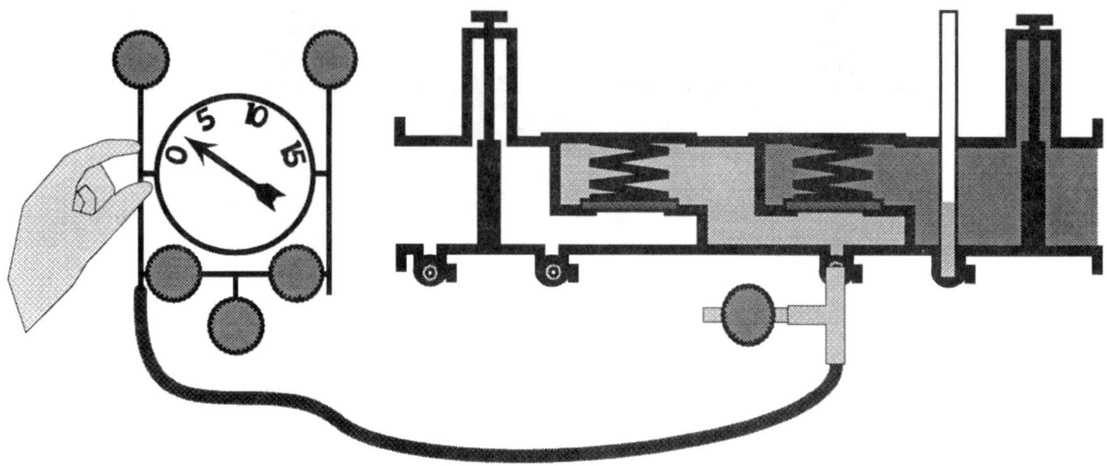

e. Should the water at test cock No. 4 recede, there is a leaking No. 2 shutoff valve. Move the gage to the centerline of the assembly and record the gage reading as the pressure differential across the No. 2 check valve.

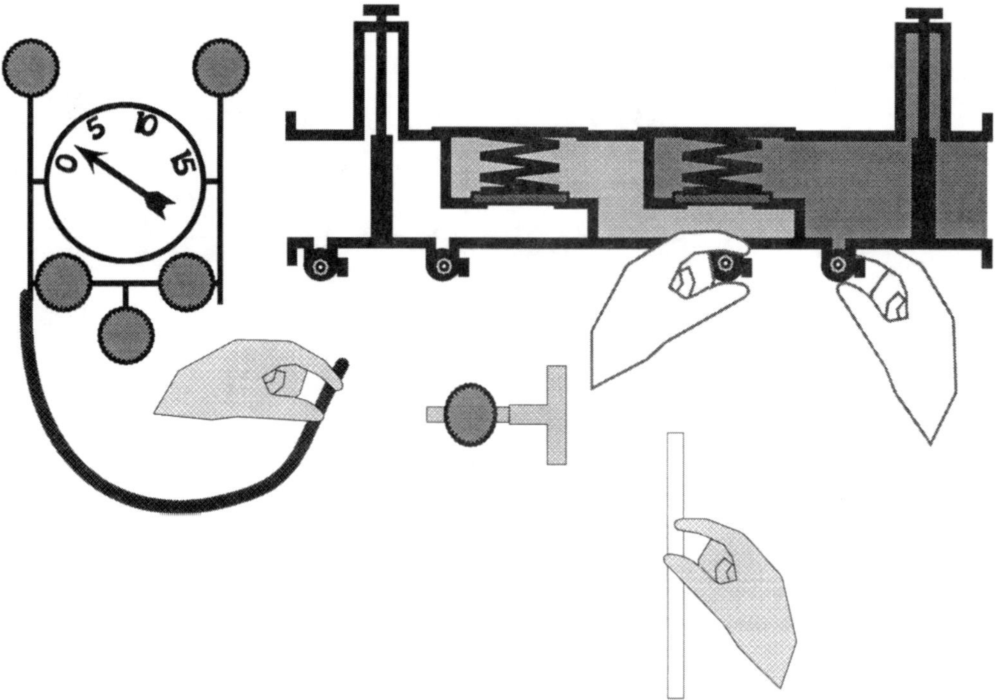

f. Close all test cocks, remove all test equipment.

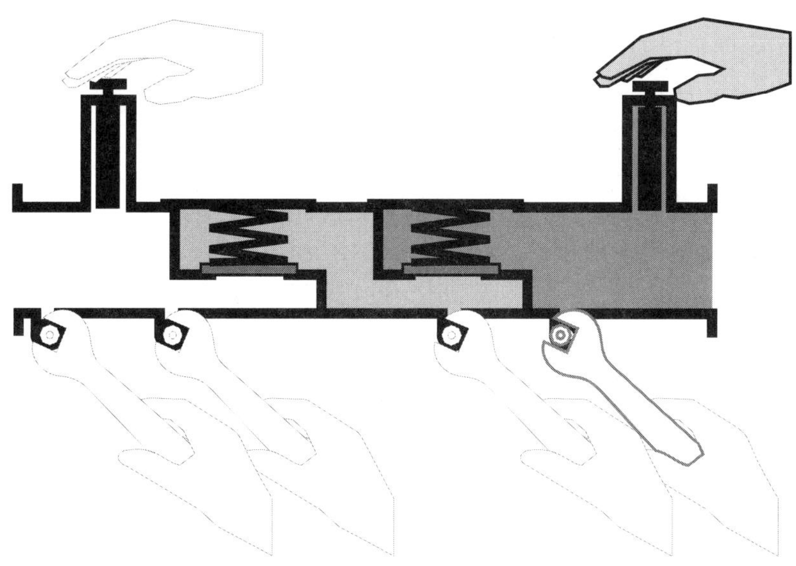

g. Remove fittings. Open shutoff valve No. 1, then slowly open shutoff valve No. 2.

9.3.3 Troubleshooting

9.3.3.1 Location of Gage Equipment

During this test it is important to keep the gage equipment and unused hoses at the appropriate elevation. A solid downstream reference point is needed for this test. If the downstream test cock is at the highest point of the body, then this can be used as the reference point. However, if the downstream test cock is below the top of the body, then a piece of pipe or tubing must be attached to the downstream test cock so that it rises above the top of the body. *See Fig. 9.6.*

To record the correct differential pressure reading the gage must be held at the same elevation as the test cock on the downstream side of the check valve being tested. If a tube is attached to the downstream test cock, then the gage must be maintained at the same height as the water in the tube. If this is not done properly, then values which are either too high or too low may be recorded.

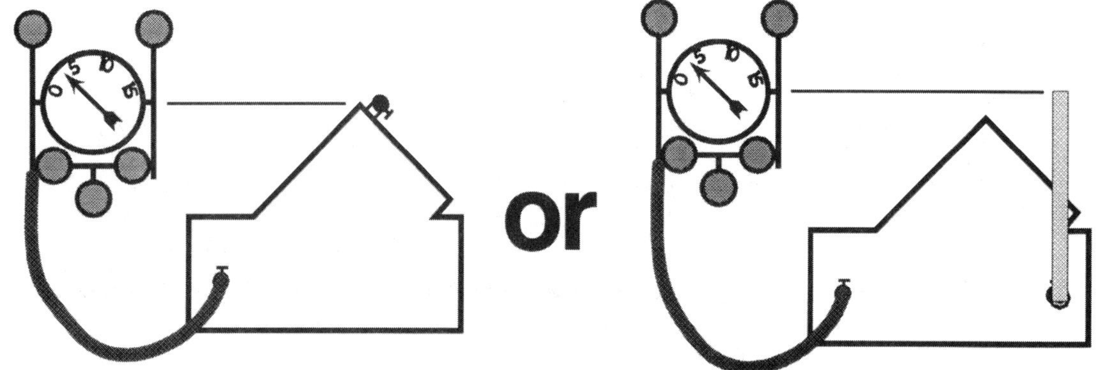

Fig. 9.6
Location of Gage Relative to Downstream Test Cock

9.3.3.2 Leaking Shutoff Valves - Water Discharges from No. 3 Test Cock

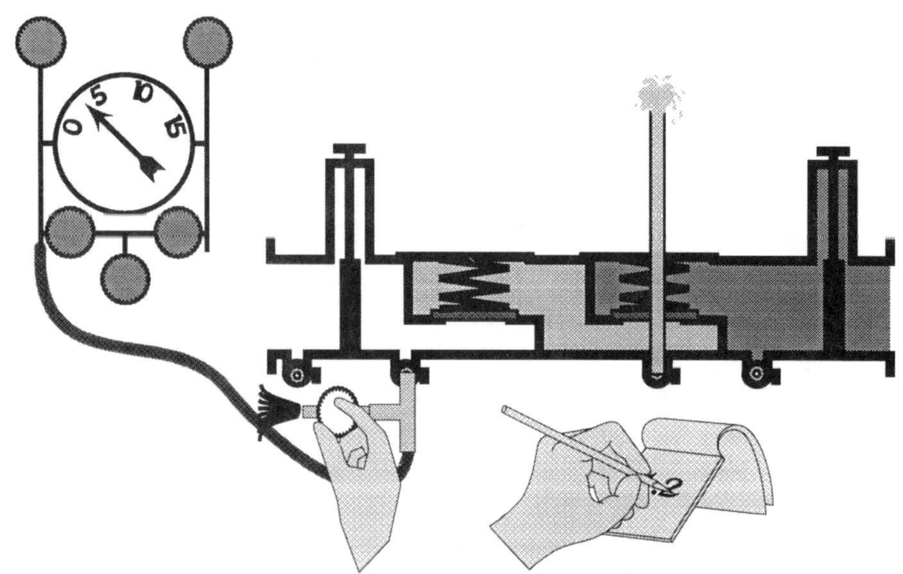

T1. If water continues to flow from test cock No. 3, then a shutoff valve may be leaking. Observe the reading on the gage, but do not record at this time. Open the bleed-off valve.

- If the water continues to flow from the bleed-off valve, and the bleed-off valve can be adjusted so there is a slight drip from the No. 3 test cock. Go to step ***T2*** below; or,
- If it is not possible to adjust the bleed-off valve to allow a slight drip at the No. 3. test cock, the No. 1 shutoff valve is leaking and must be repaired before the test may be completed. If the same condition occurs after repairing the No. 1 shutoff valve, repair the No. 1 check valve, No. 2 check valve and the No. 2 shutoff valve.
- If the water does not continue to flow from the bleed-off valve go to ***T3*** in this section.

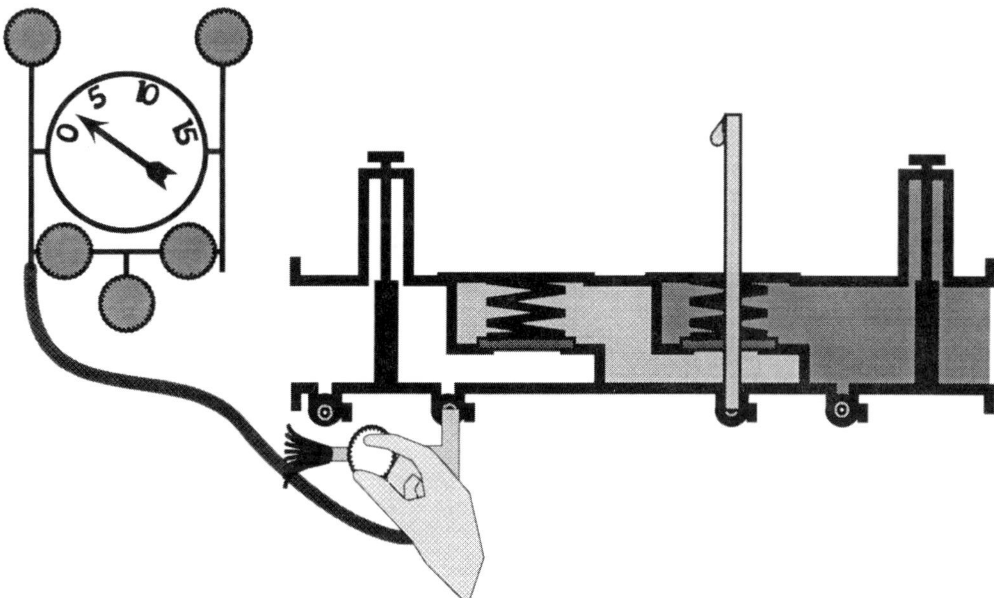

T2. After adjusting the bleed-off valve so that there is a slight drip at the No. 3 test cock, record the reading on the gage as the static pressure drop across the No. 1 check valve. This reading should be greater than or equal to 1.0 psi. Return to Test No. 1 Step ***h.*** If the reading is less than 1.0 psi, the No. 1 check valve must be repaired and retested before proceeding to test No. 2.

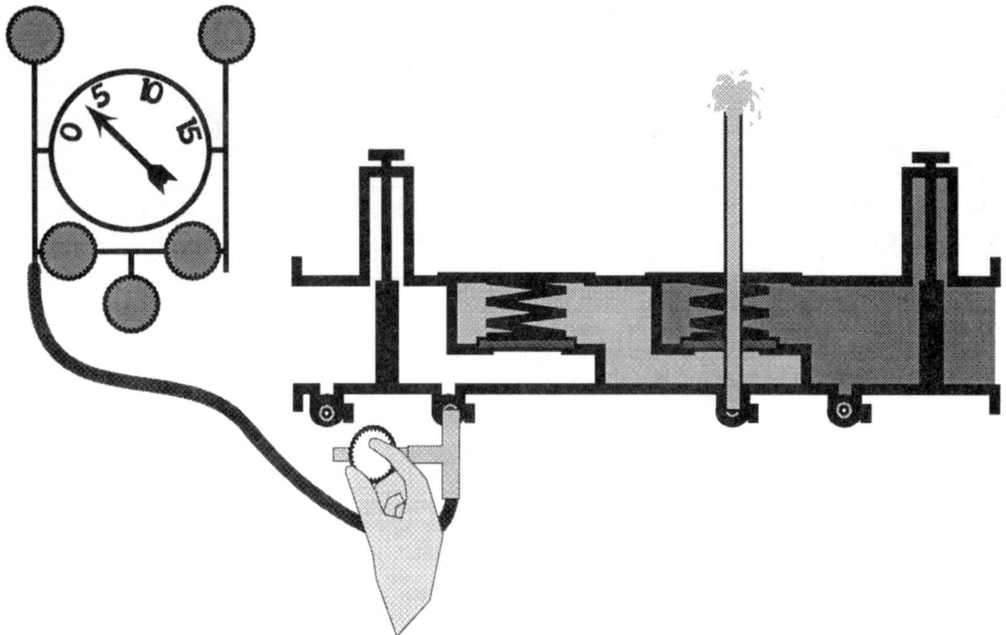

T3. If water does not continue to flow from the bleed-off valve with water still flowing from the No. 3 test cock, record the observed reading at Step ***T1*** as the static pressure drop across the No. 1 check valve. Also record the No. 2 check valve as leaking, and the No. 2 shutoff valve is leaking with backpressure. This concludes the test, proceed to Test No. 2 Step *f;* appropriate repairs must be made.

9.3.3.3 Leaking Shutoff Valves - Water Recedes from No. 3 Test Cock

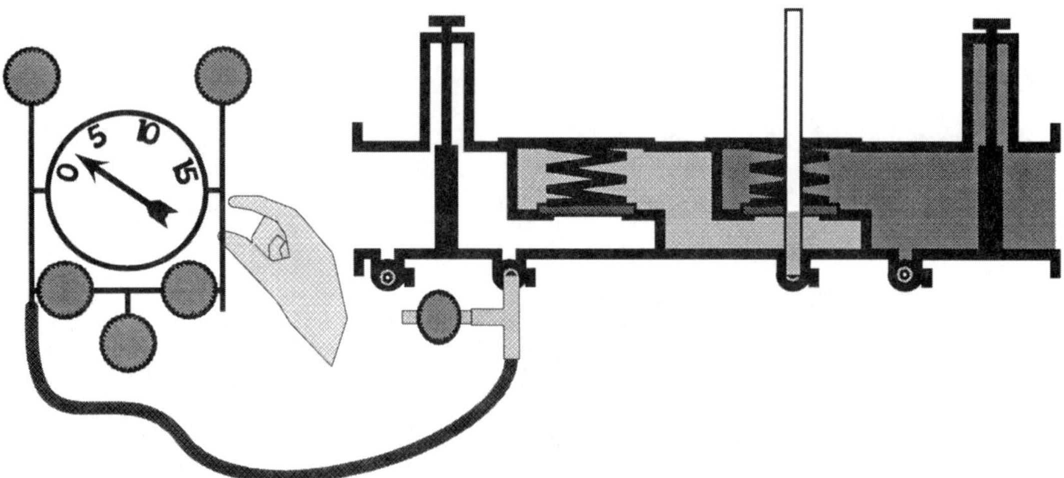

T4. If water recedes from test cock No. 3, the level of the gage should be moved to the centerline of the assembly. Record the gage reading as the static pressure drop across the No. 1 check valve. The No. 2 check valve should be recorded as leaking and the No. 2 shutoff valve as leaking. Proceed to Test No. 2 Step *f.*

9.3.3.4 Second Check Test - Leaking Shutoff Valves

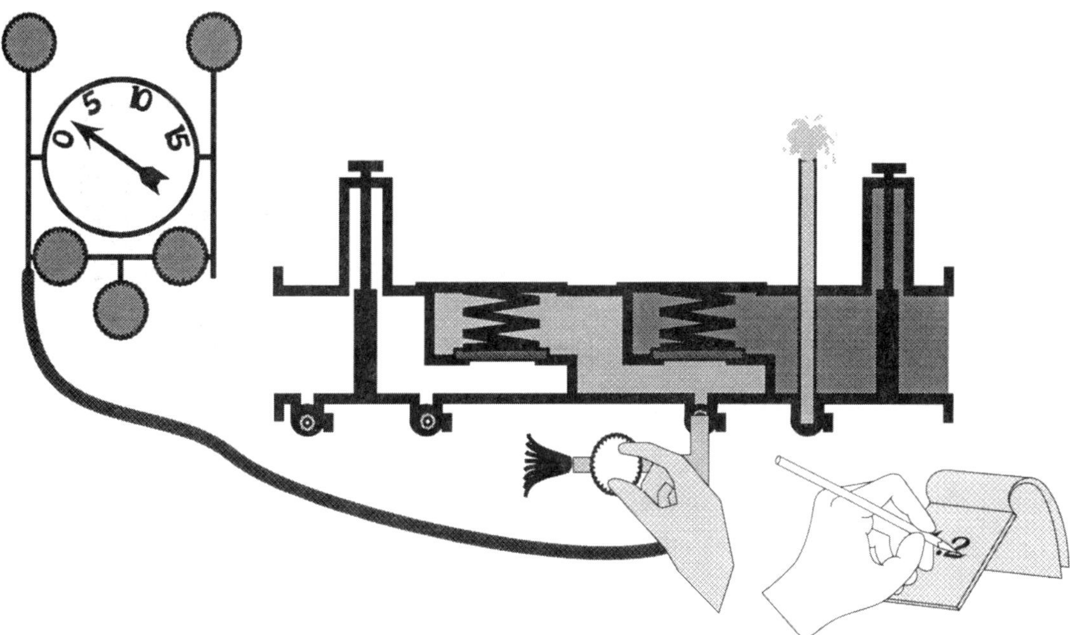

T5. If at Test No. 2 Step ***d*** water continues to flow from the No. 4 test cock, one of the shutoff valves is leaking. Observe the reading on the gage, but do not record it at this time. Open the bleed-off valve.

- If the water does not continue to flow from the bleed-off valve go to ***T6*** below.
- If the water continues to flow from the bleed-off valve, and the bleed-off valve can be adjusted so there is a slight drip from the No. 4 test cock. Go to step ***T7*** in this section; or,
- If it is not possible to adjust the bleed-off valve to allow a slight drip at the No. 4 test cock, the No. 1 shutoff valve should be checked to make sure it is closed tight. Then proceed to step ***T8.***

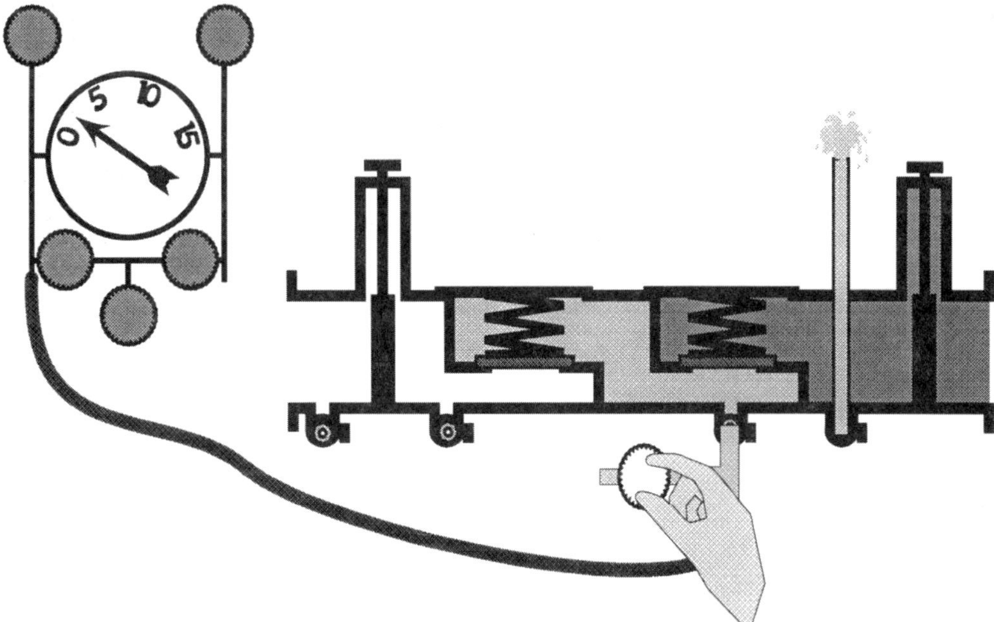

T6. If water does not continue to flow from the bleed-off needle valve with water still flowing from the No. 4 test cock, record the observed reading at Step ***T5*** as the static pressure drop across the No. 2 check valve. Also the No. 2 shutoff valve is leaking with backpressure. This concludes the test, proceed to Test No. 2 Step ***f;*** appropriate repairs must be made.

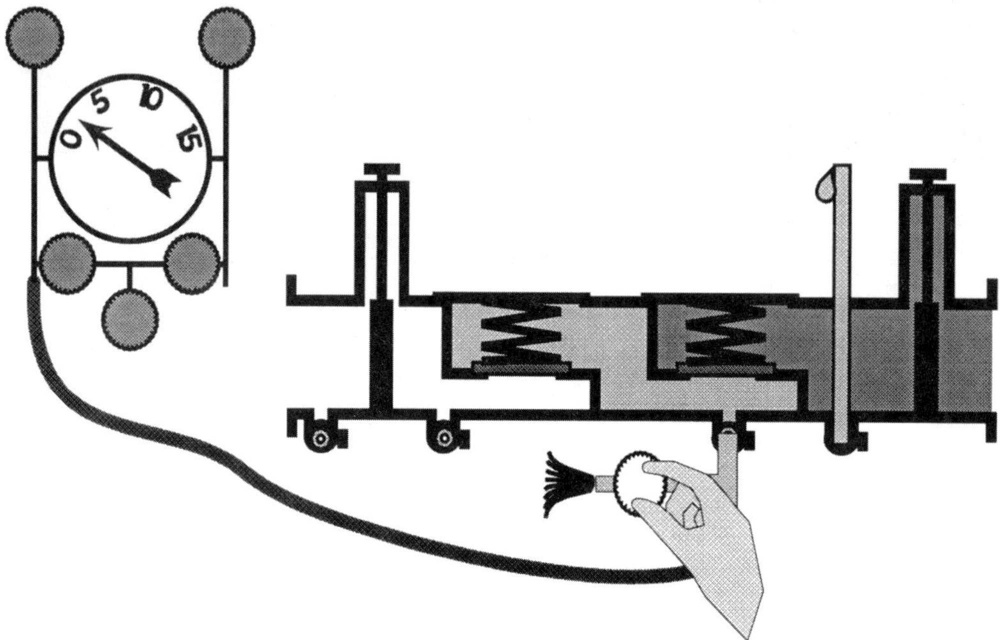

T7. After adjusting the bleed-off valve so that there is a slight drip at the No. 4 test cock record the reading on the gage as the static pressure drop across the No. 2 check valve. This reading should be greater than or equal to 1.0 psi. Return to Test No. 2 Step *f.*

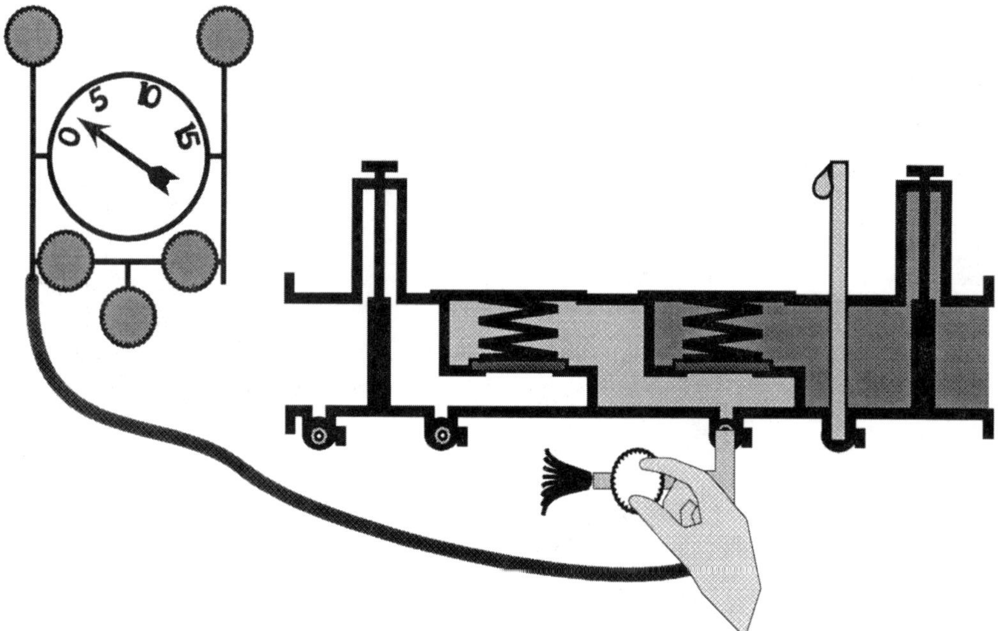

T8. If, after checking the tightness of the No. 1 shutoff valve, it is possible to adjust the bleed-off valve so there is a slight drip from the No. 4 test cock, record the reading on the gage as the static pressure drop across the No. 2 check valve and return to Test No. 2 step *f.* If it is not possible to adjust the bleed-off valve so that the water flowing at the No. 4 test cock is a slight drip, proceed to step ***T9.***

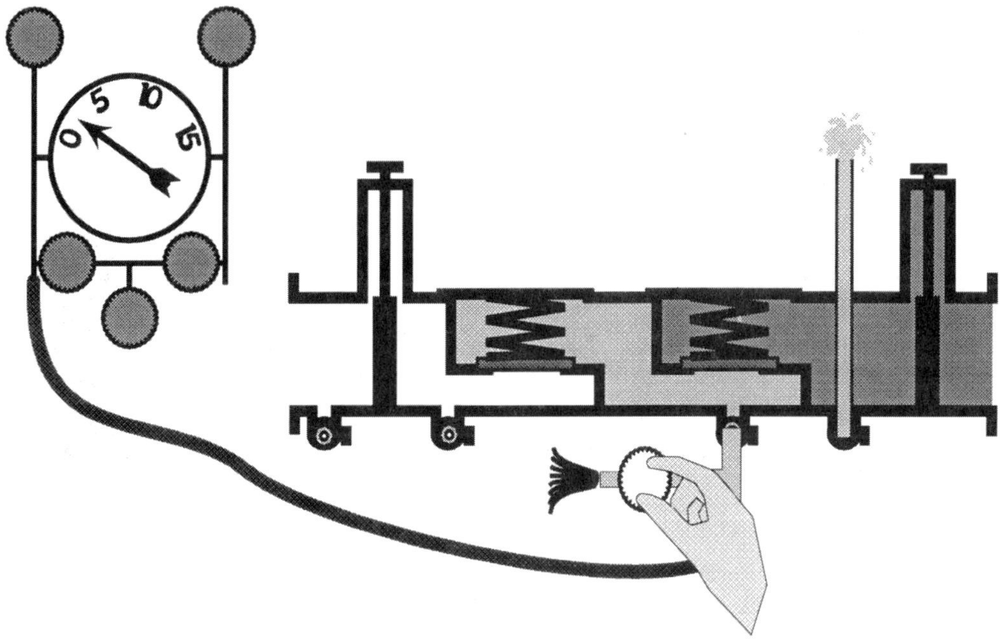

T9. If it is not possible to adjust the bleed-off valve so that the water flowing from the No. 4 test cock is a slight drip, and if check valve No. 1 was holding less than 1.0 psi in Test No. 1, the No. 1 check valve must be repaired before testing the No. 2 check valve. Then return to Test No. 1 step ***a.*** Otherwise go to step ***T10.***

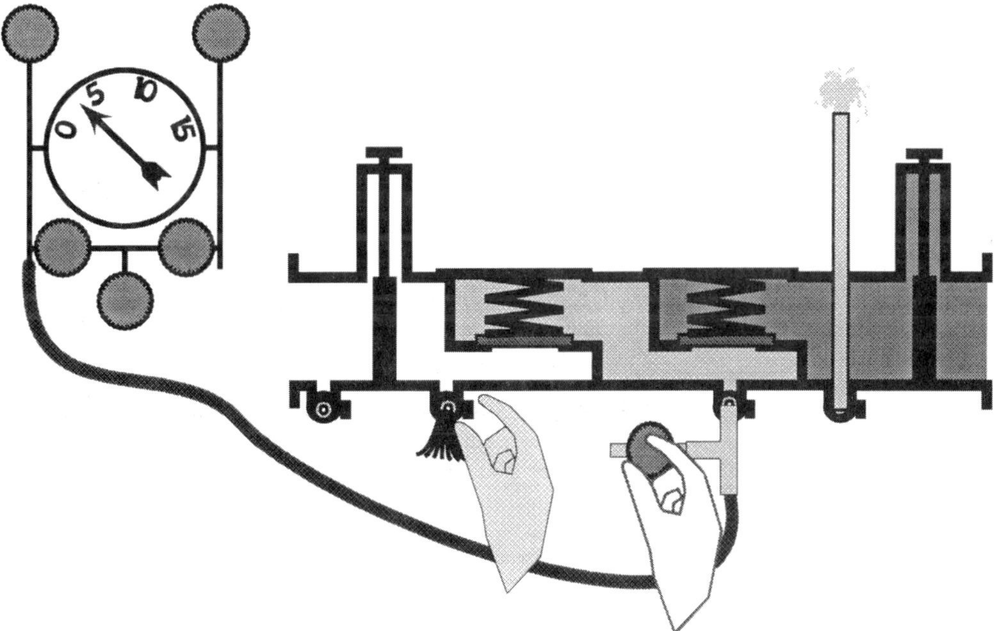

T10. If check valve No. 1 was holding 1.0 psi or more in Test No. 1, close the bleed-off valve and open the No. 2 test cock. Record the reading on the gage as the static pressure drop across the No. 2 check valve and return to Test No. 2 Step *f.*

9.4 Pressure Vacuum Breaker Backsiphonage Prevention Assembly (PVB)

Note: This procedure is for pressure vacuum breakers with loaded air inlet valves.

9.4.1 Equipment Required:

a) Differential Pressure Gage—Minimum range 0 -15 PSID (0.1 or 0.2 psid graduations)

b) One 6 ft. length—minimum 1/4"∅ high pressure hose

c) 1/4" Needle valves, for fine control of flows

d) Two 1/4" IPS x 45° SAE flare connectors—brass

e) Adapter fittings for each test cock size—brass 1/8" x 1/4"

f) Bleed-off valve (see Appendix Section A.4.1—Bleed-off Valve Arrangement)

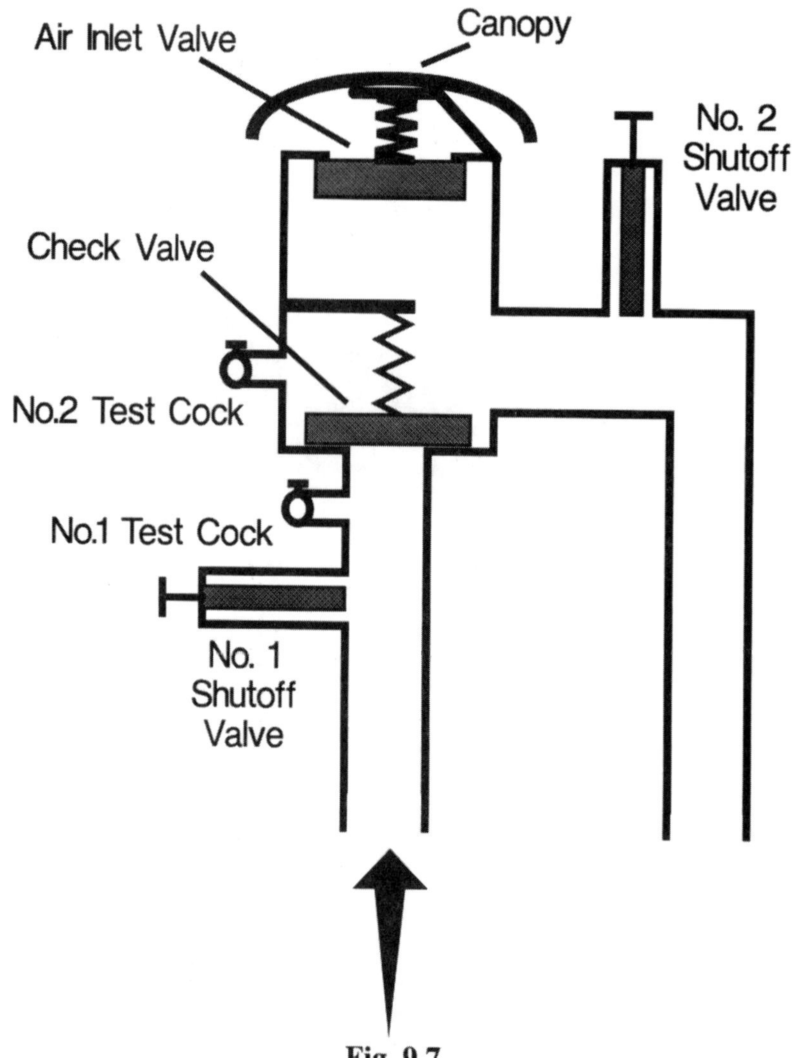

Fig. 9.7
Pressure Vacuum Breaker

NOTE: For both of the following tests the differential pressure gage must be held at the same level as the assembly being tested. Be sure that hoses not being used are also kept at this level.

9.4.2 Field Test Procedure

TEST NO. 1 ***Air Inlet Valve Opening Point***

Purpose: To determine the pressure in the body when the air inlet valve opens.

Requirement: The air inlet valve shall open when the pressure in the body is no less than 1.0 psi above atmospheric pressure. And, the air inlet valve shall be fully open when the water drains from the body.

Steps:

Follow all preliminary steps detailed in Section 9.1

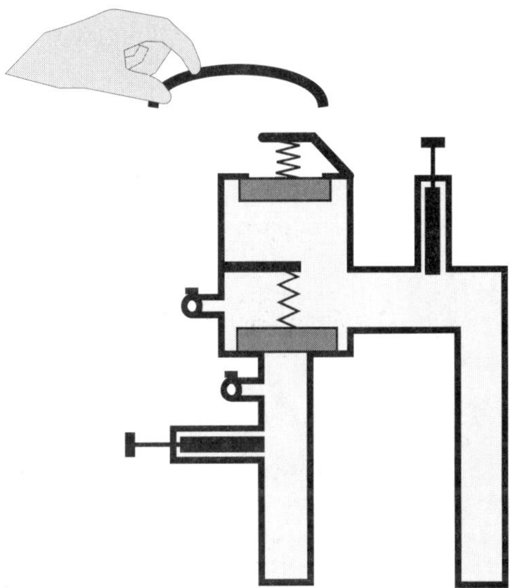

a. Remove air inlet valve canopy.

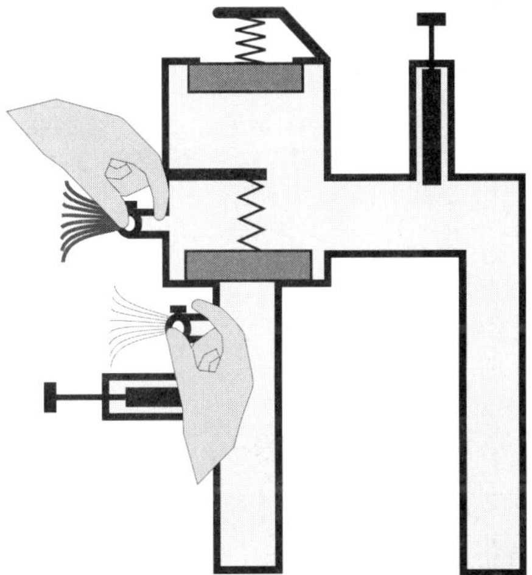

b. Bleed water through both test cocks to eliminate foreign material.

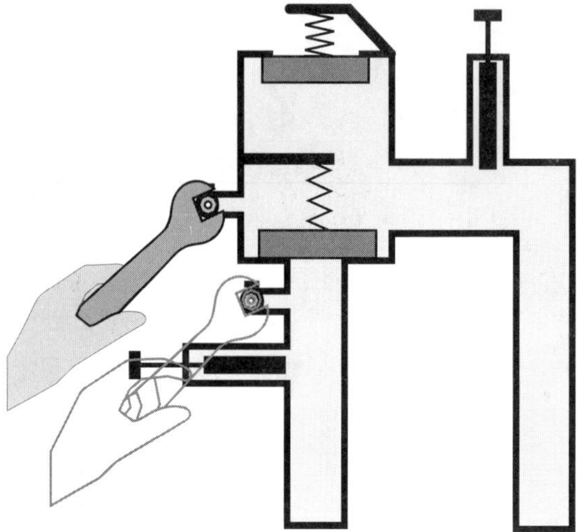

c. Install appropriate fittings to test cocks.

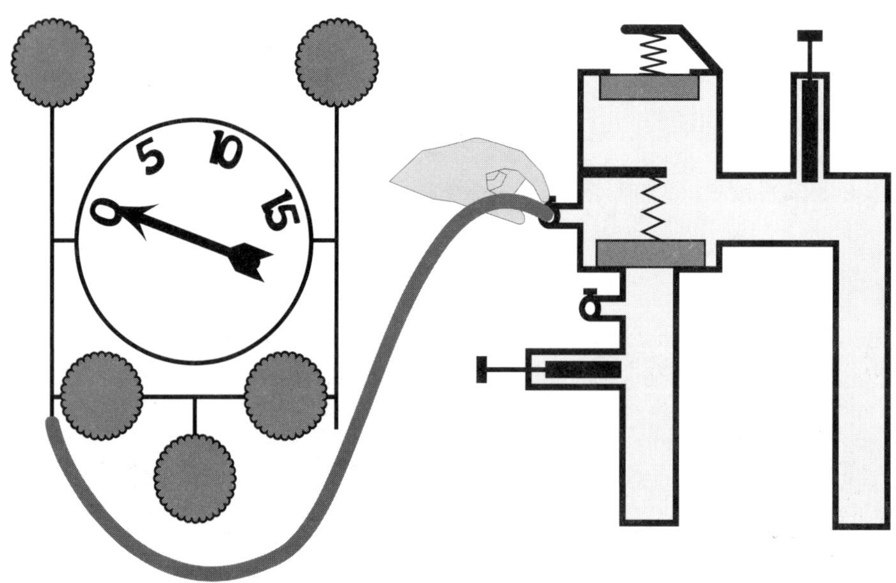

d. Attach the high side hose of the differential pressure gage to test cock No. 2, open test cock No. 2.

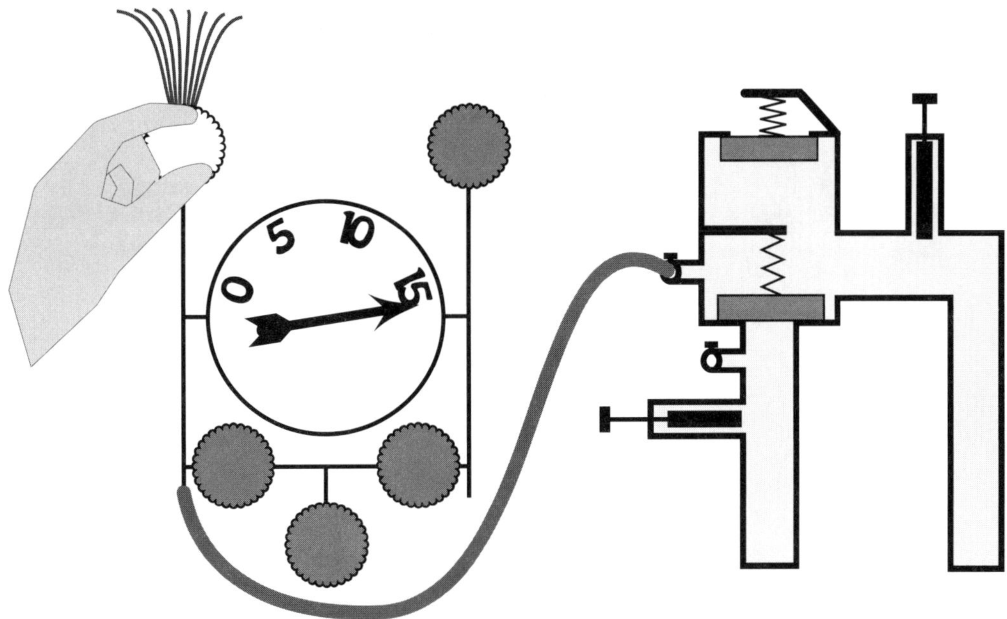

e. Bleed air from the hose and gage by opening the high side bleed needle valve. Close the high side bleed needle valve.

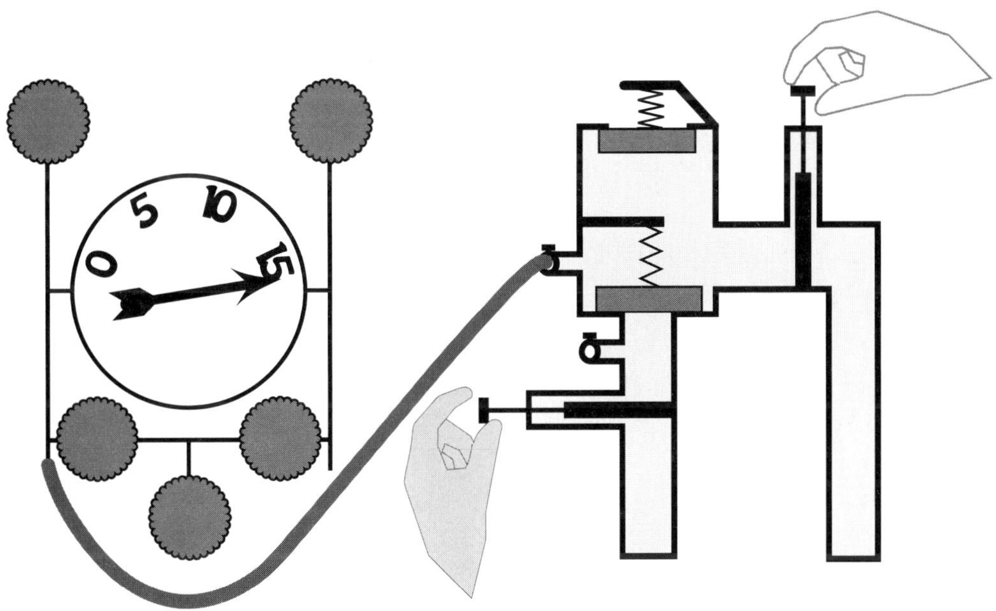

f. Close No. 2 shutoff valve, then close No. 1 shutoff valve.

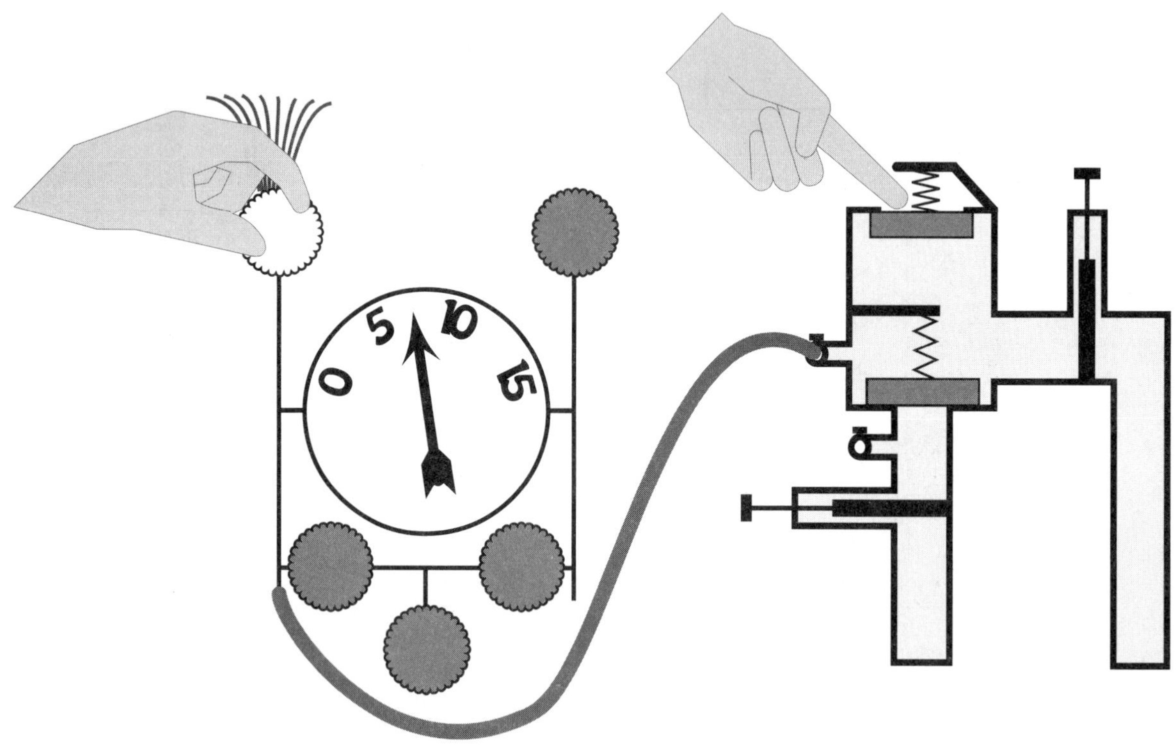

g. Slowly open the high side bleed needle valve no more than one-quarter (1/4) turn, being careful not to drop the differential pressure reading on the gage too fast. Record the differential pressure reading on the gage when the air inlet valve opens. The reading must be 1.0 psi or greater. If the high side bleed needle valve must be opened more than one-quarter (1/4) turn to lower the pressure reading in the body, see *Troubleshooting Section 9.4.3.1 - Leaking No. 1 Shutoff Valve*. Open the high side bleed needle valve to drain water from the body. Observe that the air inlet valve has opened to its fully open position.

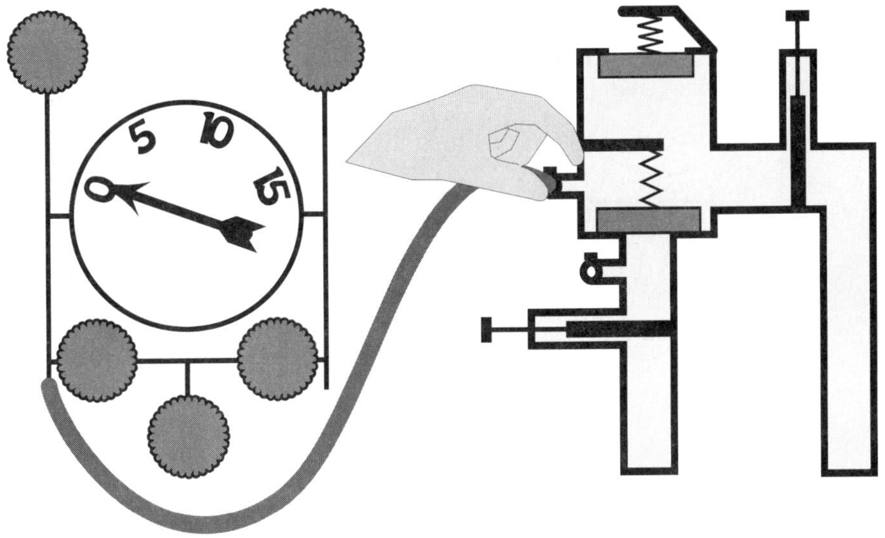

h. Close test cock No. 2.

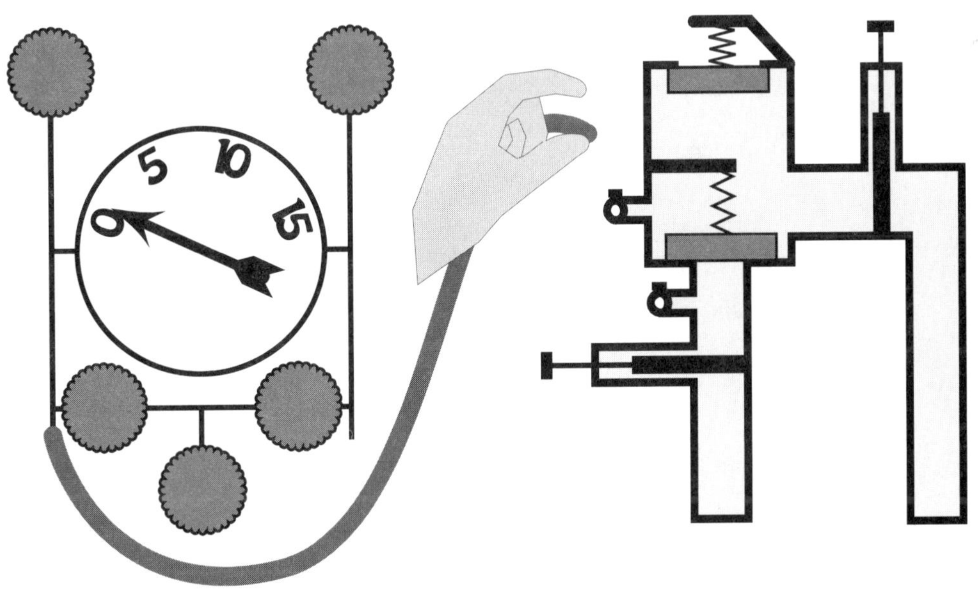

i. Remove equipment.

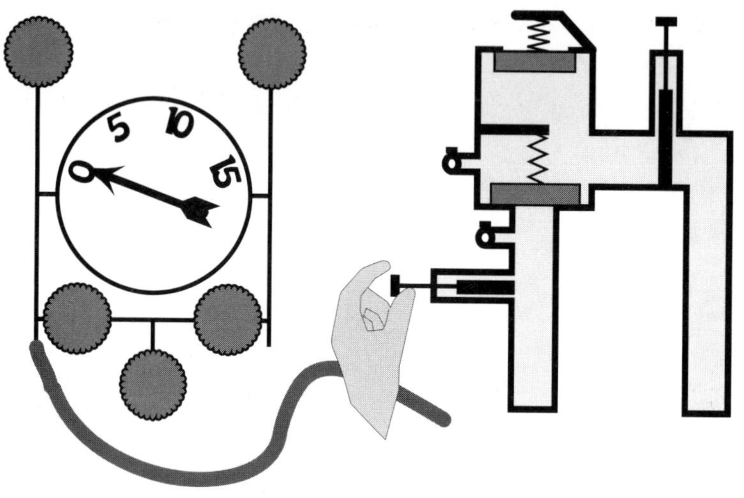

j. Open No. 1 shutoff valve.

TEST NO. 2 ***Check Valve Closing Point***

Purpose: To determine the static pressure drop across the check valve.

Requirement: The static pressure drop across the check valve shall be at least 1.0 psid.

Steps:

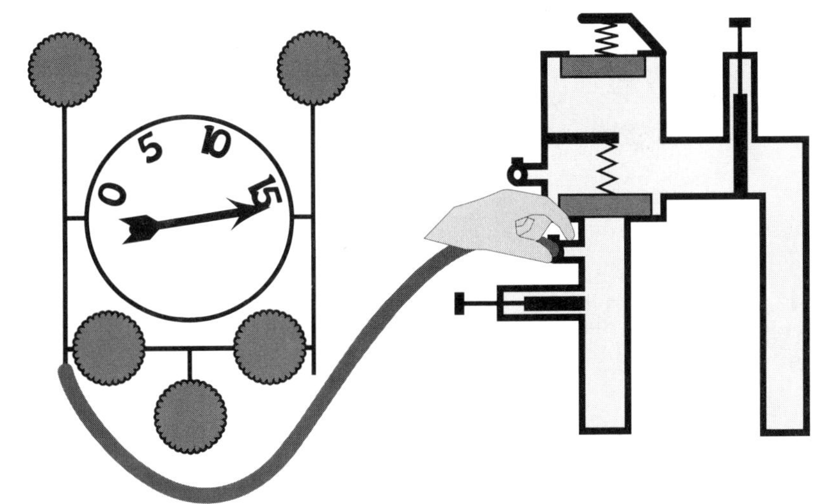

a. Attach high side hose of the differential pressure gage to test cock No. 1, open test cock No. 1.

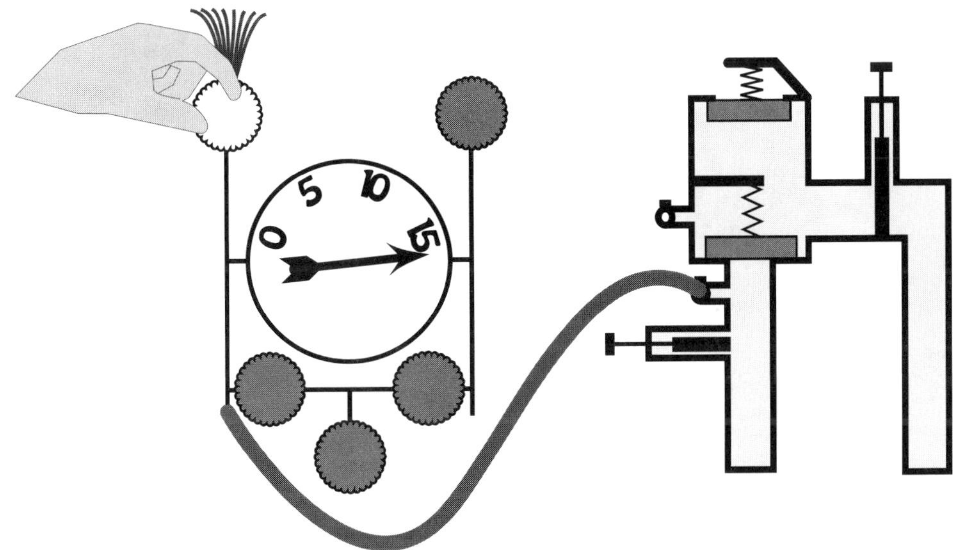

b. Bleed all air from the hose and gage by opening high side bleed needle valve. Close high side bleed needle valve.

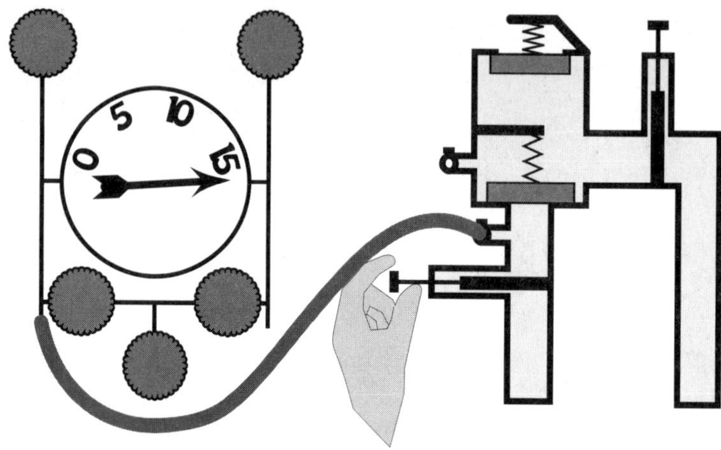

c. Close No. 1 shutoff valve (*No. 2 shutoff valve remains closed from Test No. 1*).

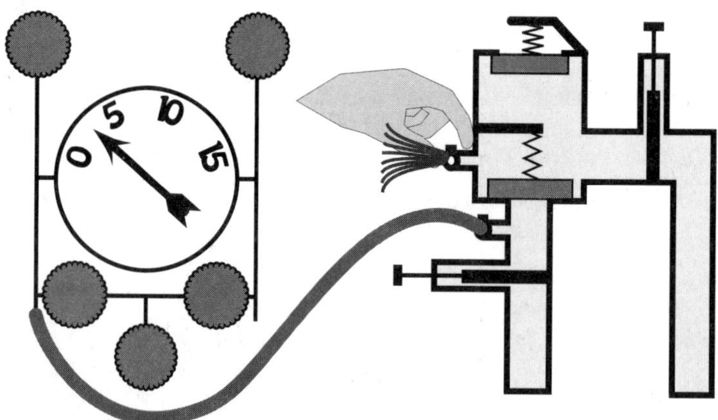

d. Open test cock No. 2. The water in the body will drain out through test cock No. 2. When this flow of water stops, and the differential pressure reading indicated by the gage settles, the gage reading will be the pressure drop across the check valve, and recorded as such. This gage reading must be 1.0 psid or greater. Record this value. If water continues to flow out of test cock No. 2, *see Troubleshooting Section 9.4.3.2 - Leaking No. 1 shutoff valve.*

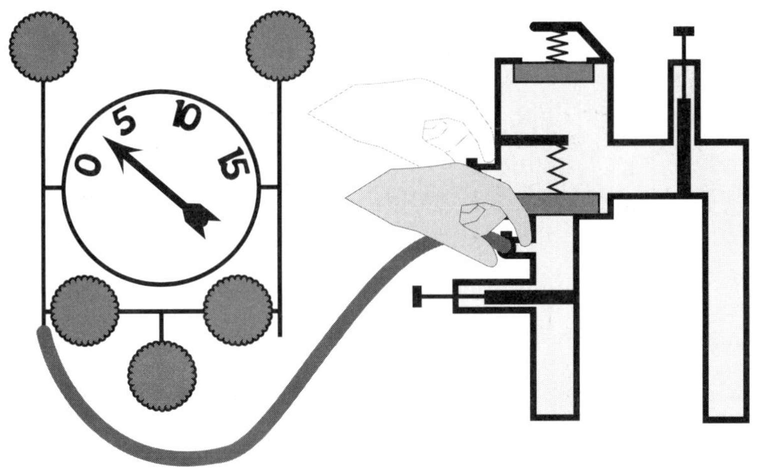

e. Close test cocks No. 1 and No. 2.

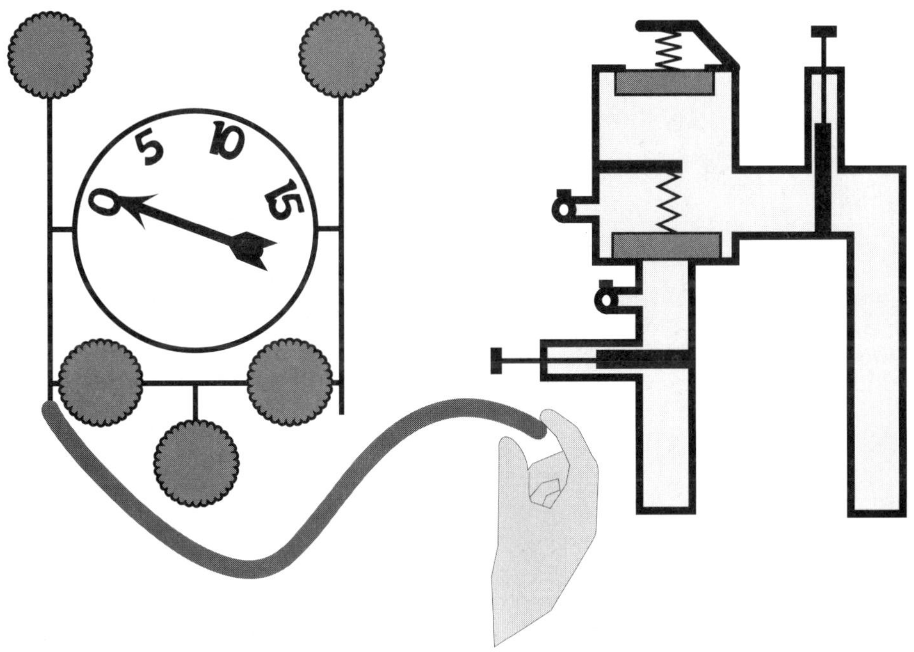

f. Remove equipment.

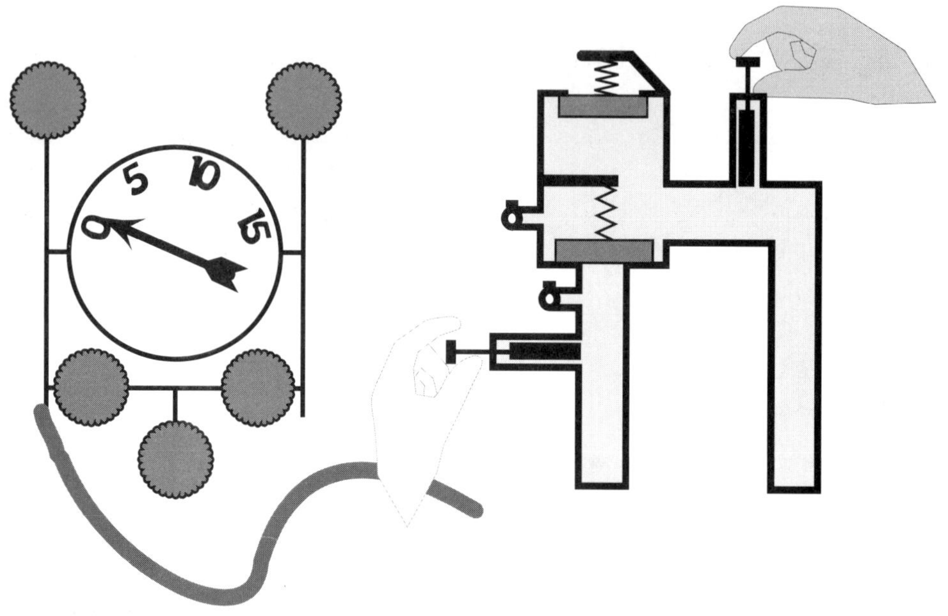

g. Open No. 1 shutoff valve, then No. 2 shutoff valve.

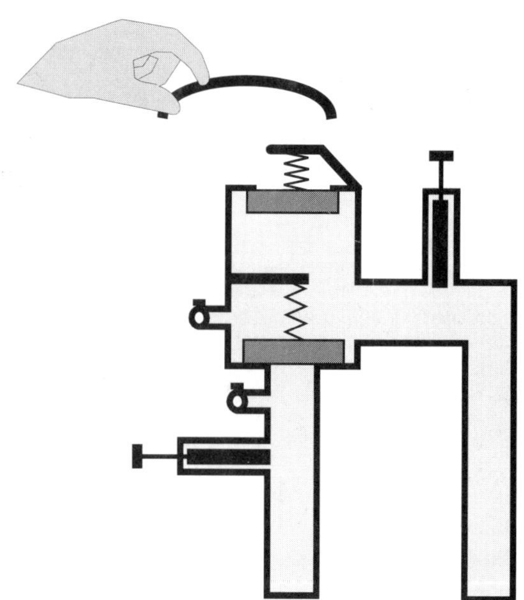

h. Replace air inlet valve canopy.

9.4.3 Troubleshooting

9.4.3.1 Leaking No. 1 Shutoff Valve - Air Inlet Valve Test (See Fig. 9.8)

Should the high side bleed needle valve need to be opened more than one-quarter (1/4) turn, it is likely that the No. 1 shutoff valve is leaking. The No. 1 shutoff valve should be re-opened and closed in an effort to a get a better seal. This may particularly occur when testing units which do not have resilient seated shutoff valves. Should the leak persist then the leak must be diverted so that the air inlet valve can be tested. Open the No. 1 test cock slowly to divert the leakage from the No. 1 shutoff valve, monitoring the gage while this is done. Once the leakage has been diverted to the No. 1 test cock, continue with step ***g*** Test No. 1. If the shutoff valve leak exceeds the limit of the No. 1 test cock, then the test can not be completed until the No. 1 shutoff valve is repaired or replaced.

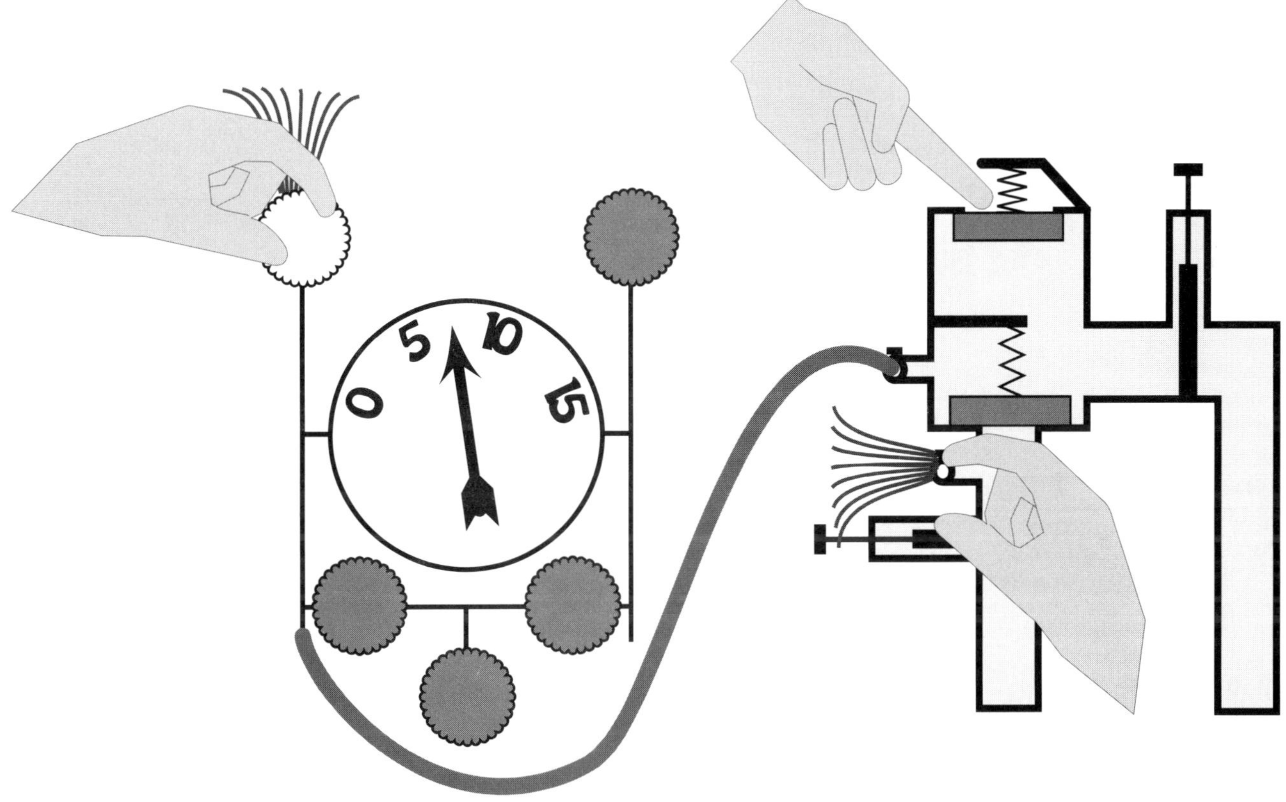

Fig. 9.8
Leaking No. 1 Shutoff valve

9.4.3.2 Leaking No. 1 Shutoff Valve - Check Valve Test (See Fig. 9.9)

Should water continue to flow out of test cock No. 2 during Test No. 2, this indicates that the No. 1 shutoff valve is leaking. Install bleed-off valve to test cock No. 1 and attach high side hose to bleed-off valve. Open test cock No. 1. Slowly open the bleed-off valve until the flow of water from test cock No. 2 is a slight drip. Record the reading on the differential gage as the static pressure drop across the check valve. If the flow of water from test cock No. 2 can not be eliminated by opening the bleed-off valve, then the tightness of the check valve can not be determined. The shutoff valve No.1 must be repaired or replaced.

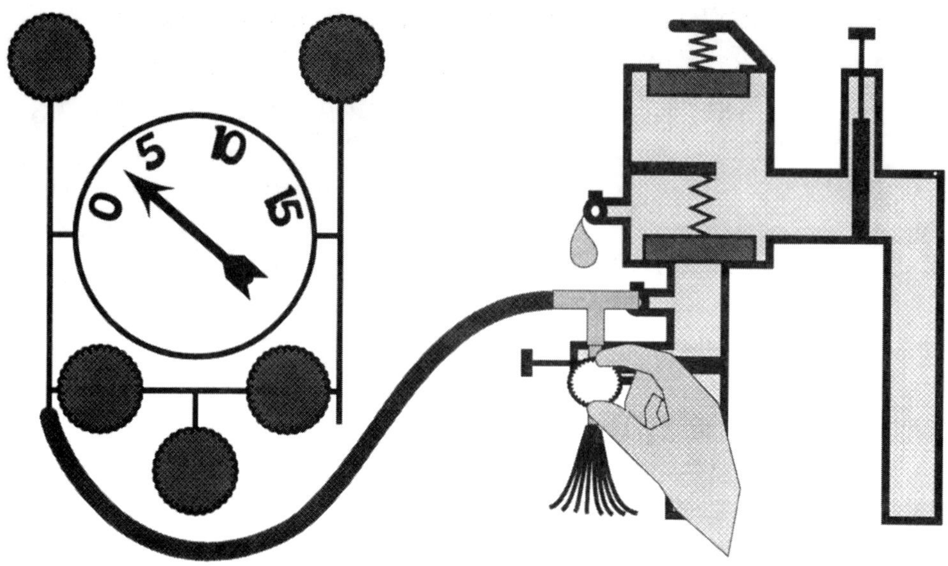

Fig. 9.9
Leaking No. 1 Shutoff Valve

9.4.3.3 Troubleshooting Summary

NOTE: Many problems can be corrected by cleaning the internal components. Carefully observe condition of components.

Problem	May be caused by
Air inlet valve does not open, as gage drops to 0.0 psid	1. Air inlet disc stuck to seat 2. Broken or missing air inlet spring 3. Old style" pressure vacuum breaker (non-loaded air inlet valve) 4. SVB—See Section 9.5
Air inlet valve does not open, and differential on gage will not drop	1. Leaky No. 1 shutoff valve 2. Parallel installation with leaky No. 2 shutoff valve.
Air inlet opens below 1.0 psid	1. Dirty or damaged air inlet disc 2. Scale buildup on seat
Check valve below 1.0 psid	1. Dirty or damaged check disc 2. Damaged seat
Water runs continuously from test cock No. 2 (Test No.2)	1. Leaky No. 1 shutoff valve

Repair note: Lubricants shall only be used to assist with the reassembly of components, and shall be non-toxic.

9.5 Spill-Resistant Pressure Vacuum Breaker Backsiphonage Prevention Assembly (SVB)

9.5.1 Equipment Required:

a) Differential Pressure Gage—Minimum range 0 -15 PSID (0.1 or 0.2 psid graduations)

b) One 6 ft. length—minimum 1/4"∅ high pressure hose

c) 1/4" Needle valves, for fine control of flows

d) One 1/4" IPS x 45° SAE flare connector—brass

e) Adapter fitting for test cock size—brass 1/8" x 1/4"

f) Bleed-off valve (see Appendix Section A.4.1—Bleed-off Valve Arrangement)

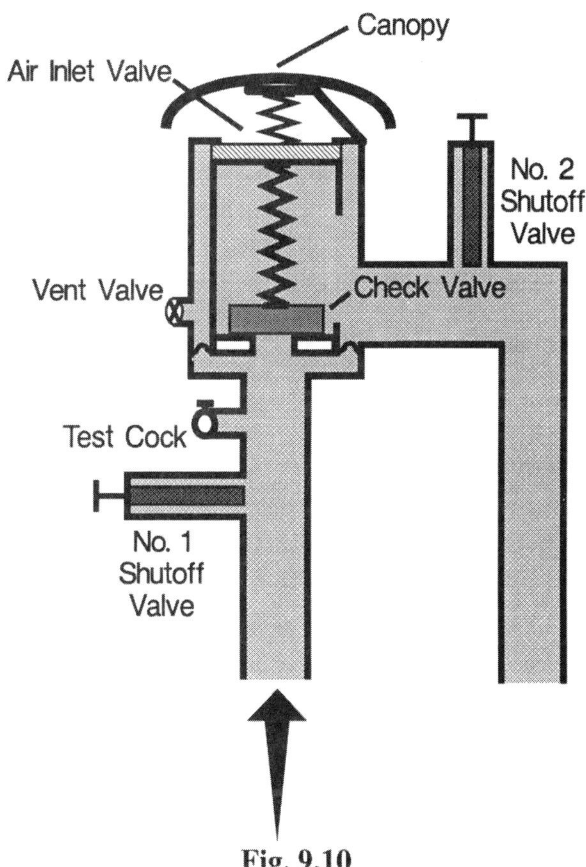

Fig. 9.10
Spill-Resistant Pressure Vacuum Breaker (SVB)

NOTE: For both of the following tests the differential pressure gage must be held at the same level as the assembly being tested. Be sure that hoses not being used are also kept at this level.

9.5.2 Field Test Procedure

TEST NO. 1 ***Air Inlet Valve Opening Point***

Purpose: To determine the inlet pressure when the air inlet valve opens.

Requirement: The air inlet valve shall open when the inlet pressure is no less than 1.0 psi above atmospheric pressure and the outlet pressure is atmospheric pressure. The air inlet valve shall be fully open when the inlet pressure is atmospheric.

Steps:

Follow all preliminary steps detailed in Section 9.1

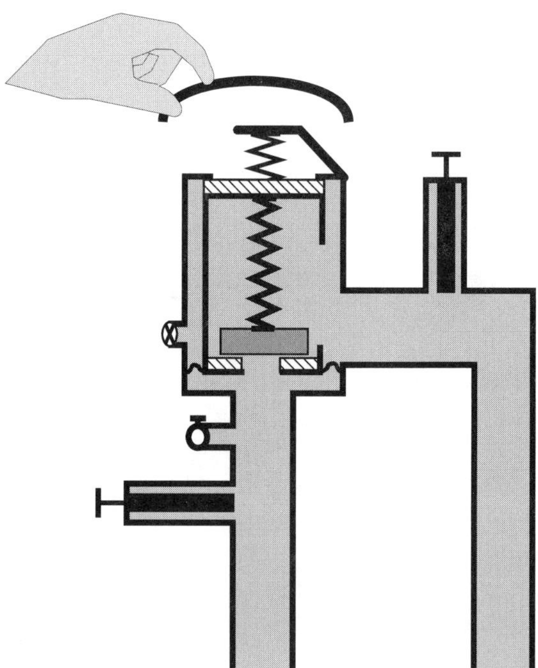

a. Remove air inlet valve canopy.

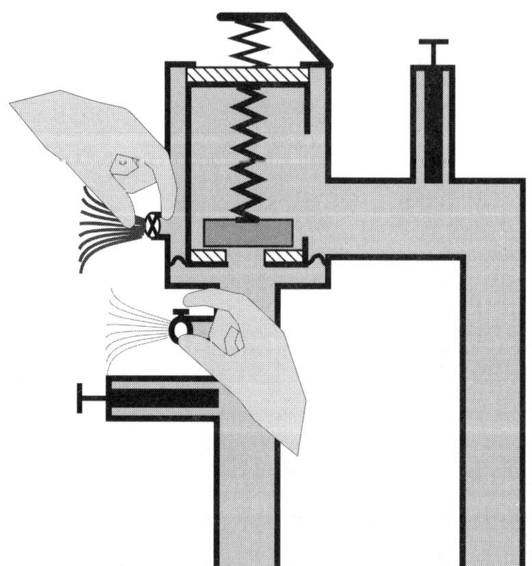

b. Bleed water through the test cock and vent valve to eliminate foreign material.

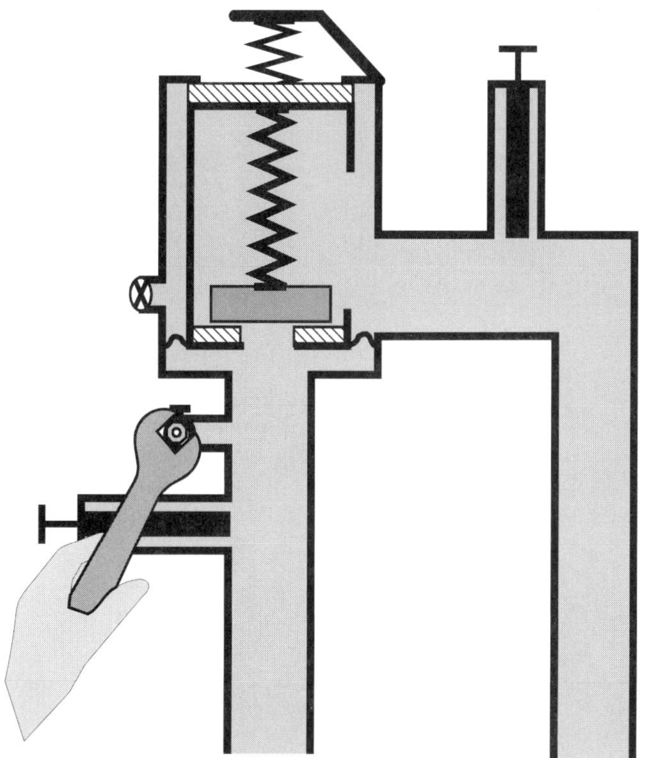

c. Install appropriate fitting to test cock.

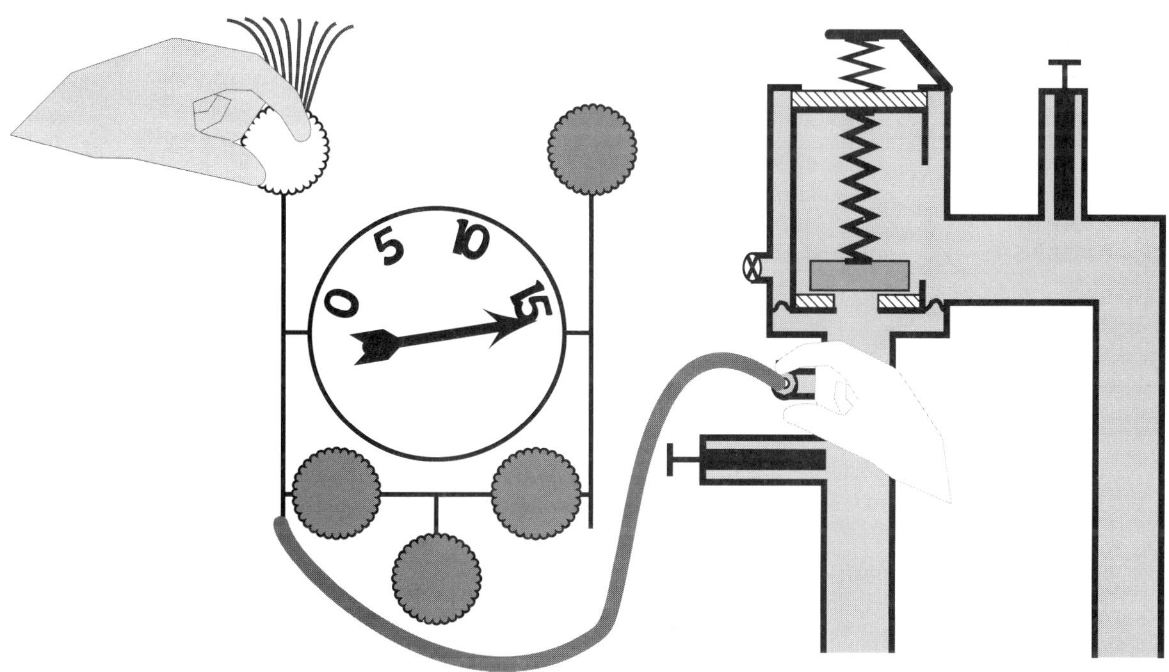

d. Attach the high side hose of the differential pressure gage to the test cock, open the test cock, and bleed air from the hose and gage by opening the high side bleed needle valve. Close the high side bleed needle valve.

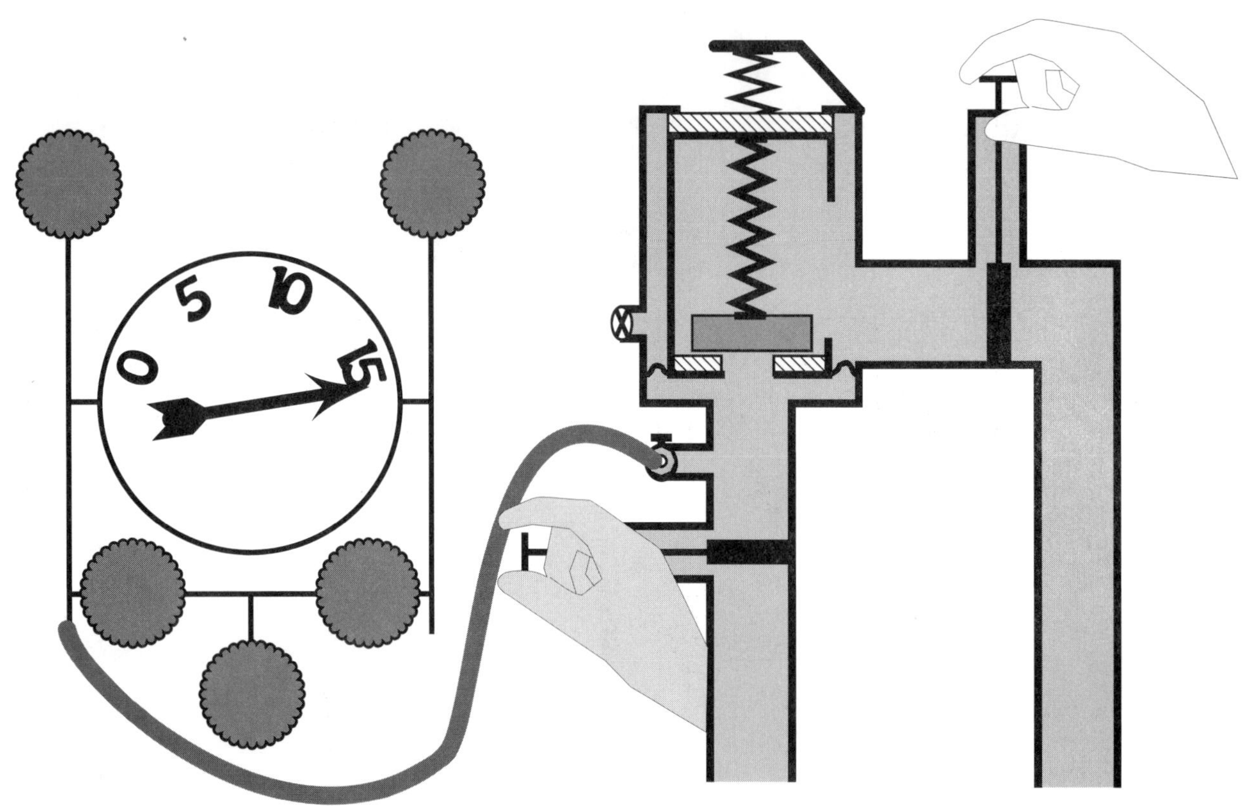

e. Close No. 2 shutoff valve, then close No. 1 shutoff valve.

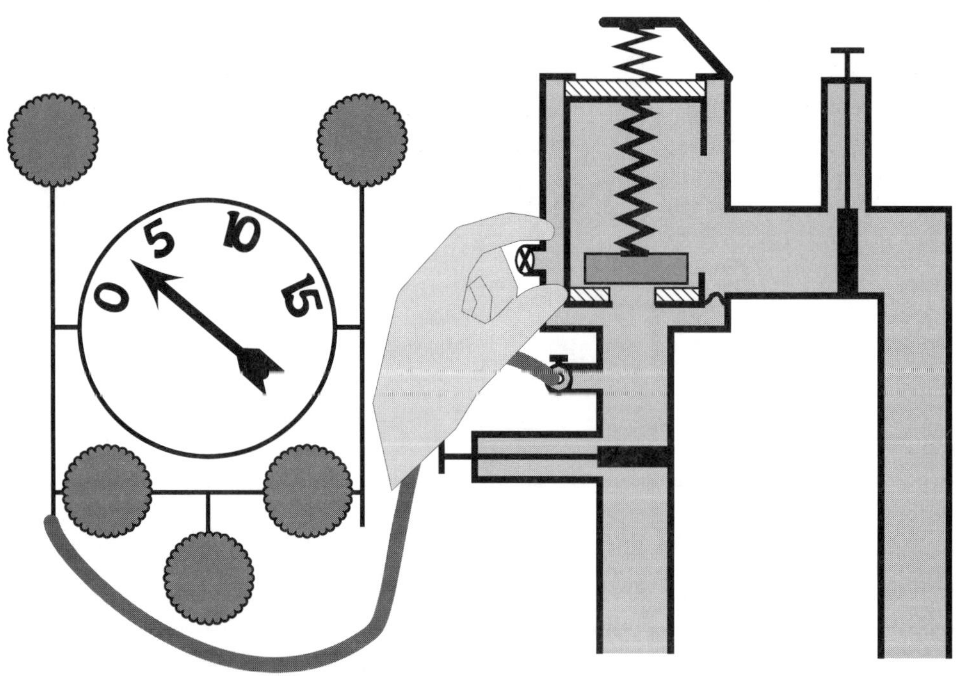

f. Open vent valve to lower outlet pressure to atmospheric. *(Observe gage reading and record value should air inlet valve open at this point.)*

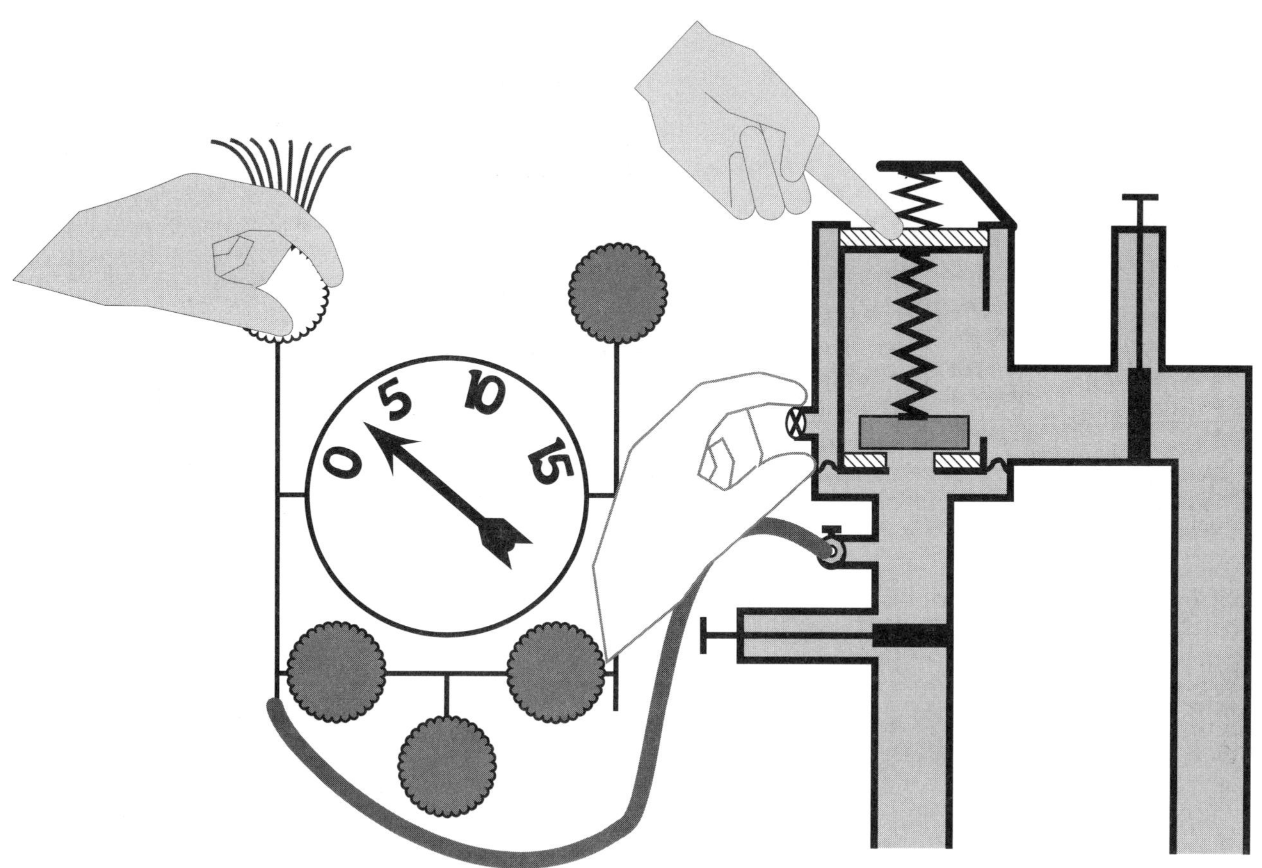

g. Slowly open the high side bleed needle valve no more than one-quarter (1/4) turn, being especially careful not to drop the pressure reading on the gage too fast. Record the pressure reading on the gage when the air inlet valve opens. The reading must be 1.0 psi or greater. If the high side bleed needle valve must be opened more than one-quarter (1/4) turn to lower the pressure reading in the body, see *Troubleshooting Section 9.5.3.1 - Leaking No. 1 shutoff valve*. Open high side bleed needle valve to lower pressure to atmospheric - 0.0 psid. Observe that air inlet valve has opened to its fully open position. Close vent valve. Close high side bleed needle valve.

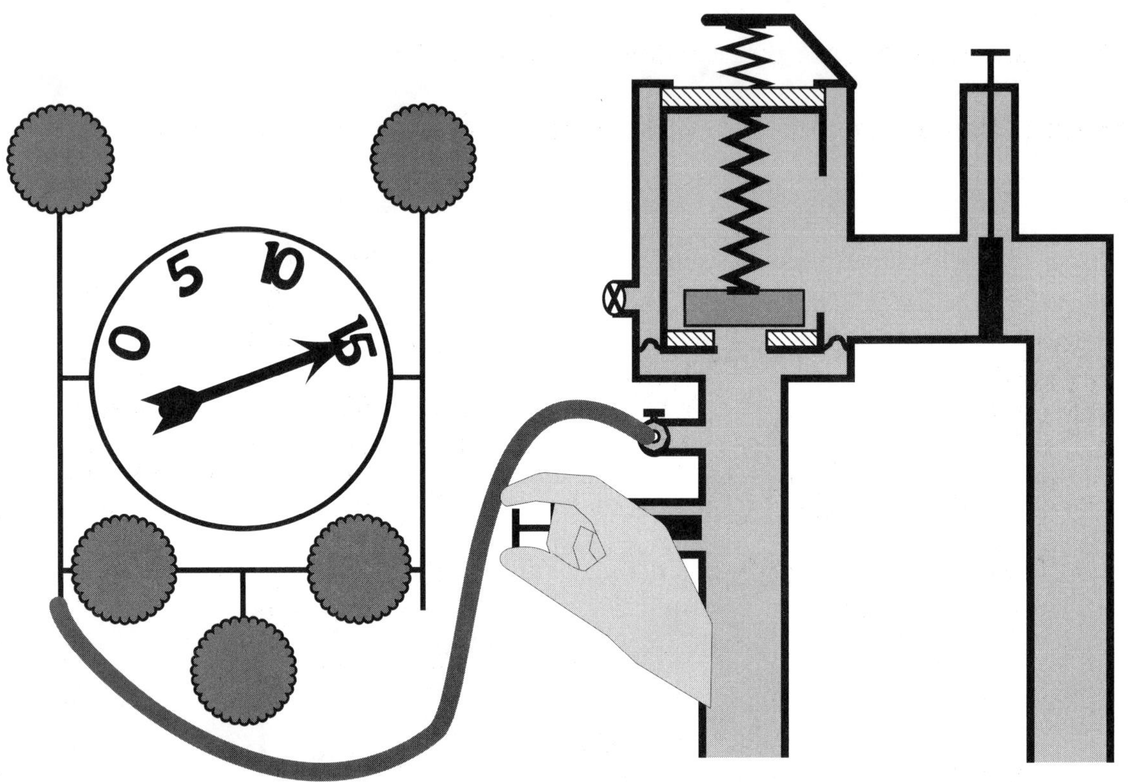

h. Open No. 1 shutoff valve slowly.

TEST NO. 2 ***Check Valve Closing Point***

Purpose: To determine the static pressure drop across the check valve.

Requirement: The static pressure drop across the check valve shall be at least 1.0 psid.

Steps:

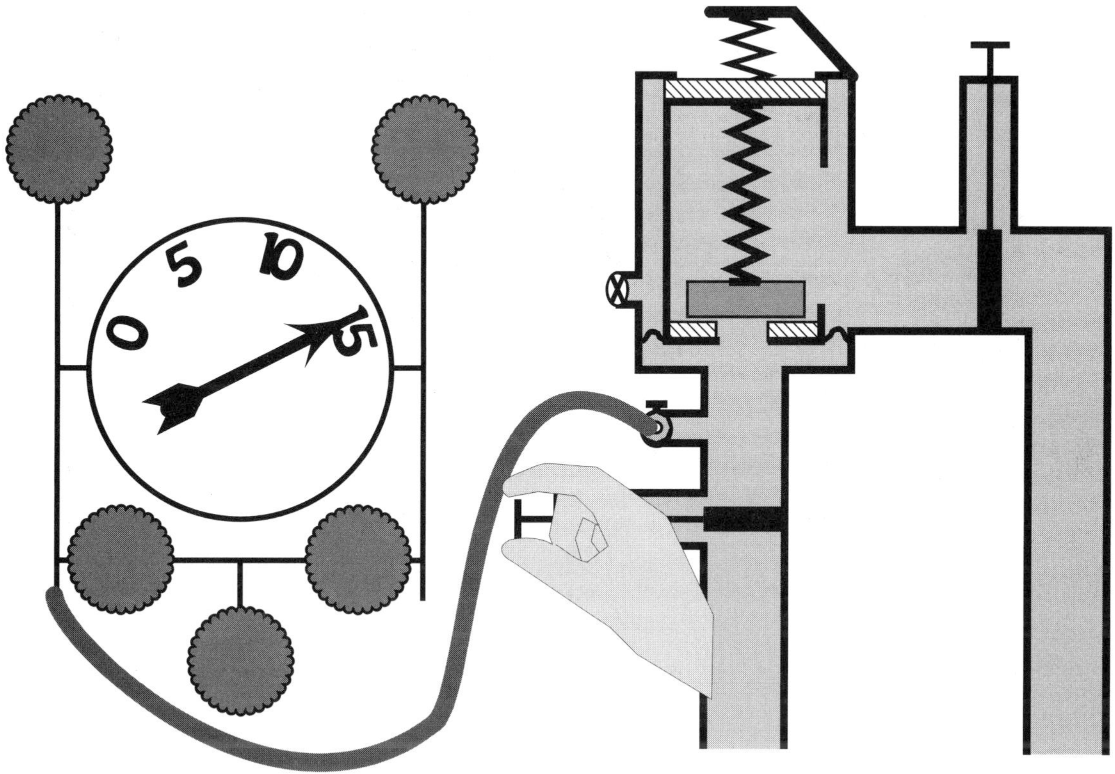

a. Close No. 1 shutoff valve (*No. 2 shutoff valve remains closed from Test No. 1*).

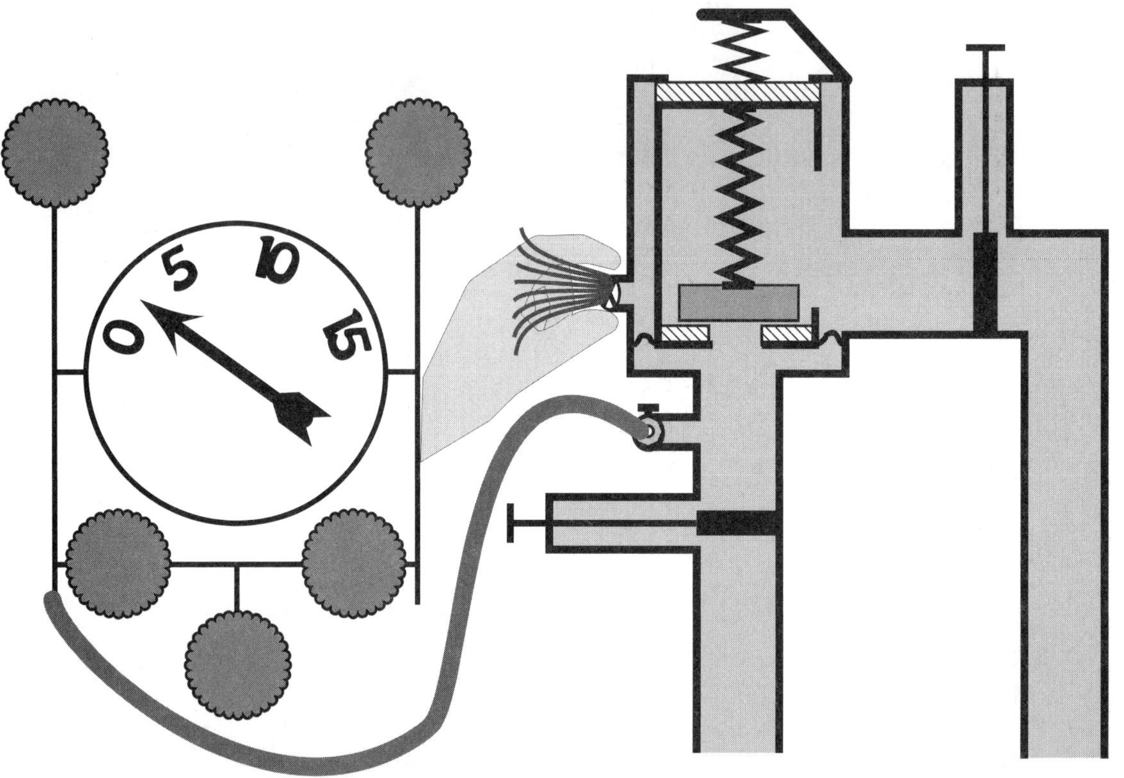

b. Open vent valve. The water in the body will drain out through the vent valve. When this flow of water stops, and the pressure reading indicated by the gage settles, the gage reading will be the static pressure drop across the check valve, and recorded as such. This gage reading must be 1.0 psid or greater. Record this value. If water continues to flow out of vent valve, *see Troubleshooting Section 9.5.3.2 - Leaking No. 1 shutoff valve.*

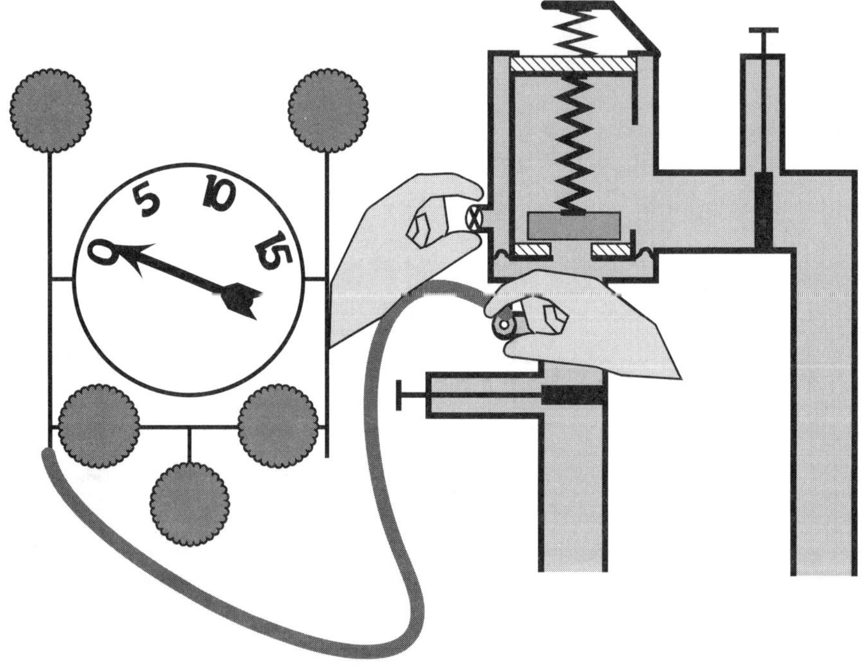

c. Close test cock and vent valve.

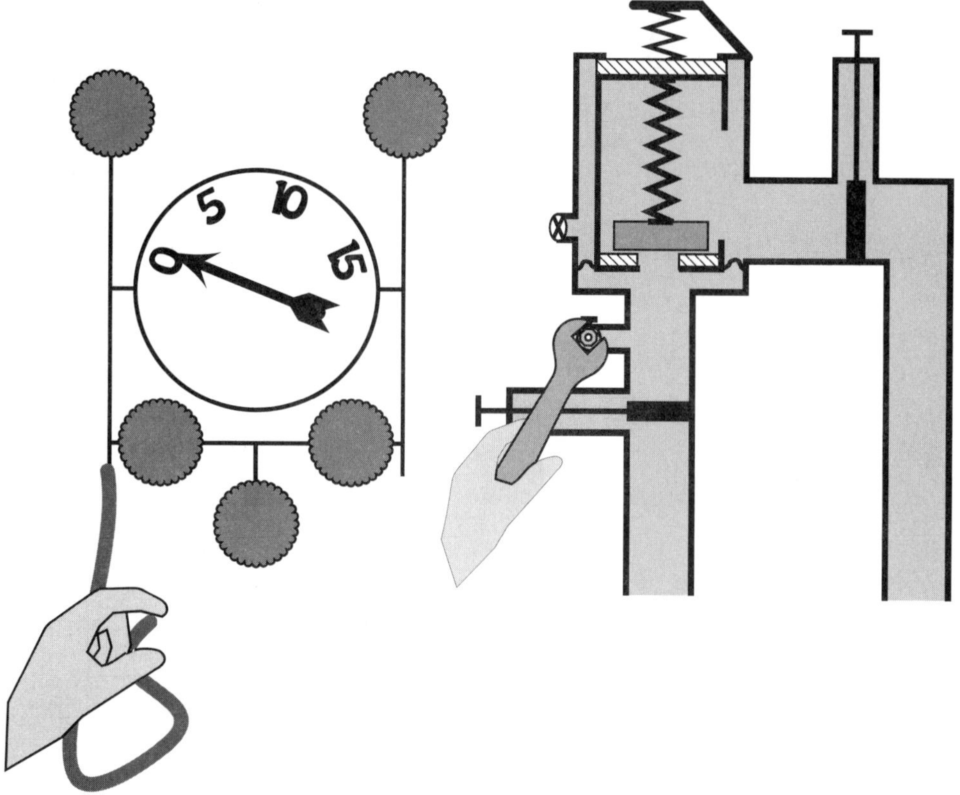

d. Remove equipment.

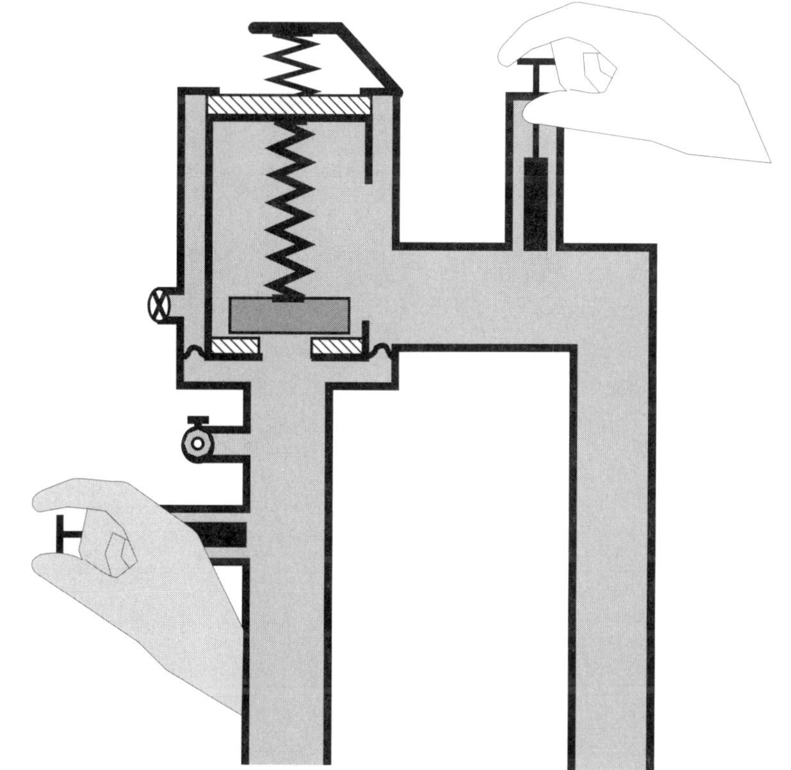

e. Open No. 1 shutoff valve, then No. 2 shutoff valve.

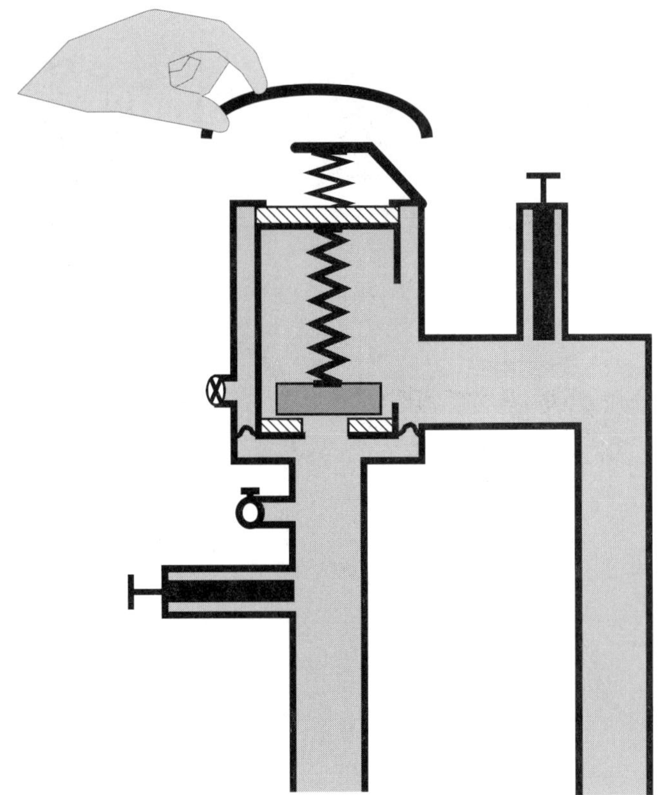

f. Replace air inlet valve canopy.

9.5.3 Troubleshooting

9.5.3.1 Leaking No. 1 Shutoff Valve - Air Inlet Valve Test (See Fig. 9.11)

Should the high side bleed needle valve need to be opened more than one-quarter (1/4) turn, it is likely that the No. 1 shutoff valve is leaking. The No. 1 shutoff valve should be re-opened and closed in an effort to a get a better seal. This may particularly occur when testing units which do not have resilient seated shutoff valves. Should the leak persist then the leak must be diverted so that the air inlet valve can be tested. Close the test cock. Attach a bleed-off valve to the test cock. Attach the high side hose to the bleed off valve. Open the test cock. Bleed air from the gage again by opening the high side bleed needle valve. Open the bleed-off valve slowly to divert the leakage from the No. 1 shutoff valve, monitoring the gage while this is being done. If the reading on the gage can be stabilized at approximately 10 psid, continue with step ***g*** Test No. 1. If the shutoff valve leak exceeds the limit of the test cock, then the test can not be completed until the No. 1 shutoff valve is repaired or replaced.

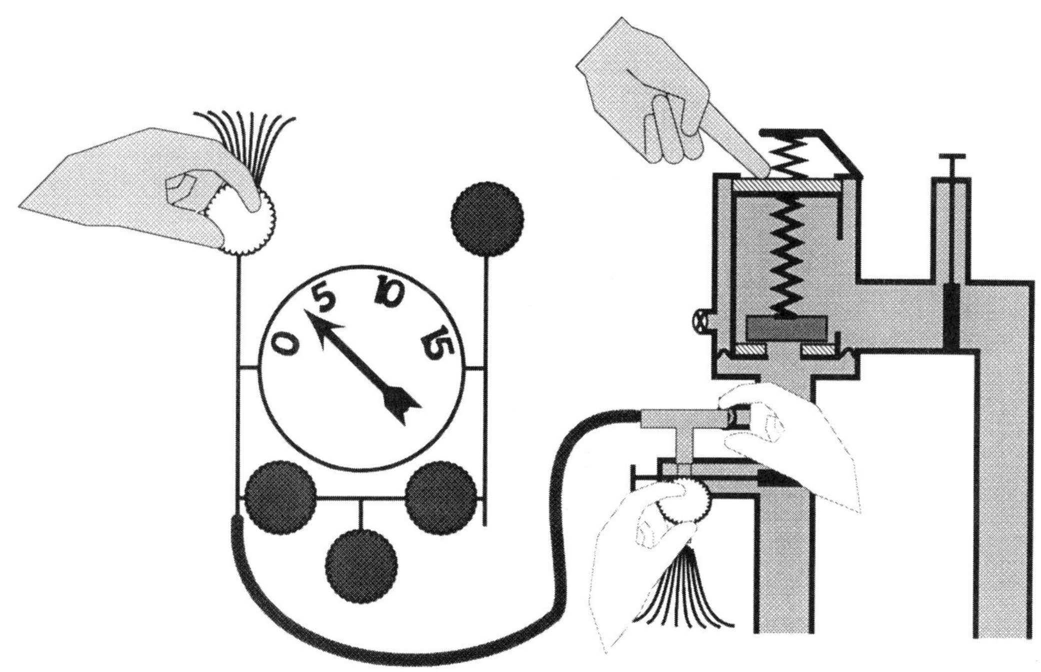

Fig. 9.11
Leaking No. 1 Shutoff Valve

9.5.3.2 Leaking No. 1 Shutoff Valve - Check Valve Test (See Fig. 9.12)

Should water continue to flow out of vent valve during Test No. 2, this indicates that the No. 1 shutoff valve is leaking. Install bleed-off valve to the test cock and attach high side hose to bleed-off valve. Open the test cock. Slowly open the bleed-off valve until the flow of water from vent valve is a slight drip. Record the reading on the differential gage as the static pressure drop across the check valve. Continue with Test No. 2 step ***c.*** If the flow of water from the vent valve can not be eliminated by opening the bleed-off valve, then the tightness of the check valve can not be determined. The shutoff valve No.1 must be repaired or replaced.

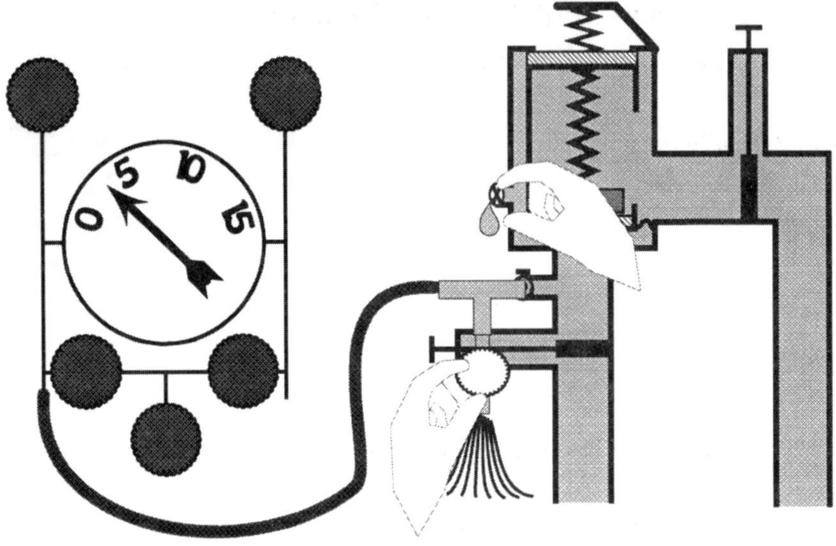

Fig. 9.12
Leaking No. 1 Shutoff Valve

9.5.3.3 Troubleshooting Summary

NOTE: Many problems can be corrected by cleaning the internal components. Carefully observe condition of components.

Problem	May be caused by
Air inlet valve does not open, as gage drops to 0.0 psid	1. Air inlet disc stuck to seat 2. Broken or missing air inlet spring 3. Parallel installation with Leaky No. 2 Shutoff valve 4. PVB not SVB
Air inlet valve does not open, and differential on gage will not drop	1. Leaky No. 1 shutoff valve
Air inlet opens below 1.0 psid	1. Dirty or damaged air inlet disc
Check valve below 1.0 psid	1. Dirty or damaged check disc 2. Damaged seat 3. Leaking Diaphragm/Seal
Water runs continuously from vent valve No. 2 (Test No.2)	1. Leaky No. 1 shutoff valve

Repair note: Lubricants shall only be used to assist with the reassembly of components, and shall be non-toxic.

9.6 Reduced Pressure Principle-Detector Backflow Prevention Assembly (RPDA)

9.6.1 Equipment Required:

a) Differential Pressure Gage—Minimum range 0 -15 PSID (0.1 or 0.2 psid graduations)

b) Three 6 ft. lengths—minimum 1/4"∅ high pressure hose

c) 1/4" Needle valves, for fine control of flows

d) Three 1/4" IPS x 45° SAE flare connector—brass

e) Adapter fittings for each test cock size—brass 1/8" x 1/4", 1/4" x 1/2", 1/4" x 3/4"

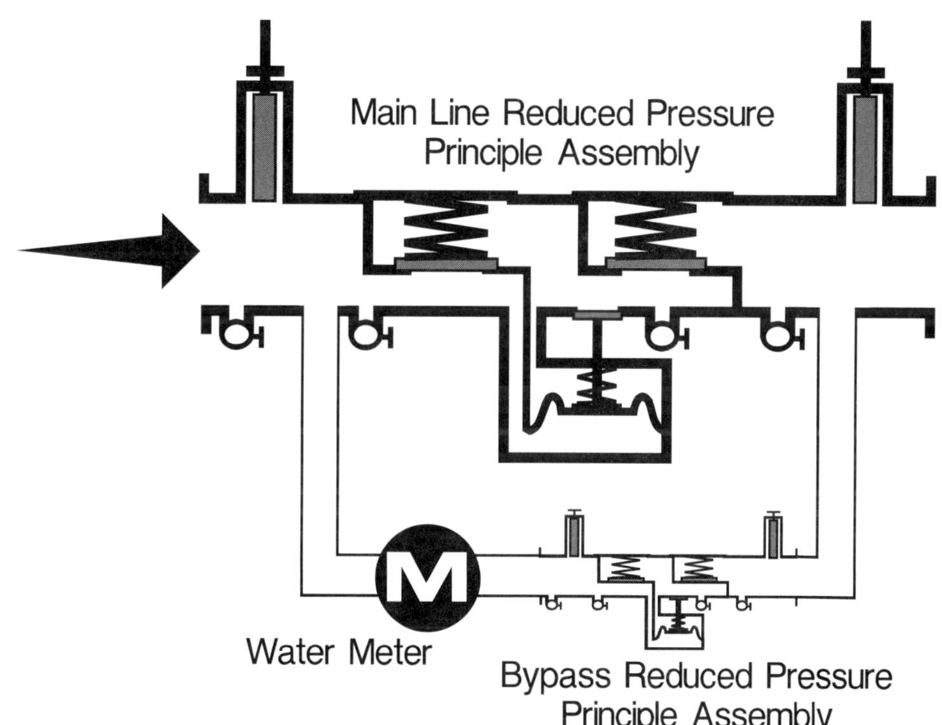

Fig. 9.13
Reduced Pressure Principle-Detector Backflow Prevention Assembly (RPDA)

9.6.2 Field Test Procedure

Test No. 1 - *Test of Main-line Assembly*

Purpose: To test the operation of the main-line reduced pressure principle backflow prevention assembly.

Requirement: The main-line reduced pressure principle backflow prevention assembly shall comply with field test requirements of Section 9.2.

Steps:

> **NOTE:** *Follow all preliminary steps detailed in Section 9.1. The tester shall request that the owner or occupant notify the authority having jurisdiction, the fire department, if required, and the alarm receiving facility before shutting down a fire sprinkler system or its water supply. The notification prior to testing shall include the purpose for the shutdown of the system, the component(s) involved, and the estimated time required.*

a. Close No. 2 shutoff valve of bypass assembly

b. Perform field test procedure per Section 9.2 for main-line reduced pressure principle backflow prevention assembly.

c. Maintain No. 2 shutoff valve of main-line assembly in closed position.

Test No. 2 - *Test of Bypass Assembly*

Purpose: To test the operation of the bypass reduced pressure principle backflow prevention assembly.

Requirement: The bypass reduced pressure principle backflow prevention assembly shall comply with field test requirements of Section 9.2.

Steps:

a. Perform field test procedure per Section 9.2 for bypass reduced pressure principle backflow prevention assembly.

b. Open all shutoff valves in RPDA.

9.7 Double Check -Detector Backflow Prevention Assembly (DCDA)

9.7.1 Equipment required:

a) Differential Pressure Gage—Minimum range 0 -15 PSID (0.1 or 0.2 psid graduations)

b) One 6 ft. lengths—minimum 1/4"∅ high pressure hose

c) 1/4" Needle valves, for fine control of flows

d) One 1/4" IPS x 45° SAE flare connector—brass

e) Adapter fittings for each test cock size—brass 1/8" x 1/4", 1/4" x 1/2", 1/4" x 3/4"

f) Street ell, pipe nipple or tube (for length - see Section 9.3.3)

g) Bleed-off valve (see Appendix Section A.4.1—Bleed-Off Valve Arrangement)

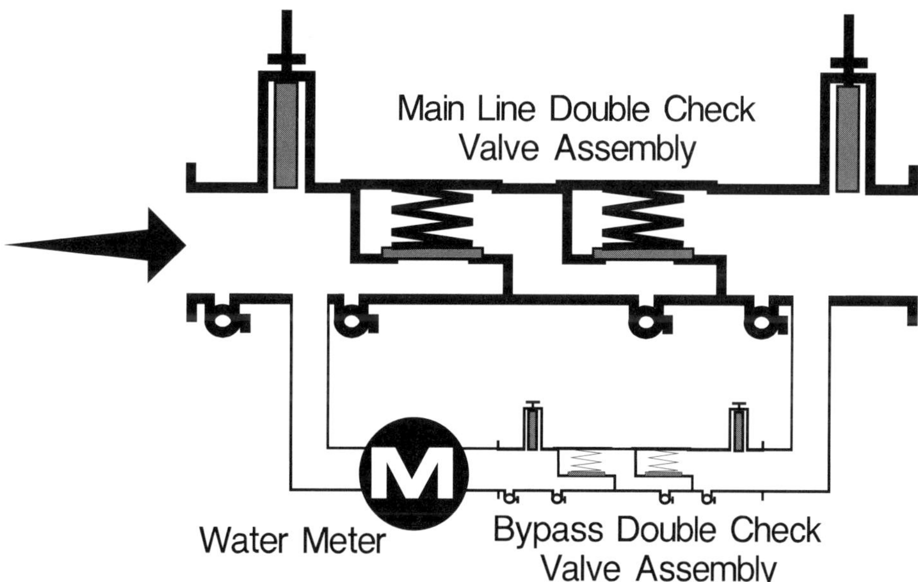

NOTE: For both of the following tests the differential pressure gage must be held at the same level as the assembly being tested. Be sure that hoses not being used are also kept at this level

Fig. 9.14
Double Check-Detector Backflow Prevention Assembly (DCDA)

9.7.2 Field Test Procedure

Test No. 1 - *Test of Bypass Assembly*

Purpose: To test the operation of the bypass double check valve backflow prevention assembly.

Requirement: The bypass double check valve backflow prevention assembly shall comply with field test requirements of Section 9.3.

Steps:

> ***NOTE:*** *Follow all preliminary steps detailed in Section 9.1 The tester shall request that the owner or occupant notify the authority having jurisdiction, the fire department, if required, and the alarm receiving facility before shutting down a fire sprinkler system or its water supply. The notification prior to testing shall include the purpose for the shutdown of the system, the component(s) involved, and the estimated time required.*

a. Perform field test procedure per Section 9.3 for bypass double check valve backflow prevention assembly.

b. Leave both shutoff valves of the bypass assembly closed.

Test No. 2 - *Test of Main-line Assembly*

Purpose: To test the operation of the main-line double check valve backflow prevention assembly.

Requirement: The main-line double check valve backflow prevention assembly shall comply with field test requirements of Section 9.3.

Steps:

a. Perform field test procedure per Section 9.3 for main-line double check valve backflow prevention assembly.

b. Open all shutoff valves in the DCDA.

9.8 Gage Accuracy Verification

9.8.1 Differential Gage Accuracy Verification — Water Column (see Fig. 9.15)

9.8.1.1 Materials

1) Two transparent tubes (approx. 1-inch diameter) minimum length — 8 ft.
2) Adapters from transparent tube to gage — Rubber stoppers, nipples, ells.

9.8.1.2 Procedure

1) Attach high side hose to base of one transparent tube.
2) Attach low side hose to base of second transparent tube.
3) Fill transparent tubes with water.
4) Bleed air from gage by opening high side bleed valve, then close; then open low side bleed needle valve and close.
5) Fill or drain transparent tubes to desired height 'h'.

   27-3/4" Water = 1.0 psi
   55-1/2" Water = 2.0 psi
   83-1/4" Water = 3.0 psi
   etc.

6) Compare gage reading to water column height 'h', the two values should be within the accuracy range of ± 0.2 psid.
7) If gage requires adjustment, contact the gage manufacturer or a qualified gage calibration facility.

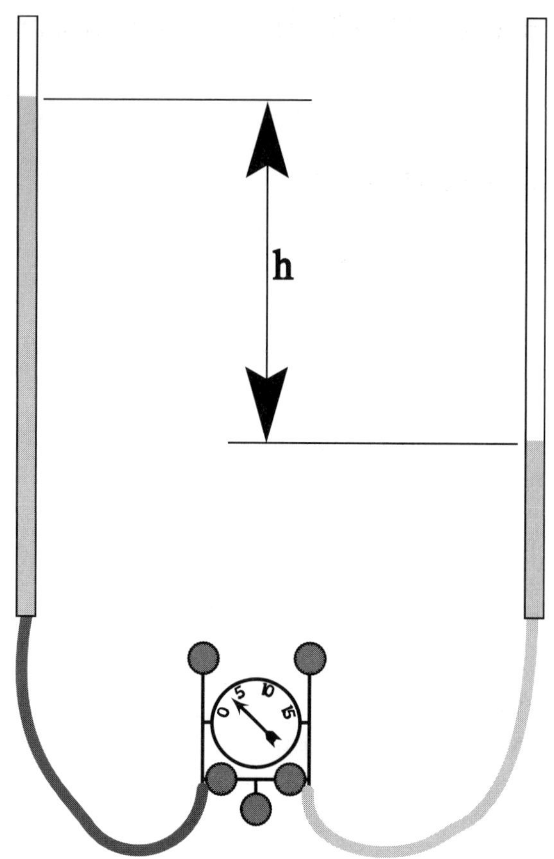

Fig. 9.15
Differential Gage Accuracy Verification - Water Column

9.8.2 Differential Gage Accuracy Verification — Manometer (see Fig. 9.16)

9.8.2.1 Materials

1) Mercury manometer — minimum length 36".

2) Miscellaneous pipe fittings — see Fig. 9.16.

3) Pressure reducing valve — proper range so that available line pressure can be reduced up to 15 psi.

CONVERSION FACTOR — Water over mercury: 1" Hg = 0.455 psi

9.8.2.2 Procedure

1) Attach hoses from mercury manometer to crosses on either side of the pressure reducing valve.

2) Attach high side hose of the differential gage to the upstream (high side) cross, and the low side hose to the downstream (low side) cross.

3) Set pressure reducer to open position (no pressure reduction).

4) Turn on inlet pressure.

5) Bleed air out of the manometer hoses by carefully opening the bleed valves at the top of each leg of the manometer.

6) Bleed air out of differential gage by bleeding through the high side hose first, then the low side hose. Both the manometer and the differential gage should be reading 0.0 psid at this time.

7) Open the low side bleed needle slightly and adjust the pressure reducing valve to establish approximately a 1.0 psid differential. Close the bleed needle valve.

8) Compare the values between the manometer and the differential gage. (i.e., measure the deflection of the mercury 'h' and multiply by 0.455 to convert the value to pounds per square inch differential — psid.) Both the manometer and the differential gage should be within the accuracy range of ±0.2 psid.

9) Repeat steps No. 7 and No. 8 with greater differential values — 2.0, 3.0, 4.0, etc., up to 15.0 psid.

10) If gage requires adjustment, contact the gage manufacturer or a qualified gage repair shop.

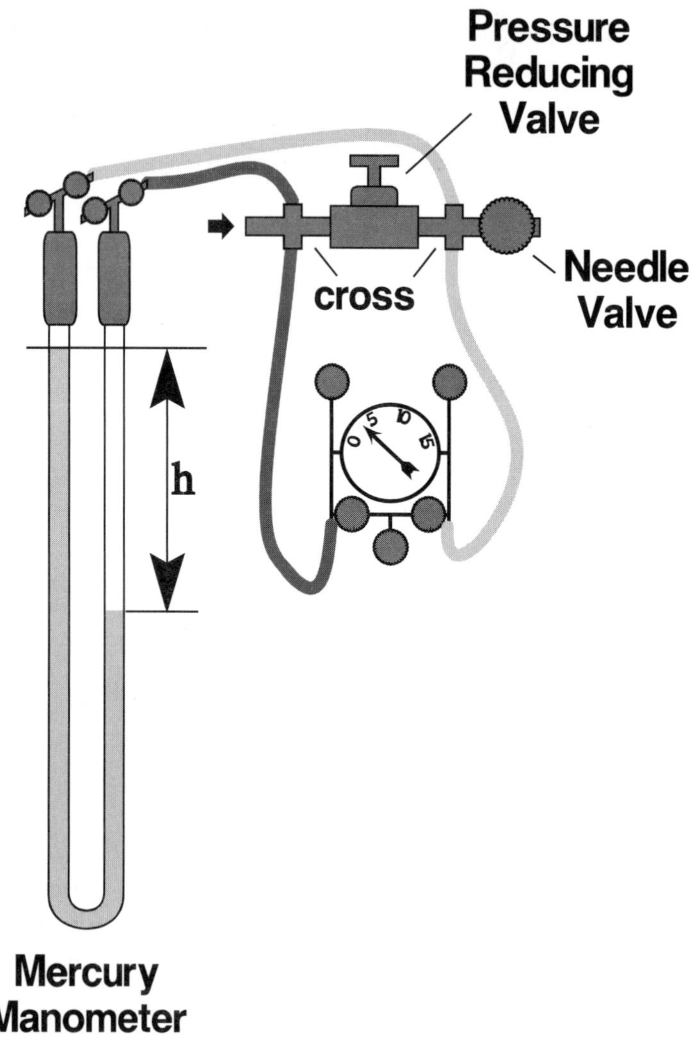

Fig. 9.16
Differential Gage Accuracy Verification
Mercury Manometer

SECTION 10

Specifications of Backflow Prevention Assemblies

10.1 GENERAL—Design and Materials Specifications and Laboratory Testing

10.1.1 General Specifications

10.1.1.1 Flow Characteristics and Pressure Loss Requirements

The flow characteristics and pressure loss requirements of these assemblies are of prime consideration in insuring their functional operation. In all cases the flow channels shall be streamlined to minimize pressure loss. The lowest possible pressure loss through the backflow prevention assembly is necessary to deal with intermittent low distribution system pressure and high in-plant pressure losses.

10.1.1.2 Rated Flow and Maximum Allowable Pressure Loss

The American Water Works Association (AWWA) has adopted values for the rated flow which must be at least met by displacement or compound meters in sizes 5/8 x 3/4 inch to 10 inch, inclusive. Those flow rates are included in Table 10-1and Table 10-5 for the above sizes together with those for 1/4, 3/8, 1/2, 1 1/4, 2-1/2, 12, 14, and 16 inch, which have been extrapolated or interpolated. For each size of backflow prevention assembly at any flow rate up to and including the rated flow, the maximum pressure loss shall not exceed the values given in Table 10-1 for Reduced Pressure Principle Assemblies and Double Check Valve Assemblies, and Table 10-5 for Pressure Vacuum Breaker Assemblies, Spill-Resistant Pressure Vacuum Breaker Assemblies, and Atmospheric Vacuum Breaker Assemblies.

As a guideline for assemblies intended for fire sprinkler service, the pressure loss at the following rates of flow are now being asked for by some agencies[1].

4-inch (101 mm)	750 gpm (47.31 L/s)
6-inch (152 mm)	1500 gpm (94.63 L/s)
8-inch (203 mm)	3000 gpm (189.25 L/s)
10-inch (254 mm)	4900 gpm (309.11 L/s)

In these cases there has been no maximum pressure loss restrictions placed on the assemblies at these rates of flow.

10.1.1.3 Standard Sizes

In this Specification the following standard sizes have been adopted for backflow prevention assemblies: 1/4, 3/8, 1/2, 5/8×3/4, 3/4, 1, 1-1/4, 1-1/2, 2, 2-1/2, 3, 4, 6, 8, 10, 12, 14, and 16 inches.

---

[1] Factory Mutual (FM), 1151 Boston-Providence Turnpike, Norwood, MA 02062 (617) 762-4300

TABLE 10-1

RATED FLOW AND MAXIMUM ALLOWABLE PRESSURE LOSS FOR VARIOUS SIZES OF BACKFLOW PREVENTION ASSEMBLIES

Size of Assembly (1)		Rated Flow (2)		Maximum Allowable Pressure Loss: Reduced Pressure Principle Assembly (3)		Double Check Valve Assembly (4)	
(inches)	(mm)	(gpm)	(L/s)	(psi)	(KPa)	(psi)	(KPa)
1/4	(6)	1*	(0.06)	24	(165.4)	10	(68.9)
3/8	(9)	3*	(0.19)	22	(151.6)	10	(68.9)
1/2	(12)	7.5*	(0.39)	22	(151.6)	10	(68.9)
5/8×3/4 ***	(16×20)	20	(1.26)	20	(137.8)	10	(68.9)
3/4	(20)	30	(1.89)	20	(137.8)	10	(68.9)
1	(25)	50	(3.15)	18	(124.1)	10	(68.9)
1 1/4	(31)	75**	(4.73)	18	(124.1)	10	(68.9)
1 1/2	(38)	100	(6.31)	16	(110.3)	10	(68.9)
2	(50)	160	(10.10)	16	(110.3)	10	(68.9)
2 1/2	(63)	225**	(14.20)	16	(110.3)	10	(68.9)
3	(75)	320	(20.19)	15	(103.4)	10	(68.9)
4	(100)	500	(31.55)	14	(96.5)	10	(68.9)
6	(150)	1000	(63.10)	14	(96.5)	10	(68.9)
8	(200)	1600	(100.96)	14	(96.5)	10	(68.9)
10	(254)	2300	(145.13)	14	(96.5)	10	(68.9)
12	(305)	3000	(189.30)	13	(89.6)	10	(68.9)
14	(355)	3700	(233.47)	13	(89.6)	10	(68.9)
16	(406)	4400	(277.64)	13	(89.6)	10	(68.9)

* Extrapolated

**Interpolated

*** 3/4"∅ inlet and outlet connections

Note: The pressure losses as shown in Column (3) and Column (4) represent the maximum permissible pressure loss at any flow rate up to and including the rated flow shown in Column (2).

All assemblies designed and constructed in sizes other than those aforementioned shall be given separate consideration.

The inlet and outlet of the assembly shall be threaded in accordance with ANSI/ASME[2,3] B1.20.1 for taper pipe connections; or ANSI B16.24 for bronze flanges; or ANSI B16.1 for iron flanges; or ANSI/AWWA[4]C606 for grooved and shoulder joints.

Metric equivalents (International System of Units — SI) have been provided wherever needed per ASTM[5] Designation: E380.

10.1.1.4 Markings

Size, model and serial number markings on backflow prevention assemblies for sizes 5/8x3/4-inch through 16-inch shall be with letters or numbers at least 1/4 inch (6 mm) in height. For backflow prevention assemblies sized 1/4, 3/8, and 1/2-inch, the size, model and serial number markings shall be with letters or numbers at least 1/8 inch (3 mm) in height. All lettering cast on a body shall be a minimum of 1/4-inch in height. All markings shall be easily read and shall be either cast or stamped on the body; or stamped, engraved, or etched on a durable nameplate permanently affixed to the body of the assembly and shall be located either; a) on both sides of the body, or b) on a top surface of the body. The nameplate shall be either brass or stainless steel and affixed with stainless steel escutcheon pins. In affixing a nameplate with escutcheon pins or stamping data in the metal of the assembly, caution shall be exercised so as not to produce an area of stress concentration.

TABLE 10-2

MARKINGS FOR BACKFLOW PREVENTION ASSEMBLIES

Make: Name or Trademark
Type of assembly *(ie., reduced pressure principle or double check valve backflow prevention assembly, pressure vacuum breaker, spill-resistant pressure vacuum breaker, double check-detector assembly, reduced pressure principle-detector assembly, or acceptable abbreviation RP, DC, PVB, SVB, DCDA, RPDA, respectively.)*
Size
Model
Direction of flow *(shown by an arrow)*
Unit serial number
Maximum rated working water pressure *(MWWP)*
Maximum rated working water temperature *(MWWT)*

For the DCDA and RPDA, the following additional information is required on the mainline assembly:
Make, Model, and Size of backflow prevention assembly in bypass.

---

[2]American National Standards Institute, 1430 Broadway, N.Y., NY 10018 (212) 354-3300
[3]American Society of Mechanical Engineers, 345 E. 47th Street, N.Y., NY 10017 (212) 705-7736
[4]American Water Works Association, 6666 Quincy Ave, Denver, CO 80235 (303) 794-7711
[5]American Society for Testing and Materials, 1916 Race Street, Philadelphia, PA 19103-1187 (215) 299-5585

The markings on a DCDA and RPDA which identify the backflow prevention assembly in the bypass shall be with letters or numbers at least 1/8-inch (3mm) in height.

The markings on a pressure vacuum breaker (PVB and SVB) shall be as noted above except that all markings shall be located either, a) on a top surface of the body, or b) on the side of the body. The removable bonnet or cover shall not be used for identification purposes.

Markings shall be permanent and not easily defaced. The markings shall include those shown in Table 10-2.

Name, model, and size markings on atmospheric vacuum breakers shall be with letters or numbers at least 1/4 inch (6 mm) in height, for line sizes 3/4-inch and larger. For line sizes 1/4, 3/8, and 1/2-inch, the name, model, and size markings shall be with letters or numbers at least 1/8 inch (3 mm) in height. The markings on an atmospheric vacuum breaker shall be as noted above except that all markings shall be located either, a) on a top surface of the body, or b) on the side of the body. The removable bonnet or cover shall not be used for identification purposes.

All markings shall be in English units (i.e., PSI, °F, Inches). Metric equivalents (i.e., KPa, °C., mm) shall be permitted in addition to the English units at the manufacturer's discretion.

10.1.1.5 Hydrostatic Test — Structural Integrity

a. All assemblies shall be pressure tested according to their designed operating pressure for use on cold water service {maximum 110°F (44.3°C)}. Normal testing shall be accomplished at a minimum of 150 psi (1034 KPa) line pressure. The hydrostatic test pressure shall be twice the maximum rated working water pressure of the assembly.

b. The entire assembly shall be subjected to the hydrostatic test both in the normal direction of flow and in the reverse direction of flow of all closed barriers with the opposite side of the barrier open to the atmosphere. There shall be no leakage across any barrier. And, the hydrostatic test pressure shall be maintained for a minimum of ten (10) minutes for each test.

c. No damage or permanent deformation of any parts of the assembly, or impairment of operation shall occur under the full hydrostatic test pressure.

10.1.1.6 General Statement of Policy Regarding Assembled Assemblies

All assemblies which consist of independent units assembled for the purpose of preventing backflow shall comply with the material, the operational and other specifications as required for backflow prevention assemblies. In order to ensure proper installations, all backflow prevention assemblies shall be delivered for installation completely assembled by the original manufacturer with all components as approved. Resilient seated shutoff valves and test cocks are considered integral parts of the assembly.

10.1.2 Design Specifications

10.1.2.1 Policy Regarding Design

In the design of any backflow prevention assembly, prime consideration shall be given to the construction of a trouble-free unit. All materials used shall be of the best quality. The water way shall be as free as possible from obstructions and pockets which could interfere with the free flow of water. All moving parts shall be designed to operate up to the rated flow without chatter or vibration. The moving parts shall have adequate clearance to prevent binding or galling, or to prevent the assembly from becoming inoperative by being thrown out of balance, by being distorted, by having one part interfere with another, or by becoming encrusted with lime, rust, or scale deposits.

All foundry and machine work shall be performed in accordance with the best modern practice for the class of work involved. All parts shall conform, with allowable tolerances, to the manufacturer's specifications and shall be free from injurious defects. All flanged joints shall be faced true and machined at right angles to their respective axes; while threaded joints must be concentric and accurately cut. All joints shall be watertight where subject to water pressure. All ferrous parts receiving a bronze or other mounting shall be finished to fit. Such handwork as required in finishing shall produce a neat, workmanlike, well-fitting and smoothly operating product. All replaceable parts of assemblies of the same size and model shall be interchangeable with the original parts.

Assemblies intended for cold water service {i.e., maximum of 110°F (43.3°C)} shall be so designed and tested as to function satisfactorily up to a minimum temperature of 140°F (60°C) at the maximum rated working water pressure.

Assemblies intended for elevated temperature service above the normal cold water designation of 110°F (43.3°C), shall be so designed and tested as to function satisfactorily over the specified temperature and pressure ranges. And further, these assemblies are to carry a distinctive name plate (different from the normal cold water service assembly name plate) which specifies the higher temperature range of the assembly.

Any changes of design, materials or coating, require full disclosure by the manufacturer to the approving laboratory for approval by the approving laboratory prior to implementation.

10.1.2.2 Removability of Major Components from the Line

A backflow prevention assembly shall be designed so that each principle component (i.e., check valve if a separate body, and the differential pressure relief valve) of the assembly shall be permitted to be removed and reinstalled individually from the line.

10.1.2.3 Accessibility of Internal Parts

A backflow prevention assembly shall be provided with one or more openings through which the internal parts may be removed, repaired, or inspected without having to remove the body of the assembly from the line.

10.1.2.4 Interdependence of Components

In the double check valve assembly there shall be no mechanical linkage between the two check valves. Each check valve shall be free to operate independently through its entire movement. The movement of either the first or second check valve through its full limit of travel shall not affect the operation of the other check valve.

In the reduced pressure principle backflow prevention assembly each of the check valves shall likewise be free of any coupling mechanical linkage and shall be free to operate independently through its entire movement. Further, the relief valve shall be mechanically independent of both check valves yet hydraulically dependent upon the pressure differential across the first check valve. The movement of either the first check valve, second check valve, or differential pressure relief valve through their respective full limits of travel shall not affect the operation of the other components.

In the pressure vacuum breaker (PVB and SVB) the check valve shall be free to operate independently from the air inlet valve. The movement of either the check valve or the air inlet valve through their full limits of travel shall not affect the operation of the other.

10.1.2.5 Differential Pressure Relief Valve

The differential pressure relief valve of a reduced pressure principle backflow prevention assembly shall be located so that its valve seat(s) and port(s) to the atmosphere are below the lowest point of the first check valve so as to preclude backsiphonage. The water passageway from the differential pressure relief valve seat(s) to the discharge port(s) to the atmosphere shall not cause any backsiphonage air to pass through a pocket of water that is in or been discharged from the differential pressure relief valve.

For designs utilizing a separate differential pressure relief valve body, the relief valve shall not be attached to the body of the assembly with standard tapered pipe thread, pipe flange, or grooved joint sizes as detailed in Section 10.1.1.3 . To reduce the possibility of unauthorized field modifications the shape of the connection or boss on the body shall be such that no piping, hose connection, flange, or plate can easily be attached to it.

The differential pressure relief valve discharge port(s) to the atmosphere shall not be of a size that can be threaded for NPT threads or adapted to tubing internally or externally. Further, the shape of the port or its boss shall be such that no piping, hose connection, flange, or plate can easily be attached to it.

If a drain funnel is available as an accessory to an assembly, the drain funnel shall provide adequate backflow protection between the discharge port(s) of the relief valve and the drain funnel. (See Section 10.2.2.3.9) The outlet of the drain funnel shall be provided with connections as detailed in Section 10.1.1.3 or connections adaptable to hubless connectors complying with ASTM C564. The drain funnel shall be rigidly attached to the assembly.

10.1.2.6 Design of Waterway

a. Area of Waterway — Backflow prevention assemblies shall be so designed that the minimum waterway area normal to the direction of flow shall not adversely affect the overall pressure loss.

b. Obstructions — The waterways of the assembly shall be so designed as to minimize cavitation and eliminate all cavities which could entrap foreign materials.

c. Turbulence — The turbulence within or created by the assembly shall not be excessive for any flow rate up to the rated flow conditions.

10.1.2.7 Clearance

a. Between guide stems and guides, valve stems and guides, hinge pins and bushings and other similar parts the clearances shall be adequate to prevent sticking or binding.

b. Binding or clogging of parts might occur between the body of a valve or hinge arm, clapper, counterweight, poppet valve or other similar free-moving part. Wherever such binding or clogging of parts might occur which would prevent the assembly from being free to operate normally, there shall be provided minimum clearances as follows: for ferrous-body assemblies of all sizes, 1/2 inch (12.7 mm) clearance; for bronze-body assemblies in sizes up through 1-inch (25.4 mm), 1/8 inch (3.1 mm) clearance; for bronze assemblies over 1 inch (25.4 mm), 1/4 inch (6.35 mm) clearance.

c. The elastomer disc for all sizes of poppet type valves shall be supported by an adequate bearing surface such that the elastomer disc will comply with all tests contained in Section 10.2. Furthermore, the elastomer disc must be wide enough to overhang the seating surface of the seat ring at least 1/16 inch (1.58 mm) on the outside. A minimum of 1/16 inch (1.58mm) shall be provided between the inside edge of the seat ring and the facing clamping member.

d. The elastomer disc for all sizes of swing or clapper type valves shall be supported by an adequate bearing surface such that the elastomer disc will comply with all tests contained in Section 10.2. Furthermore, the elastomer disc must be wide enough to overhang the seating surface of the seat ring at least 1/8 inch (3.18 mm) on sizes up to 2 inches (50 mm) and 1/4 inch (6.35 mm) on sizes 2 inches (50 mm) and larger. A minimum of 1/16 inch (1.58 mm) shall be provided between the inside edge of the seat ring and the facing ring clamp.

e. All bushings shall project inside the body of the assembly as follows:

   1/8 inch (3.18 mm) for sizes up through 1 inch (25.4 mm)
   1/4 inch (6.35 mm) for sizes over 1 inch (25.4 mm)

10.1.2.8 Body and Bonnet

When a body and bonnet are bolted together, the bolts shall be of such design that the maximum stress at the root diameter shall not exceed a calculated 7,500 psi, based on the maximum rated working water pressure.

Further, the clearance between the bolt circle flange or the valve body flange and the adjacent body of the assembly shall be sufficient so that no special tools, slotted bolt holes or non-standard bolts or nuts are required.

In addition, a notch shall be provided in the bonnet or a lug on the bonnet and body by means of which the bonnet may be loosened from the body without seriously damaging the seal or sealing surfaces.

10.1.2.9 Tapping and Threading

Valve bodies shall be provided with an adequate boss at each location where a tapped hole is required. Female pipe threaded connections shall be tapped into bosses in the body and shall be so constructed that it shall not be possible to thread a pipe into them far enough to interfere with internal components.

All bolts and cap screws used internally and/or externally are to be of one standard — either English or SI.

10.1.2.10 Test Cocks

Double check valve assemblies and reduced pressure principle backflow prevention assemblies shall be equipped with resilient seated test cocks located as follows:

a. On the upstream side of the No. 1 shutoff valve
   Exception: Not required on 1/4" or 3/8" reduced pressure principle backflow prevention assemblies.

b. Between the No. 1 shutoff valve and the No. 1 check valve

c. Between the check valves

d. Between the No. 2 check valve and the No. 2 shutoff valve.

(See Figs. 10-1 and 10-4.)

Pressure vacuum breaker backsiphonage prevention assemblies shall be equipped with resilient seated test cocks located as follows:

a. Between the No. 1 shutoff valve and the check valve

b. Between the check valve and the air inlet valve

(See Fig. 10-5.)

Spill-resistant pressure vacuum breaker backsiphonage prevention assemblies shall be equipped with a resilient seated test cock and a resilient seated vent valve located as follows:

Test cock: Between the No. 1 shutoff valve and the check valve.

Vent valve: Between the check valve and the air inlet valve.

(See Fig. 10- 9.)

The sizes of these test cocks shall be as given in Table 10-3. The body port into which the test cock is threaded shall be rated size for the test cock and the waterway through the test cock itself shall be at minimum rated size. All 1/4", 1/2", and 3/4" test cocks shall have IPS female ends on the discharge side in accordance with ANSI standards for the standard pipe sizes mentioned in Table 10-3. The 1/8" test cock shall be permitted to be either IPS female threads or male 1/4" flared type connection per SAE[6] J513.

The operating stem of a ball valve type test cock shall be designed so that it is inserted from inside the body, *i.e.*, blow out proof stem. The operator on quarter turn valves must clearly show if the valve is open or closed, even if the handle is removed.

The test cocks shall withstand all hydraulic testing of the assembly, in both the open and closed positions, without leakage, damage, or permanent deformation.

10.1.2.11 Control Piping and Diaphragms

All control piping or passageways shall be of corrosion-resistant material or protected with a suitable protective coating in accordance with section 10.1.3.14 The pipes or passageways shall be sized sufficiently large to prevent clogging. The pipe or passageways shall be located in a manner as to prevent entrapment of foreign materials or air. Furthermore, all external control piping shall be placed in such a manner that it is not readily damaged when the assembly is located in unprotected areas.

TABLE 10-3
SIZES OF TEST COCKS

Size of Assembly in inches	Minimum Equivalent Orifice Size of Test Cock in inches
1 (25mm) and below	1/8 Ø (3.2 mm)
1 1/4 (31mm) to 2 (50mm) inclusive	1/4 Ø (6.4 mm)
2 1/2 (63mm) to 4 (100mm) inclusive	1/2 Ø (12.7 mm)
6 (150mm) and larger	3/4 Ø (19.1 mm)

[6]Society of Automotive Engineers, 400 Commonwealth Drive, Warrendale, PA 15096-0001 (412) 776-4841

When diaphragms or bellows are used as barriers in control piping which bypasses one or more check valves in the backflow prevention assembly, such a diaphragm or bellows shall be installed in such a manner that its failure shall produce visible evidence of such failure.

10.1.2.12 Air Release

Provision shall be made for bleeding trapped air from the highest point of the assembly when the normal flow of water will not displace it.

10.1.2.13 Valve Seats

The DC, RP, DCDA, RPDA, PVB, and SVB backflow prevention assemblies shall be fitted with replaceable valve seats (i.e., check valve, differential pressure relief valve, and air inlet valve seats) which shall be provided with a means of in situ insertion and removal.

10.1.2.14 Alignment

The clapper or poppet elastomer disc, clamping ring and the securing bolt, stud or cap screw each shall be concentric to an axis which is normal to the face of the elastomer disc. A means shall be provided to prevent clamping the elastomer disc so tight as to deform its face.

Provision shall be made to prevent the clapper from tipping and catching under the seat ring or any faulty action which would prevent true alignment. All parts shall be constructed and supported in such a manner as to preclude distortion or misalignment.

10.1.2.15 Chatter, Vibration

All moving parts shall be designed to operate from static to the rated flow condition in a positive manner without chatter or vibration.

10.1.2.16 Material Selection

When a relative motion is to be allowed between mating parts the materials shall be sufficiently different so that there will be no scuffing or galling.

10.1.2.17 Lubrication of Components

Proper operation of an assembly shall not be dependent upon a particular type or class of lubricant.

10.1.2.18 Shutoff Valves

The shutoff valves used as an integral component of approved backflow prevention assemblies are typically separated into two categories. The 2-inch (50 mm) and smaller assemblies generally use bronze bodied shutoff valves, whereas the 2-1/2-inch (63 mm) and larger assemblies generally use ferrous bodied shutoff valves. But this is not a restriction of use.

The specifications for resilient seated gate valves on 3-inch through 12-inch units are taken from the American Water Works Association (AWWA) Standard #C509 with an additional requirement for the use of a polymerized coating on all ferrous surfaces. The 2-1/2, 14, and 16-inch shutoff valves are to be constructed per the applicable requirements of AWWA #C509 Standard, even though they are not specifically listed in this standard.

All ferrous bodies shall be coated with a holiday-free polymerized coating of suitable thickness to comply with AWWA Standard #C550. The protective coating shall be composed of materials deemed acceptable in the Food and Drug Administration Document, Title 21 of the Federal Regulations on Food additives, Sec.175.300, entitled "Resinous and Polymeric Coatings." The manufacturer shall provide documentation showing suitability for potable water use.

The AWWA #C550 Standard is listed as a minimum requirement.

Each shutoff valve shall be marked with the following:

a. Manufacturer's or private labeler's name or identifying symbol.
b. Nominal size of valve.
c. Model number with designation of resilient seat: such as "RS" or "R" must be cast, molded, or affixed onto the body or bonnet of the valve.
d. Working pressure.

Below are the requirements for the resilient seated shutoff valves which do not come under the AWWA Standard #C509.

The inlet of the upstream shutoff valve and the outlet of the downstream shutoff valve body shall be threaded, flanged, or grooved in accordance with the ANSI Standards for the standard pipe sizes mentioned in Section 10.1.1.3. The valve body shall be provided with an adequate boss at each location where a tapped hole is required for the No. 1 test cock.

For threaded valve bodies, the hole in the wall of the No. 1 shutoff valve for the No. 1 test cock shall not enter into the body where it can be covered by the inlet piping. A thread stop must be provided so that the inlet piping does not obstruct the test cock sensing hole. The sensing hole must be fully sized according to the test cock sizes in Table 10-3.

The operating stem in a ball valve shall be designed so that it is inserted from inside the body, and is seated with an adjustable stem packing nut on the outside. This blow-out-proof type of stem allows in-line service of the stem packing. The operating stem and handle on quarter turn shutoff valves must clearly show if the valve is open or closed, even if the handle is removed. Stops must be provided to limit the valve actuation to ninety (90) degrees. The handle must be in line with the piping for the open position, and at ninety (90) degrees to the piping in the closed position. The operating handle must be suitably protected from corrosion.

The water passageway shall be at least equal in area to the rated size of the unit, and shall not adversely affect the overall pressure loss. The water passageway shall be permitted to be less than the rated size of the unit providing that the connections between the shutoff valves and the check valve body(s) are not standard NPT, flange, or groove per Section 10.1.1.3., and comply with all performance requirements of Section 10.2.

Each valve shall be capable of withstanding a hydrostatic pressure of twice the working pressure of the assembly for a minimum of ten (10) minutes, in both the open and closed positions, in both directions, with no leakage, damage, or permanent deformation.

10.1.2.19 Manifold Assembly

A manifold assembly shall be comprised of two (2) or more backflow prevention assemblies (double check valve assemblies, or reduced pressure principle backflow prevention assemblies) in parallel with a single inlet and outlet connection per Section 10.1.1.3. The size of the manifold assembly shall be determined by the inlet and outlet connections. (See Fig. 8.10.) The manifold assembly shall comply with the following requirements of this Section.

10.1.2.19.1 For each size of manifold assembly at any flow rate up to and including the rated flow, the maximum allowable pressure loss shall not exceed the values given in Table 10-1.

10.1.2.19.2 The manifold assembly shall comply with Hydrostatic Test requirements per Section 10.1.1.5.

10.1.2.19.3 Each of the individual backflow prevention assemblies contained in the manifold assembly shall be approved backflow prevention assemblies.

10.1.2.19.4 Each of the individual backflow prevention assemblies contained in the manifold assembly shall be of the same manufacturer, model, and size.

10.1.2.19.5 The manifold adaptor fittings on both the inlet and outlet of the manifold assembly are considered integral components and shall meet all of the material specifications in Section 10.1.3; and shall be marked per the marking requirements in Section 10.1.1.4 with the manifold assembly model designation to identify the adaptor fittings as integral components of the manifold assembly.

10.1.2.19.6 Each of the individual backflow prevention assemblies contained in the manifold assembly shall be flow tested at the rated flow of the manifold assembly. See Sections 10.2.2.3.2 or 10.2.3.3.2.

10.1.2.20 Repair Tools

Tools not commercially available for the repair/maintenance of a backflow prevention assembly shall be made readily available from the backflow prevention assembly manufacturer.

10.1.3 Material Specifications

10.1.3.1 Statement of Policy

The following material specifications and current ASTM (American Society for Testing Materials) and/or ANSI (American National Standards Institute) designations shall be adhered to at all times except where a manufacturer desires to use an equivalent or better material. In such cases the substitute specifications shall be submitted for approval prior to the inclusion of such materials in an assembly. In the subsequent list of materials no attempt has been made to bar the use of alloys, rubbers, plastics, or other materials which may be adaptable and which will give at least the equivalent trouble free service. All materials shall be non-toxic . The manufacturer shall provide documentation showing suitability for potable water use.

10.1.3.2 Dissimilar Metals

In the construction of backflow prevention assemblies a minimum of dissimilar metals shall be used. In all cases where it is impossible to use similar metals, steps shall be taken, insofar as it is practicable, to prevent the formation of galvanic electrolytic couples. To this end, when two dissimilar metals must come close to each other, the metals chosen shall be as nearly as possibly electrolytically similar and shall be insulated wherever possible.

10.1.3.3 Corrosion

Ferrous metals contained in backflow prevention assemblies shall be protected to resist corrosion.

10.1.3.4 Body and Cover

Materials to be used in construction of these parts of the assembly shall be either valve bronze which conforms to ASTM Designation: B61 or B62 or B584 UNS number C84400; or gray iron which conforms to ASTM Designation: A126, Class B or Class C; or ductile iron which conforms to ASTM Designation A536, Grade 65-45-12; or stainless steel which conforms to ASTM Designation: A276 or A296 either UNS No. S30400, S30500, S31600 or Schedule 40 steel pipe and flanges — suitably protected against corrosion (see 10.1.3.14 Protective Coatings); or engineered plastic.

10.1.3.5 Seat Rings

The seat rings shall be constructed of valve bronze which conforms to ASTM Designation: B61 or B62 or B584 UNS number C84400; or stainless steel which conforms to ASTM Designation: A276 either UNS No. S30400, S30500 or S31600; or engineered plastic.

10.1.3.6 Check Valve

The clapper, poppet or similar check valve shall be constructed of valve bronze which conforms to ASTM Designation: B61 or B62 or B584 UNS number C84400; or stainless steel which conforms to ASTM Designation: A276 either UNS No. S30400, S30500, or S31600; or engineered plastic.

10.1.3.7 Elastomer Disc(s)

a. The check valve elastomer disc shall be composed of molded natural rubber or a synthetic elastomer. It must be of even thickness, smooth-faced and with a Shore "A" hardness (ASTM D2240-86 Rubber Property - Durometer Hardness) of between 35 and 45 inclusive.

b. The differential pressure relief valve elastomer disc(s) shall be composed of a molded natural rubber or a synthetic elastomer. It must be of even thickness, smooth-faced and with a Shore "A" hardness of between 55 and 65 inclusive.

c. The air inlet valve elastomer disc shall be composed of a molded natural rubber or a synthetic elastomer. It must be of even thickness, smooth-faced and with a Shore "A" hardness of between 65 and 75 inclusive.

d.. For elastomer discs, a) ,b) and c) above, the material must be capable of withstanding the rated temperature and pressure of the assembly it is being used in with no permanent change in characteristics per performance evaluation in Section 10.2. It is suggested that the manufacturer add a means of identification to the elastomer discs to identify them as original manufactured parts.

e. The manufacturer shall supply the ASTM Designation: D2000 for all elastomers used in the assembly, and material properties such as tensile strength, tear resistance, and compression set for identification purposes.

The above noted Shore "A" hardness figures shall be used only for disc thicknesses of 1/4-inch or more. For thinner disc thicknesses the equivalent hardness in Micro Hardness Shore "M" degrees shall be used per ASTM D1415 - Rubber Property - International Hardness.

10.1.3.8 Swing Arm

The swing arm shall be made of valve bronze which conforms to ASTM Designation: B61 or B62 or B584 UNS number C84400; or of stainless steel which conforms to ASTM Designation: A276, UNS numbers S30400, S30500 or S31600; or engineered plastic.

10.1.3.9 Swing Pin and Guide Stem

The swing pin or guide stem shall be made either of phosphor bronze which conforms to ASTM Designation: B139, UNS numbers C51000, C52100, or C52400; or of stainless steel which conforms to ASTM Designation: A276, UNS numbers S30400, S30500 or S31600; or engineered plastic.

10.1.3.10 Bushings

Bushings shall have corrosion resistance at least equal to ASTM Designation B584 UNS number C84400.

10.1.3.11 Counterbalance

The counterbalance shall be weighted with a corrosion resistant material and/or protected from corrosion with a suitable protective coating in accordance with Section 10.1.3.14.

10.1.3.12 Springs

All springs shall be made of either stainless steel which conforms to ASTM Designation: A313; or of phosphor bronze which conforms to ASTM Designation: B159, UNS numbers C52100 or C52400; or equal.

10.1.3.13 Diaphragms

The diaphragm material shall be of a natural rubber, synthetic elastomer, or thermoplastic elastomer.

10.1.3.14 Protective Coatings

All ferrous bodies and parts shall be coated with a holiday-free polymerized coating per AWWA #C550. The use of synthetic protective coatings on ferrous bodies and parts shall be subject to the individual approval not only of the specific use but of the coating and process of application.

The protective coating shall be composed of materials deemed acceptable in the Food and Drug Administration Document, Title 21 of the Federal Regulations on Food additives, Sec. 175.300, entitled "Resinous and Polymeric Coatings." The manufacturer shall provide documentation showing suitability for potable water use.

Machining of a protective coating after the coating has been applied in order to facilitate the assembly of the assembly or to provide dimensional control is to be avoided. Extra care is to be taken with areas around ports or threaded surfaces such that focal points for corrosion are eliminated.

10.1.3.15 Studs, Bolts, and Cap Screws

Such parts for internal use or bonnet or closure plates shall be made of naval bronze rod which conforms to ASTM Designation: B21; or stainless steel which conforms to ASTM Designation: F593 UNS numbers S30400, S30500 or S31600. Bolts for line flanges may be cadmium or zinc plated steel.

10.1.3.16 Control Piping

Control piping shall have corrosion resistance at least equal to ASTM Designation B584 UNS number C84400.

10.1.3.17 Test Cocks

The test cocks on backflow prevention assemblies are required so that the assembly may be tested periodically, in situ. Any leakage of a closed test cock constitutes an undesirable nuisance. Hence, a test cock that will not shutoff tight or that is easily damaged by frequent use is to be avoided. The test cock shall be resilient seated and have a full flow characteristic.

Materials to be used in construction of the body and internal wetted components of the test cock shall be either valve bronze which conforms to ASTM Designation: B61, B62, or B584 UNS number C84400; or stainless steel which conforms to ASTM Designation: A276, UNS numbers S30400, S30500, or S31600; or suitable engineered plastic.

In ball valves, the solid ball shall be stainless steel which conforms to ASTM Designation: A276, UNS numbers S30400, S30500, or S31600; or hard chromium plated brass per Federal Specification QQ-C-320B — Chromium Plating (Electrodeposited).

In plug valves, the plug shall be stainless steel which conforms to ASTM Designation: A276, UNS numbers S30400, S30500, or S31600; or valve bronze which conforms to ASTM Designation: B61, B62, or B584 UNS number C84400; or hard chromium plated brass per Federal Specification QQ-C-320B — Chromium Plating (Electrodeposited).

Material to be used for the resilient seals shall be of a suitable elastomer or polymer, and capable of withstanding the action of line fluids and operation under long term service at rated conditions.

An approved backflow prevention assembly shall utilize test cocks as detailed above and in Section 10.1.2.10.

10.1.3.18 Shutoff Valves

The shutoff valves on backflow prevention assemblies are required so that the assembly may be tested and/or maintained periodically, in situ. Any internal leakage of a closed shutoff valve may eliminate the possibility of accurate testing. Hence, a shutoff valve that will not shut off tight or that is easily damaged by frequent use is to be avoided. The shutoff valves shall be resilient seated and shall have a full flow characteristic.

Below are the requirements for the resilient seated shutoff valves which do not come under the AWWA C509 Standard as detailed in Section 10.1.2.18.

Materials to be used in construction of the body of the shutoff valve shall be either valve bronze which conforms to ASTM Designation: B61, B62, or B584 UNS number C84400; or gray iron which conforms to ASTM Designation: A126, Class B or Class C; or ductile iron which conforms to ASTM Designation: A536 Grade 65-45-12; or stainless steel which conforms to ASTM Designation:A276, UNS numbers S30400, S30500, or S31600; or engineered plastic.

Materials to be used in construction of the internal wetted components of the shutoff valve shall be either valve bronze which conforms to ASTM Designation: B61, B62, or B584 UNS number C84400; or stainless steel which conforms to ASTM Designation: A276, UNS numbers S30400, S30500, or S31600; or engineered plastic.

In ball valves, the solid ball shall be stainless steel which conforms to ASTM Designation: A276, UNS numbers S30400, S30500, or S31600; or hard chromium plated brass per Federal Specification QQ-C-320B — Chromium Plating (Electrodeposited).

In plug valves, the plug shall be stainless steel which conforms to ASTM Designation: A276, UNS numbers S30400, S30500, or S31600; or valve bronze which conforms to ASTM Designation: B61, B62, or B584 UNS number C84400; or hard chromium plated brass per Federal Specification QQ-C-320B — Chromium Plating (Electrodeposited).

Material to be used for the resilient seals shall be of a suitable elastomer or polymer, and capable of withstanding the action of line fluids and operation under long term service at rated conditions.

An approved backflow prevention assembly shall utilize shutoff valves as detailed above and in Section 10.1.2.18.

10.1.3.19 Lubricants

The lubricant shall be suitable for lubricating parts for assembly purposes only. The lubricant shall be non-toxic, shall not support the growth of bacteria, and shall have no deteriorating effects on any component. Manufacturer must supply documentation stating that lubricant is suitable for contact with potable water.

10.2 — Evaluation of Design and Performance

10.2.1 General

The specifications set forth in this Manual supersede the specifications of previous editions of this Manual as well as those of Paper No. 5 and USCEC Report No. 48-101.

10.2.1.1 Drawings and Specifications

A full set of working drawings and specifications of all materials shall be furnished with each make, model, and size of assembly that is submitted for evaluation.

10.2.1.2 Assemblies Required for Evaluation

a. Laboratory Evaluation - One or more assemblies, selected at random from the manufacturer's stock, of each size and model shall undergo a complete laboratory evaluation of its design and operating characteristics under the supervision of the Foundation for Cross-Connection Control and Hydraulic Research or an approved testing laboratory. For the sizes up to and including 2-inch (50 mm), the manufacturer shall submit a minimum of three assemblies. For the sizes 2½-inch (63 mm) and larger, the manufacturer shall submit a minimum of one assembly.

   Repair/maintenance tools necessary to remove or install all components of the assembly shall be identified by the manufacturer. Tools not commercially available, per Section 10.1.2.20, shall be supplied with the assembly(s) submitted.

   The Laboratory Evaluation shall be performed in the orientation for which each assembly is submitted. Horizontal, vertical (flow up and/or down), and axial rotation orientations shall be identified by the manufacturer at the time of submittal.

b. Field Evaluation - Upon completion of a satisfactory laboratory evaluation, a minimum of three (3) assemblies, selected at random from the manufacturer's stock, of each size and model shall undergo a satisfactory simultaneous twelve (12) consecutive months field performance under the supervision of the Foundation for Cross-Connection Control and Hydraulic Research or an approved testing laboratory.

   The Field Evaluation shall be performed in each of the orientation(s) successfully completing the Laboratory Evaluation.

10.2.1.3 Selection of Field Locations

The manufacturer shall be responsible for locating and installing all assemblies in acceptable field evaluation sites, however, the Foundation for Cross-Connection Control and Hydraulic Research or an approved testing laboratory shall be permitted to reject any field evaluation site submitted by the manufacturer. In general the following conditions will govern site selection; but, these conditions will not constitute the only site selection criteria.

a. Reduced pressure principle assemblies shall not normally be sited where conditions would require the use of this type of assembly.

b. The sites shall provide as wide a range of water conditions (i.e., line pressure, flow rates) and types of use as practical.

c. No more than one site per model and size shall be permitted on a non-flowing static service (e.g., fire sprinkler system).

   For the purposes of the Field Evaluation in vertical orientation(s), the manufacturer shall be permitted to install *no more than* ***two (2)*** *sites per size on a non flowing static service* providing that:

   1. The assembly is currently Approved by the Foundation for Cross-Connection Control and Hydraulic Research in the horizontal orientation, or

   2. The assembly is currently undergoing the Field Evaluation in the horizontal orientation.

d. Parallel installations for field evaluation are not normally recognized; but, if such a site is accepted the two assemblies shall be considered to be only one of the required sites.

e. All sites shall be freely accessible during normal working hours to the Foundation's or approved testing laboratory's field evaluation personnel.

f. The owner of the property (or the manager), the water agency and/or health agency having cognizance must all supply written acknowledgment prior to the installation of the assembly(s) to the Foundation for Cross-Connection Control and Hydraulic Research or approved testing laboratory that the site will be an acceptable field evaluation location. (see sample letter Section 8.17)

g. Each field site shall be suitably protected from freezing conditions or vandalism.

h. The manufacturer shall submit the Field Site Application Form (see sample letter Section 8. 18) for each proposed field site prior to the installation of the assembly.

10.2.1.4 Period of Field Evaluation

After the manufacturer has submitted the Field Site Application Form for each of the specific field site locations, and the form has been reviewed and accepted by the Foundation for Cross-Connection Control and Hydraulic Research or approved testing laboratory, the initial in situ field evaluation test will be conducted. When a minimum of three (3) assemblies for each size and model have successfully complied with the initial field test by the Foundation for Cross-Connection Control and Hydraulic Research or approved testing laboratory, the field evaluation period shall be started. The initial field test shall include:

a) Recording of field test data per Section 9 of this Manual, and
b) Recording of static differential pressure reading across each check valve, and
c) Recording of inlet line pressure, and
d) Disassembly and physical inspection of all components, reassembly, and
e) Recording of field test data per Section 9 of this Manual

A minimum of three assemblies undergoing field evaluation for each size and model of assembly shall provide simultaneous trouble-free operation for a minimum period of twelve (12) consecutive months in conformance with the Field Evaluation Specifications as set forth herein for each type of assembly. If more than three (3) assemblies are utilized for the Field Evaluation, and two assemblies fail to comply with the performance requirements in a similar manner, then this shall be cause for rejection. Before a specific size and model of assembly shall be released from the Field Evaluation it shall be inspected to determine that all of the working parts continue to conform to the design and performance specifications. The concluding field test at the end of the Field Evaluation shall include:

a) Recording of field test data per Section 9 of this Manual, and
b) Recording of static differential pressure reading across each check valve, and
c) Recording of inlet line pressure, and
d) Disassembly and physical inspection of all components, reassembly, and
e) Recording of field test data per Section 9 of this Manual

Should a make, model, and size assembly fail to comply with the specifications set forth in this Manual, the following offices, having cognizance of the field test locations, in addition to the manufacturer shall be immediately notified:

a) the health agency
b) the plumbing authority
c) the water purveyor

Should a model and size assembly not complete the Field Evaluation and not gain Approval, then the manufacturer shall be responsible for replacing the field test assemblies in all field evaluation sites with currently Approved assemblies.

During the entire Field Evaluation program only the Foundation for Cross-Connection Control and Hydraulic Research or approved testing laboratory shall be permitted to test, repair, or inspect an assembly undergoing Field Evaluation. The Foundation for Cross-Connection Control and Hydraulic Research or approved testing laboratory shall be permitted to reject a field site should any testing, repair, or inspection of an assembly(s) be performed by unauthorized personnel.

10.2.1.5 Approval

In addition to the satisfactory completion of the Laboratory and Field Evaluation, the manufacturer shall supply to the Foundation for Cross-Connection Control and Hydraulic Research or approved testing laboratory the sales literature and installation/maintenance literature for the model and size assembly being evaluated under these specifications. Before consideration for Approval these materials shall be reviewed for:

a) Installation detailing,
   including orientation and clearances
b) General operating parameters of the assembly
   MWWP, MWWT, Pressure loss vs flow rate curves
c) Repair/Maintenance Procedures
   Tool requirements per Section 10.1.2.20
d) Field test procedures.

Upon the completion of a satisfactory Laboratory and Field Evaluation, and acceptable review of the manufacturer's literature, the manufacturer shall supply to the approving agency the serial number of the first assembly manufactured in the Approved configuration. The Foundation for Cross-Connection Control and Hydraulic Research or an approved testing laboratory shall then grant an Approval for the specific size and model evaluated under these specifications. This Approval shall be valid for a period of no more than three (3) years — and may be rescinded for cause before that time. The approving agency shall issue a list of all assemblies that are currently approved by said agency.

Should only original design replacement parts (see Section 10.2.1.9) be available from the manufacturer, and not the entire assembly, then the particular assembly shall be identified with the Greek letter Ψ (psi) on the list. When the original manufacturer ceases to maintain a supply of original design replacement parts for an approved model and size of assembly that particular model and size shall have its Approval rescinded.

10.2.1.6 Renewal of Approval

Continuing verification of compliance with these specifications and field performance shall be accomplished at a maximum of once every three (3) years to the satisfaction of the original approving agency. The original approving agency, at its own discretion retains the right of determining the extent of re-evaluation required before renewal is granted. The manufacturer's current sales, installation and maintenance literature shall be submitted for review at this time. Acceptable re-evaluation, literature, past performance of the assembly under field operating conditions, and spare parts availability shall be considered before the renewing of an Approval. Failure to meet these requirements shall result in the automatic rescinding of the approval for that size and model of assembly. The latest listing of approved assemblies shall reflect the current status of a particular model and size assembly.

10.2.1.7 Change of Design, Materials or Operation

The original approving agency shall be notified by the manufacturer in writing of any proposed change in design, materials or operation of an approved assembly. The original approving agency, at its own discretion, shall be permitted to require another laboratory and/or field evaluation. Failure to notify the original approving agency of any changes of design, materials or operation prior to implementation shall be cause for the rescinding of the Approval for the size and model of assembly involved.

10.2.1.8 Prior Approval

a. On the publication date of this Manual all makes, models, and sizes of assemblies having been previously approved shall henceforth be designated by reference to the specifications under which each assembly was approved. The list of approved assemblies will show compliance with either Paper No. 5*, USCEC 48-101**, or Editions 1, 2, 3, 4, 5, 6, 6 (Revised), 7. or 8 of this Manual***.

b. Backflow prevention assemblies approved under the above paragraph (10.2.1.8a) shall be approved for installation as set forth herein.

c. Continuing verification of compliance with the specification under which a size and model of assembly was originally approved shall be accomplished at least once every three (3) years from the date of the most recent renewal to the satisfaction of the original approving agency. The original approving agency, at its own discretion retains the right of determining the extent of re-evaluation required before renewal is granted. Acceptable current literature and past performance of the assembly under field operation conditions shall be considered before the renewing of an Approval. Failure to meet this requirement shall result in the automatic rescinding of the Approval for that make, model, and size of assembly.

d. A Certificate of Full Approval or Approval granted under conditions of prior approval (10.2.1.8a) or a Certificate of Approval granted under the conditions of this 9th Edition of the Manual shall be permitted to be rescinded at any time by the Foundation for Cross-Connection Control and Hydraulic Research or an approved testing laboratory for cause.

10.2.1.9 Replacement Parts

The replacement parts of a backflow prevention assembly shall include all control piping, internal components, and access covers for internal components of the assembly, excluding the primary assembly body(s) {i.e., main pressure containing vessel(s)}.

---

* Paper No. 5: "Objectives, General Testing Procedure, Specifications, Results of Tests;" Foundation for Cross-Connection Control Research, University of Southern California; April, 1948.

** USCEC Report 48-101 "Definitions and Specifications of Double Check Valve Assemblies and Reduced Pressure Principle Backflow Prevention Devices;" Foundation for Cross-Connection Control Research, University of Southern California; January 30, 1959.

*** Manual of Cross-Connection Control: Foundation for Cross-Connection Control and Hydraulic Research, University of Southern California; 1st Ed., 1960; 2nd ed., 1965; 3rd Ed., 1966; 4th Ed., 1969; 5th Ed., 1974; 6th Ed., 1979; Revised 6th Ed., 1982; 7th Ed., 1985, 8th Ed., 1988.

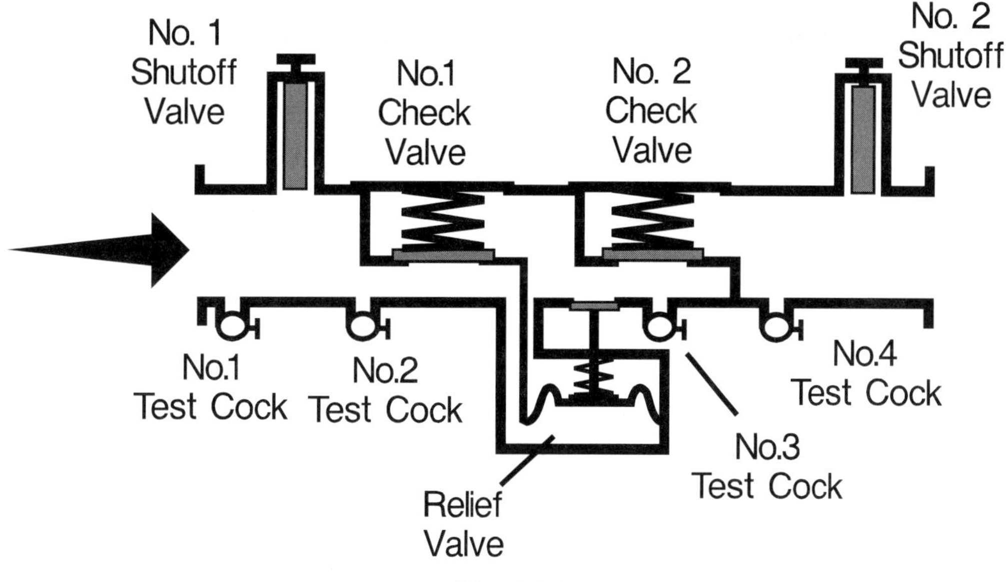

Fig. 10.1
Reduced pressure principle backflow prevention assembly — showing location of shutoff valves, differential pressure relief valve, and test cocks.

10.2.2 DESIGN, OPERATIONAL AND EVALUATION SPECIFICATIONS FOR REDUCED PRESSURE PRINCIPLE BACKFLOW PREVENTION ASSEMBLIES (RP)

10.2.2.1 Design and Operational Specifications

a. This assembly shall include two approved, independently operating check valves with an automatically operating, mechanically independent, hydraulically dependent pressure differential relief valve located between the two check valves. The assembly shall include a tightly closing resilient seated shutoff valve on each end of the body and each assembly shall be fitted with four properly located resilient seated test cocks (see Fig. 10.1). During normal flow and at the cessation of normal flow, the pressure in the zone, i.e., the zone between these two check valves, shall be at least 2 psi (13.78 KPa) less than the upstream (or supply) pressure. The pressure drop across the first check valve at static (no flow) conditions shall be essentially a constant value for all line pressures from 20 psi (137.8 KPa) through the maximum working water pressure (MWWP), but not less than 150 psi (1034 KPa).

b. With no flow from the upstream side when the pressure on the supply side drops to 2 psi (13.78 KPa) above the zone pressure the relief valve shall discharge water to maintain the zone at 2 psi below the supply pressure. When the pressure on the upstream side of the first check valve drops below 2 psi above the zone pressure the relief valve shall be open.

c. Under a backflow condition, when the upstream pressure is 2 psi (13.78 KPa) up to maximum working water pressure, the relief valve shall discharge from the zone to atmosphere the quantities of backflowing water given in Table 10-4, and the pressure in the zone shall be at least 1/2 psi (3.44 KPa) below the upstream pressure.

d. The second check valve shall be internally loaded and shall at all times be drip tight in the normal direction of flow with the inlet pressure at least 1 psi (6.89 KPa) and the outlet under atmospheric pressure.

e. When the upstream prcssurc is atmospheric the differential pressure relief valve shall discharge water from the zone to atmosphere with the rate of discharge corresponding to the data shown in Table 10-4; and the pressure in the zone shall not exceed 1 1/2 psi (10.34 KPa).

f. The differential pressure relief valve shall open and close positively and quietly. Further, it shall not spit water under fluctuating line conditions when the upstream pressure is 3.0 psi (20.68 KPa) or more above the differential pressure required to open the relief valve.

g. The differential pressure relief valve shall be located so that the valve seat(s) and the discharge port(s) are below the lowest portion of the No. 1 check valve so as to preclude backsiphonage.

h. The differential pressure relief valve and the No. 1 check valve shall be so designed that their characteristic opening points shall not drift unduly as a function of the line pressure.

i. The assembly shall operate at, and withstand operating parameters up to its rated maximum working water temperature (MWWT), maximum working water pressure (MWWP), and flow rate specified in Table 10-7 for at least 100 hours with no damage or permanent deformation.

j. The differential pressure relief valve shall not discharge while flowing up to and including 200 percent of the rated flow of the assembly. This includes both increasing and decreasing rates of flow.

k. For assemblies intended for fire sprinkler service, the differential pressure relief valve shall not discharge while flowing up to and including the rates of flow detailed in Section 10.1.1.2.

l. The differential pressure relief valve shall not discharge water when the test cocks are fully opened one at a time.

TABLE 10-4

MINIMUM FLOW RATES AND SIZE OF MINIMUM AREA OF DIFFERENTIAL PRESSURE RELIEF VALVE OPENING

Size of Assembly		Minimum Flow Rate Through Relief Valve		Minimum Equivalent Diameter of Relief Valve Porting	
(inches)	(mm)	(gpm)	(L/s)	(inches)	(mm)
1/4	6	0.5	0.03	3/16	5
3/8	9	1	0.06	1/4	6
1/2	12	3	0.19	3/8	10
5/8×3/4	16×20	5	0.32	3/8	10
3/4 — 1	20 — 25	5	0.32	1/2	12
1 1/4 — 1 1/2	31 — 38	10	0.63	3/4	19
2	50	20	1.26	1	25
2 1/2	63	20	1.26	1	50
3	75	30	1.89	1 1/4	31
4	100	40	2.52	1 1/2	38
6	150	40	2.52	1 1/2	38
8	200	60	3.78	2	50
10	254	60	3.78	2	50
12	305	75	4.73	2 1/2	64
14	355	90	5.68	3	75
16	406	100	6.31	3	75

10.2.2.2 Laboratory Evaluation

One assembly shall be inspected and evaluated in the following order. Unless otherwise noted, tests shall be performed at ambient temperature.

a. Conformance to the general, design and material requirements outlined in Section 10.1.

b. Conformance to the operational requirements outlined in Section 10.2.

c. Conformance to the working drawings and materials specifications.

d. Hydrostatic tests — the test assembly shall be subjected to the conditions of Section 10.1.1.5.

e. Pressure loss characteristics for flow rates up to the rated conditions.

f. Differential pressure relief valve opening point test.

g. Sensitivity of differential pressure relief valve to opening of test cocks.

h. Static closing point of check valve No. 1.

i. Static closing point of check valve No. 2.

j. Differential pressure relief valve discharge capacity.

k. The assembly shall be evaluated for effectiveness when simultaneous backsiphonage and backpressure conditions are applied.

l. Differential pressure relief valve drain funnel backsiphonage test.

m. Thermal Test — the assembly shall be evaluated at the greater of 140°F (60°C) or the maximum working water temperature (MWWT), maximum working water pressure (MWWP), and specified flow rate.

n. Life cycle test.

10.2.2.3 Evaluation Procedure

10.2.2.3.1

Purpose: To determine the capability of the assembly to withstand the required hydrostatic test pressure.

Requirement: All components of the assembly shall withstand a hydrostatic pressure of twice (2×) the maximum working water pressure (MWWP), as stated by the manufacturer, for a minimum period of ten (10) minutes without any damage, permanent deformation or impairment of operation.

Steps:

a. With both shutoff valves closed and all of the air bled from the assembly, a hydrostatic pressure of twice (2×) the maximum working water pressure shall be supplied through the No. 2 test cock for a minimum period of ten (10) minutes. Any evidence of leakage shall be cause for rejection.

b. The required hydrostatic pressure shall be supplied through test cock No. 4 with test cock No. 3 open to atmosphere for a minimum period of ten (10) minutes. Any evidence of leakage shall be cause for rejection.

c. The required hydrostatic pressure shall be supplied through test cock No. 3 with test cock No. 2 open to atmosphere for a minimum period of ten (10) minutes. Any evidence of leakage shall be cause for rejection.

NOTE: A reverse hydrostatic pressure cannot be imposed on the No. 1 check valve without sealing the differential pressure relief valve, which is an abnormal condition. The backpressure test on the No. 1 check valve must be performed with the differential pressure relief valve isolated from the assembly, or mechanically held in the closed position.

d. The assembly shall be disassembled and inspected for any damage, permanent deformation, or impairment of operation. Evidence of such shall be cause for rejection.

10.2.2.3.2

Purpose: To determine the overall pressure loss of the complete assembly as a function of the rate of flow.

Requirement: At any rate of flow from zero (static) up to and including the rated flow the overall pressure loss shall not exceed the values shown in Table 10-1. The assembly shall withstand a minimum of 200% of the rated flow without any damage, permanent deformation or impairment of operation.

Steps:

a. Install the assembly in a suitable hydraulic test line having a flow capacity in excess of the rated flow conditions (see Table 10-1) and an available line pressure of at least 150 psi (1034 KPa).

b. Install in the supply line at a suitable location upstream from the assembly a piezometer ring; and, install a similar piezometer ring at a suitable location downstream from the assembly (located as indicated in ANSI/ISA Standard S75.02). Then install a mercury manometer, or other suitable means of measuring pressure differentials between the two piezometer rings to measure the observed overall pressure loss at various rates of flow.

c. Record differential pressure between piezometer rings at static condition. Increase flow rate and record steady state differential pressure at a sufficient number of increments to fully define flow curve characteristic. Data shall be taken for both increasing and decreasing flow conditions. The flow rate shall then be increased to 200% of the rated flow for five (5) minutes, then returned to static.

   Manifold Assembly (Section 10.1.2.19) Only: Close all shutoff valves so that only one of the individual assemblies of the manifold assembly is open. Gradually increase the flow of water through the individual assembly until the rated flow of the manifold assembly is reached, then return to static condition. Repeat flow test for each individual assembly contained in the manifold assembly.

d. Remove the assembly from the line and couple the downstream piping directly to the supply piping. Then observe data for a friction head correction curve for the pipe fittings required to adapt the test assembly to the supply line between the existing piezometer rings. The data of this correction curve is then subtracted from the observed overall pressure loss data of the assembly from the upstream face of the upstream shutoff valve to the downstream face of the downstream shutoff valve.

e. Exceeding the maximum allowable pressure loss listed in Table 10-1 for any rate of flow up to and including rated flow, or discharge from the differential pressure relief valve, shall be cause for rejection. The assembly shall be disassembled and inspected for any damage, permanent deformation, or impairment of operation. Evidence of such shall be cause for rejection.

10.2.2.3.3

Purpose: To test operation of differential pressure relief valve.

Requirement: The differential pressure relief valve shall operate to maintain the zone between the two check valves at least 2 psi (13.78 KPa) less than the supply pressure for all line pressures from 20 psi (137.8 KPa) up to the maximum working water pressure, but not less than 150 psi (1034.1 KPa) line pressure.

Steps:

a. Connect the high side hose of the differential pressure gage (or manometer) to test cock No. 2.

b. Connect the low side hose of the differential pressure gage (or manometer) to test cock No. 3.

c. Open test cocks No. 2 and No. 3 and bleed air from the gage (or manometer).

d. Close No. 2 shutoff valve.

e. Slowly (maximum rate of 0.2 psid per second) equalize the pressure between the high side and low side hoses, noting the gage (or manometer) reading at the initial opening of the differential pressure relief valve.

f. Repeat step ***e*** for each 10 psi (68.9 KPa) increment between 20 psi (137.8 KPa) and the maximum working water pressure (MWWP), but not less than 150 psi (1034.1 KPa) line pressure.

g. Failure of the differential pressure relief valve to open at or before 2.0 psid (13.78 KPa) is reached, for all pressures as in step ***f***, shall be cause for rejection.

10.2.2.3.4

Purpose: To determine if the differential pressure relief valve shall discharge when the test cocks are opened one at a time.

Requirement: The differential pressure relief valve shall not discharge water when the test cocks are fully opened one at a time.

Steps:

a. Install the assembly in a suitable hydraulic test line which is capable of maintaining an inlet pressure of at least 150 psi (1034 KPa) during the following steps.

b. Close the No. 2 shutoff valve while maintaining the No. 1 shutoff valve fully open.

c. Slowly open (4 seconds ± 1 second) test cock No. 1 until fully open. Then slowly close (4 seconds ± 1 second) test cock No. 1 .

d. Repeat step ***c*** with test cocks No. 2, No. 3, and No. 4.

e. Evidence of discharge from the differential pressure relief valve shall be cause for rejection.

10.2.2.3.5

Purpose: To determine the static pressure drop across check valve No. 1.

Requirement: The static pressure drop across check valve No. 1 shall be at least 3.0 psi (20.7 KPa) greater than the pressure differential between inlet line pressure and the zone required to open the differential pressure relief valve, for all line pressures from 20 psi (137.8 KPa) up to maximum working water pressure (MWWP), but not less than 150 psi (1034.1 KPa).

Steps:

a. With a differential pressure gage or manometer connected as in Section 10.2.2.3.3, flow a sufficient amount of water through the No. 2 shutoff valve of the assembly to re-establish the normal pressure gradient across check valve No. 1. Record the static pressure differential across check valve No. 1.

b. Repeat step ***a*** for each 10 psi (68.9 KPa) increment between 20 psi (137.8 KPa) and the maximum working water pressure (MWWP), but not less than 150 psi (1034.1 KPa) line pressure.

c. Failure of the first check valve to maintain a static pressure differential of at least 3.0 psi (20.7 KPa) greater than the differential pressure relief valve opening point at the corresponding line pressure, shall be cause for rejection.

10.2.2.3.6

Purpose: To determine the static pressure drop across check valve No. 2.

Requirement: No. 2 check valve shall be drip-tight in the normal direction of flow with the inlet pressure at least 1 psi (68.9KPa) and the outlet pressure at atmospheric.

Steps:

a. Install a vertical transparent tube at least six feet (1.83 m) long at test cock No. 3. If test cock No. 4 is not at the highest point of the check valve body, then another vertical transparent tube must be installed on test cock No. 4 so that it just rises above the top of the check valve body. See Fig. 10.2.

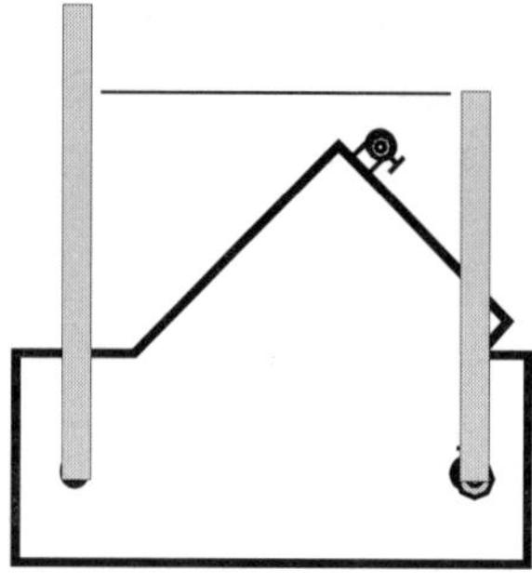

Fig. 10.2
Location of vertical transparent tubes

b. Open test cock No. 3 and fill the tube with water, then close test cock No. 3. If a tube is attached to test cock No. 4, open test cock No. 4 to fill the tube so that the water level is above the highest point of the check valve body, then close test cock No. 4.

c. Close the No. 2 shutoff valve, then close the No. 1 shutoff valve.

d. Open test cock No. 4, then open test cock No. 3.

e. Water from the tube on test cock No. 3 will flow through check valve No. 2 until closure of check valve No. 2 is attained. When no further fall of the water in the tube on test cock No. 3 is observed, record the height of the water in the tube on test cock No. 3 above the water level at test cock No. 4. The difference of height of water in the transparent tube(s) must be greater than 27 3/4 inches (704.85 mm).

f. Failure to maintain a minimum level of 27 3/4 inches (704.85mm) shall be cause for rejection.

10.2.2.3.7

Purpose: To determine the discharge capacity of the differential pressure relief valve.

Requirement: For the backpressure requirements specified by Sections 10.2.2.1.c and 10.2.2.1.e, the differential pressure relief valve discharge capacity shall be a minimum of the rates shown in Table 10-4. These rates shall be achieved with and without the manufacturer's drain funnel attached to the assembly.

Steps:

a. Either cause line pressure to be piped into the downstream piping, or, install the assembly so that the supply line feeds directly into the No. 2 shutoff valve (i.e., assembly installed backwards)

b. Remove the No. 2 check valve moving member (i.e., poppet or clapper).

c. Locate a weigh tank directly beneath the relief valve discharge port to measure the rate of discharge from the relief valve, or provide other suitable means to measure rate of flow.

d. Install a vertical transparent tube on the No. 3 test cock so that the back pressure in the zone can be observed.

e. By means of a control valve located a minimum of 5 pipe diameters from the No. 2 shutoff valve (maintain the No. 2 shutoff valve in the fully open position), control the flow into the zone so that pressure in the zone is maintained at 1-1/2 psi (10.34 KPa). The inlet of the No. 1 shutoff valve and No. 2 test cock shall be open to atmosphere.

f. Record the discharge rate of the differential pressure relief valve while the zone pressure is maintained as in step ***e***. Shut off control valve. Failure of the differential pressure relief valve to discharge the required flow rate shall be cause for rejection

g. Install a tee on the No. 2 test cock and a bypass hose from the supply line to the No. 2 test cock; and also, place a differential pressure gage or manometer between the No. 2 and No. 3 test cocks. Pressurize the bypass hose to the maximum working water pressure, or at least a minimum of 150 psi (1034 KPa), and bleed all of the air out of the hoses and gage, as well as the assembly being tested.

h. Open the control valve on the supply line until the pressure in the zone is maintained at 1/2 psi (3.44 KPa) below the line pressure, or the control valve is fully open. Determine the discharge rate of flow by means of a weigh tank or by other suitable means. Failure of the differential pressure relief valve to discharge the required flow rate shall be cause for rejection

i. If a drain funnel is provided by the manufacturer, repeat steps ***e*** through ***h*** with the manufacturer's drain funnel attached to the assembly. Failure of the differential pressure relief valve to discharge the required flow rate with the drain funnel attached to the assembly shall be cause for rejection. (All water need not flow through the drain funnel.) Any damage or permanent deformation of the drain funnel during steps ***e*** through ***h*** shall be cause for rejection.

10.2.2.3.8

Purpose: To determine if simultaneous backsiphonage and backpressure conditions coupled with leaking No. 1 and No. 2 check valves will permit the carry over of water from the downstream piping into the upstream piping.

Requirement: There shall be no backsiphonage of water from the downstream piping through fouled No. 1 and No. 2 check valves under conditions of up to 25 inches of mercury vacuum (16.89 KPa) in the supply piping and backpressure conditions up to the rated working pressure of the assembly.

Steps:

a. The No. 1and the No. 2 check valves shall be fouled with a wire of the size shown in Table 10-6. The fouling wires shall be placed across the seating surface in the lower quadrant of a hinged or horizontally moving poppet check valve, or at a single point of a vertically moving check valve.

b. The assembly shall be subjected to a downstream condition of from 5 inches (127 mm) water column backpressure up to the maximum working water pressure or at least 150 psi (1034 KPa) backpressure to determine the greatest discharge rate out of the relief valve port(s).

c. With the backpressure established at the level producing the greatest discharge rate per step ***b,*** the assembly shall then be subjected to an upstream condition of from 5 inches to 25 inches of mercury vacuum (84.45 KPa to 16.89 KPa absolute).

d. Each of the vacuum pressures and backpressures are to be slowly imposed and are to be maintained for at least five (5) minutes. Any amount of carry-over water shall be cause for rejection.

e. Each of the vacuum pressures is to be shock applied (by means of a quick opening valve) for five (5) cycles and any carry-over water shall be cause for rejection.

f. The assembly shall be inspected for any damage due to the above vacuum tests. Any damage or permanent deformation shall be cause for rejection. Damage or permanent deformation caused by the insertion of the fouling wire shall be disregarded.

g. If a drain funnel is provided, tests ***d*** through ***e*** shall be repeated with the manufacturer's drain funnel attached to the assembly. Evidence of carry over water shall be cause for rejection.

10.2.2.3.9

Purpose: To determine if there is adequate backflow protection between the drain funnel and the differential pressure relief valve discharge port(s).

Requirement: There shall be no backsiphonage of water from the drain funnel back into the differential pressure relief valve discharge port(s) under vacuum conditions.

Steps:

a. Attach inlet of assembly to vacuum source.

b. Remove the No. 1 check valve moving member (i.e., poppet or clapper).

c. Attach drain funnel to assembly per manufacturer's installation requirements. Plug outlet of drain funnel, then fill drain funnel to overflow with water.

d. Apply 25 inches of mercury vacuum (16.89 KPa absolute) to inlet of assembly by means of a quick opening valve.

e. Evidence of water in the drain funnel carrying over into the relief valve discharge port(s) shall be cause for rejection.

10.2.2.3.10

Purpose: To determine if all components of the assembly shall operate properly under rated temperature and pressure conditions, and flow rates as specified in Table 10-7.

Requirements: All components of the assembly shall operate at, and withstand a thermal test at the greater of 140°F (60°C) or at the maximum working water temperature (MWWT), maximum working water pressure (MWWP), and the flow rate as specified in Table 10-7, for a minimum period of one hundred (100) hours without any leakage, damage, permanent deformation or impairment of operation.

Steps:

a. Install assembly into suitable test line which can simultaneously generate the following minimum parameters:

   Maximum working water pressure *{minimum of 150 psi (1034 KPa)}*
   Maximum working water temperature *{minimum of 140°F (60°C)}*
   Flow requirements per Table 10-7

b. The dimensions and durometer hardness of all elastomer components shall be inspected and recorded.

c. Set controls to the rated temperature and pressure conditions of the assembly, as well as rate of flow as specified in Table 10-7.

d. The temperature and pressure conditions shall be continuously monitored and recorded.

e. During and after a minimum of one hundred (100) hours at rated temperature and pressure the assembly shall be tested to determine if the assembly operates satisfactorily per 10.2.2.3.3, 10.2.2.3.5, and 10.2.2.3.6. Failure to comply with these tests shall be cause for rejection.

f. The assembly shall then be disassembled and inspected for any damage or permanent deformation. Evidence of such shall be cause for rejection.

g. Once the assembly returns to ambient temperature conditions, the assembly shall then be tested per Section 10.2.2.3.1. Failure to comply with this test shall be cause for rejection.

10.2.2.3.11

Purpose: To determine if the assembly sustains any damage, permanent deformation, or impairment of operation following the specified cycles.

Requirement: The assembly shall withstand the specified cycle without leakage, damage, permanent deformation, or impairment of operation.

Steps:

a. Install assembly into suitable test line which can generate the following parameters:

Rated flow of assembly under test (per Table 10-1)
Line pressure of 60 psi ± 10 psi (414 KPa ± 69 KPa)
Temperature of 110°F $^{+30}/_{-0}$ °F (43°C. $^{+17}/_{-0}$ °C)
Backpressure of 150psi ± 10 psi (1034 KPa ± 69 KPa)
Control valves capable of closing/opening within five (5) seconds *.
*(*Resulting rise of pressure shall not exceed MWWP of the assembly.*)

b. With Valve #3 (Valve #3A open and Valve #3B closed) and Valve #4 (see Figure 10.3) in the closed position, and Valve #1 and Valve #2 in the open position, establish flow through the assembly equal to 25 percent of the rated flow per Table 10-1.

c. Maintain the flow through the assembly for a minimum of six (6) seconds. Close Valve #1.

d. After five (5) seconds, close Valve #2

e. After three (3) seconds, open Valve #4.

f. After three (3) seconds, open Valve #3. Maintain Valve #3 open for six (6) seconds. Close Valve #3.

g. After two (2) seconds, close Valve #4.

h. After three (3) seconds, open Valve #2.

i. After two (2) seconds, open Valve #1.

j. Repeat steps ***c*** through ***i*** 1250 times.

k. Raise flow through the assembly to 50 percent of rated flow. Repeat steps ***c*** through ***i*** 1250 times.

l. Raise flow through the assembly to 75 percent of rated flow. Repeat steps ***c*** through ***i*** 1250 times.

m. Raise flow through the assembly to 100 percent of rated flow. Repeat steps ***c*** through ***i*** 1250 times.

n. After a minimum of 5000 total cycles the assembly shall be tested to determine if the assembly complies with 10.2.2.3.3, 10.2.2.3.5, and 10.2.2.3.6. Failure to comply with these tests shall be cause for rejection.

o. The assembly shall then be disassembled and inspected for any damage or permanent deformation. Evidence of such shall be cause for rejection.

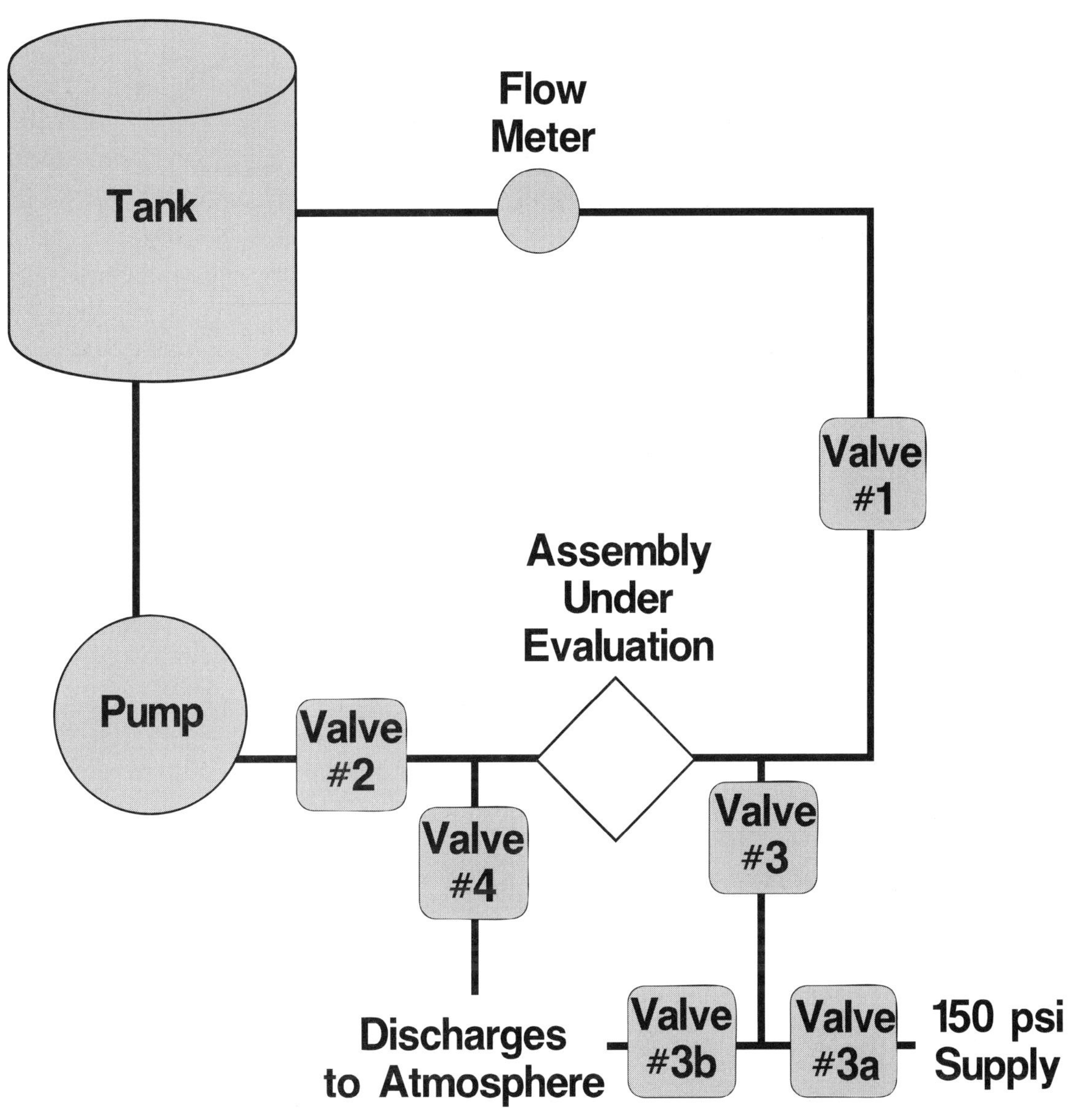

Fig. 10.3
Life cycle test system
Diagrammatic sketch showing location of system components

10.2.2.3.12

Purpose: To determine the overall compliance of the assembly with the Specifications as set forth in Section 10 of this Manual.

Requirement: All Sections of the Specifications of this Manual which are not specifically covered by the above tests shall be satisfactorily met.

Steps:

a. For each pertinent Section of the Specifications inspect, test or otherwise be assured that the assembly meets the minimum conditions of these Specifications.

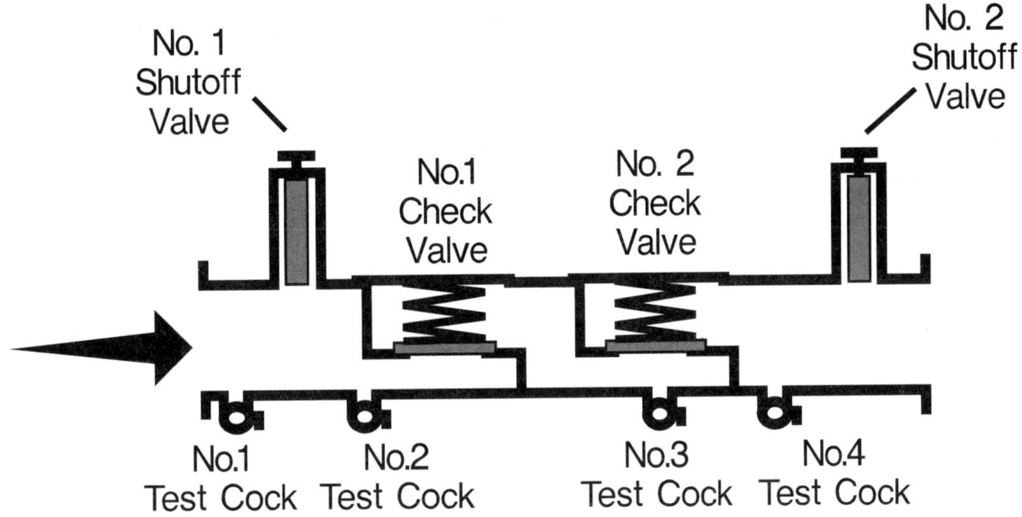

Fig. 10. 4
Double check valve backflow prevention assembly showing location of shutoff valves, and test cocks

10.2.3 DESIGN, OPERATIONAL AND EVALUATION SPECIFICATIONS FOR DOUBLE CHECK VALVE BACKFLOW PREVENTION ASSEMBLIES (DC)

10.2.3.1 Design and Operational Specifications

a. This assembly shall include two independently acting approved check valves mounted between two tightly closing resilient seated shutoff valves, and the necessary four resilient seated test cocks. (See Fig. 10.4.)

b. Each check valve shall be internally loaded and shall at all times be drip-tight in the normal direction of flow with the inlet pressure at least 1 psi and the outlet under atmospheric pressure.

c. Each check valve shall permit no leakage in a direction reverse to the normal flow under all conditions of a pressure differential.

10.2.3.2 Laboratory Evaluation

One assembly shall be inspected and evaluated in the following order. Unless otherwise noted, tests shall be performed at ambient temperature.

a. Conformance to the general design and material requirements outlined in Section 10.1.

b. Conformance to the operational requirements outlined in Section 10.2.

c. Conformance to the working drawings and materials specifications.

d. Hydrostatic tests — the test assembly shall be subjected to the conditions of Section 10.1.1.5.

e. Pressure loss characteristics for flow rates up to the rated conditions.

f. Static closing point of check valve No. 1.

g. Static closing point of check valve No. 2.

h. Thermal Test — the assembly shall be evaluated at the greater of 140°F (60°C) or the maximum working water temperature (MWWT), maximum working water pressure (MWWP), and specified flow rate.

i. Life cycle test.

10.2.3.3 Evaluation Procedure

10.2.3.3.1

Purpose: To determine the capability of the assembly to withstand the required hydrostatic test pressure.

Requirement: All components of the assembly shall withstand a hydrostatic pressure of twice (2×) the maximum working water pressure (MWWP), as stated by the manufacturer, for a minimum period of ten (10) minutes without any damage, permanent deformation or impairment of operation.

Steps:

a. With both shutoff valves closed and all of the air bled from the assembly, a hydrostatic pressure of twice (2×) the maximum working water pressure shall be supplied through the No. 2 test cock for a minimum period of ten (10) minutes. Any evidence of leakage shall be cause for rejection.

b. The required hydrostatic pressure shall be supplied through test cock No. 3 with test cock No. 2 open to atmosphere for a minimum period of ten (10) minutes. Any evidence of leakage shall be cause for rejection.

c. The required hydrostatic pressure shall be supplied through test cock No. 4 with test cock No. 3 open to atmosphere for a minimum period of ten (10) minutes. Any evidence of leakage shall be cause for rejection.

d. The assembly shall be disassembled and inspected for any damage, permanent deformation, or impairment of operation. Evidence of such shall be cause for rejection.

10.2.3.3.2

Purpose: To determine the overall pressure loss of the complete assembly as a function of the rate of flow.

Requirement: At any rate of flow from zero (static) up to and including the rated flow the overall pressure loss shall not exceed the values shown in Table 10-1. The assembly shall withstand a minimum of 200 percent of rated flow without damage, permanent deformation, or impairment of operation.

Steps:

a. Install the assembly in a suitable hydraulic test line having a flow capacity in excess of the rated flow conditions (see Table 10-1) and an available line pressure of at least 150 psi (1034 KPa).

b. Install in the supply line at a suitable location upstream from the assembly a piezometer ring; and, install a similar piezometer ring at a suitable location downstream from the assembly (located as indicated in ANSI/ISA S75.02). Then install a mercury manometer, or other suitable means of measuring pressure differentials between the two piezometer rings to measure the observed overall pressure loss at various rates of flow.

c. Record differential pressure between piezometer rings at static condition. Increase flow rate and record steady state differential pressure at a sufficient number of increments to fully define the flow curve characteristic. Data shall be taken for both increasing and decreasing flow conditions. The flow rate shall then be increased to 200% of the rated flow for five (5) minutes, then returned to static.

   Manifold Assembly (Section 10.1.2.19) Only: Close all shutoff valves so that only one of the individual assemblies of the manifold assembly is open. Gradually increase the flow of water through the individual assembly until the rated flow of the manifold assembly is reached, then return to static condition. Repeat flow test for each individual assembly contained in the manifold assembly.

d. Remove the assembly from the line and couple the downstream piping directly to the supply piping. Then observe data for a friction head correction curve for the pipe fittings required to adapt the test assembly to the supply line between the existing piezometer rings. The data of this correction curve are then subtracted from the observed overall pressure loss data of the assembly from the upstream face of the upstream shutoff valve to the downstream face of the downstream shutoff valve.

e. Exceeding the maximum allowable pressure loss listed in Table 10-1 for any rate of flow up to and including rated flow, shall be cause for rejection. The assembly shall be disassembled and inspected for any damage, permanent deformation, or impairment of operation. Evidence of such shall be cause for rejection.

10.2.3.3.3

Purpose: To determine the static pressure drop across check valve No. 1.

Requirement: The No. 1 check valve shall be drip-tight in the normal direction of flow with the inlet pressure at least 1 psi (6.89 KPa) and the outlet pressure at atmospheric.

a. Install a vertical transparent tube at least six feet (1.83 m) long at test cock No. 2. If test cock No. 3 is not at the highest point of the check valve body, then another vertical transparent tube must be installed on test cock No. 3 so that it just rises above the top of the check valve body. (See Fig. 10.2.)

b. Open test cock No. 2 and fill the tube with water, then close test cock No. 2. If a tube is attached to test cock No. 3, open test cock No. 3 to fill the tube, then close test cock No. 3.

c. Close the No. 2 shutoff valve, then close the No. 1 shutoff valve.

d. Open test cock No. 3, then open test cock No.2.

e. Water from the tube on test cock No. 2 will flow through check valve No. 1 until closure of check valve No. 1 is attained. When no further fall of the water in the tube on test cock No. 2 is observed, record the height of the water in the tube on test cock No. 2 above the water level at test cock No. 3. The difference of height of water in the transparent tube(s) must be greater than 27 3/4 inches (704.85 mm).

f. Failure to maintain a minimum level of 27 3/4 inches (704.85mm) shall be cause for rejection.

10.2.3.3.4

Purpose: To determine the static pressure drop across check valve No. 2.

Requirement: The No. 2 check valve shall be drip-tight in the normal direction of flow with the inlet pressure at least 1 psi (6.89 KPa) and the outlet pressure at atmospheric.

Steps:

a. Install a vertical transparent tube at least six feet (1.83 m) long at test cock No. 3. If test cock No. 4 is not at the highest point of the check valve body, then another vertical transparent tube must be installed on test cock No. 4 so that it just rises above the top of the check valve body. (See Fig. 10.2.)

b. Open test cock No. 3 and fill the tube with water, then close test cock No. 3. If a tube is attached to test cock No. 4, open test cock No. 4 to fill the tube, then close test cock No. 4.

c. Close the No. 2 shutoff valve, then close the No. 1 shutoff valve.

d. Open test cock No. 4, then open test cock No. 3.

e. Water from the tube on test cock No. 3 will flow through check valve No. 2 until closure of check valve No. 2 is attained. When no further fall of the water in the tube on test cock No. 3 is observed, record the height of the water in the tube on test cock No. 3 above the water level at test cock No. 4. The difference of height of water in the transparent tube(s) must be greater than 27 3/4 inches (704.85 mm).

f. Failure to maintain a minimum level of 27 3/4 inches (704.85mm) shall be cause for rejection.

10.2.3.3.5

Purpose: To determine if all components of the assembly shall operate properly under rated temperature and pressure conditions, and flow rates as specified in Table 10-7.

Requirements: All components of the assembly shall operate at, and withstand a thermal test at 140°F (60°C) or at the maximum working water temperature (MWWT) whichever is greater, maximum working water pressure (MWWP), and flow rate as specified in Table 10-7, for a minimum period of one hundred (100) hours without any damage, permanent deformation or impairment of operation.

Steps:

a. Install assembly into suitable test line which can generate the following minimum parameters:

   Maximum working water pressure *{minimum of 150 psi (1034 KPa)}*
   Maximum working water temperature *{minimum of 140°F (60°C)}*
   Flow requirements per Table 10-7

b. The dimensions and durometer hardness of all elastomer components shall be inspected and recorded.

c. Set controls to the rated temperature and pressure conditions of the assembly, as well as rate of flow as specified in Table 10-7.

d. The temperature and pressure conditions shall be continuously monitored and recorded.

e. During and after a minimum of one hundred (100) hours at rated temperature and pressure the assembly shall be tested to determine if the assembly operates satisfactorily per 10.2.3.3.3 and 10.2.3.3.4. Failure to comply with these tests shall be cause for rejection.

f. The assembly shall then be disassembled and inspected for any internal damage or permanent deformation. Evidence of such shall be cause for rejection.

g Once the assembly returns to ambient temperature conditions, the assembly shall then be tested per Section 10.2.3.3.1. Failure to comply with this test shall be cause for rejection.

10.2.3.3.6

Purpose: To determine if the assembly sustains any damage, permanent deformation, or impairment of operation following the specified cycles.

Requirement: The assembly shall withstand the specified cycle without leakage, damage, permanent deformation, or impairment of operation.

Steps:

a. Install assembly into suitable test line which can generate the following parameters:

   Rated flow of assembly under test (per Table 10-1)
   Line pressure of 60 psi ± 10 psi (414 KPa ± 69 KPa)
   Temperature of 110°F $^{+30}/_{-0}$ °F (43°C. $^{+17}/_{-0}$ °C)
   Backpressure of 150psi ± 10 psi (1034 KPa ± 69 KPa)
   Control valves capable of closing/opening within five (5) seconds *.
   *(*Resulting rise of pressure shall not exceed MWWP of the assembly.*)

b. With Valve #3 (Valve #3A open and Valve #3B closed) and Valve #4 (see Figure 10.3) in the closed position, and Valve #1 and Valve #2 in the open position, establish flow through the assembly equal to 25 percent of the maximum rated flow per Table 10-1.

c. Maintain the flow through the assembly for a minimum of six (6) seconds. Close Valve #1.

d. After five (5) seconds, close Valve #2.

e. After three (3) seconds, open Valve #4.

f. After three (3) seconds, open Valve #3. Maintain Valve #3 open for six (6) seconds. Close Valve #3.

g. After two (2) seconds, close Valve #4.

h. After three (3) seconds, open Valve #2.

i. After two (2) seconds, open Valve #1.

j. Repeat steps ***c*** through ***i*** 1250 times.

k. Raise flow through the assembly to 50 percent of rated flow. Repeat steps ***c*** through ***i*** 1250 times.

l. Raise flow through the assembly to 75 percent of rated flow. Repeat steps ***c*** through ***i*** 1250 times.

m. Raise flow through the assembly to 100 percent of rated flow. Repeat steps ***c*** through ***i*** 1250 times.

n. After a minimum of 5000 total cycles the assembly shall be tested to determine if the assembly complies with 10.2.3.3.3 and 10.2.3.3.4. Failure to comply with these tests shall be cause for rejection.

o. The assembly shall then be disassembled and inspected for any damage or permanent deformation. Evidence of such shall be cause for rejection.

10.2.3.3.7

Purpose: To determine the overall compliance of the assembly with the Specifications as set forth in Section 10 of this Manual.

Requirement: All Sections of the Specifications of this Manual which are not specifically covered by the above tests shall be satisfactorily met.

Steps:

a. For each pertinent Section of the Specifications inspect, test or otherwise be assured that the assembly meets the minimum conditions of these Specifications.

TABLE 10-5

RATED FLOW AND MAXIMUM ALLOWABLE PRESSURE LOSS FOR VARIOUS SIZES OF ATMOSPHERIC (AVB) AND PRESSURE (PVB & SVB) VACUUM BREAKERS

Size of Assembly (1)		Rated Flow (2)		Maximum Allowable Pressure Loss — Atmospheric AVB (3)		Maximum Allowable Pressure Loss — Pressure PVB & SVB (4)	
(inches)	(mm)	(gpm)	(L/s)	(psi)	(KPa)	(psi)	(KPa)
1/4	(6)	1	(0.31)	24	(165.45)	10	(68.9)
3/8	(9)	3	(0.50)	22	(151.66)	10	(68.9)
1/2 ***	(12)	15	(0.75)	22	(151.66)	10	(68.9)
1/2 ****	(12)	7.5	(0.47)	NA		10	(68.9)
3/4	(19)	30	(1.89)	20	(137.88)	10	(68.9)
1	(25)	50	(3.15)	18	(124.09)	10	(68.9)
1 1/4	(31)	75**	(4.73)	18	(124.09)	10	(68.9)
1 1/2	(38)	100	(6.31)	16	(110.30)	10	(68.9)
2	(50)	160	(10.10)	16	(110.30)	10	(68.9)
2 1/2	(63)	225	(14.20)	16	(110.30)	10	(68.9)
3	(75)	320	(20.19)	15	(103.41)	10	(68.9)
4	(100)	500	(31.55)	14	(96.51)	10	(68.9)
6	(150)	1000	(63.10)	14	(96.51)	10	(68.9)
8	(200)	1600	(100.96)	14	(96.51)	10	(68.9)
10	(254)	2300	(145.13)	14	(96.51)	10	(68.9)

* Extrapolated
** Interpolated
*** AVB and PVB only
**** SVB only

Note: The pressure losses as shown in Column (3) and Column (4) represent the maximum permissible pressure loss at any flow rate up to and including the rated flow shown in Column (2).

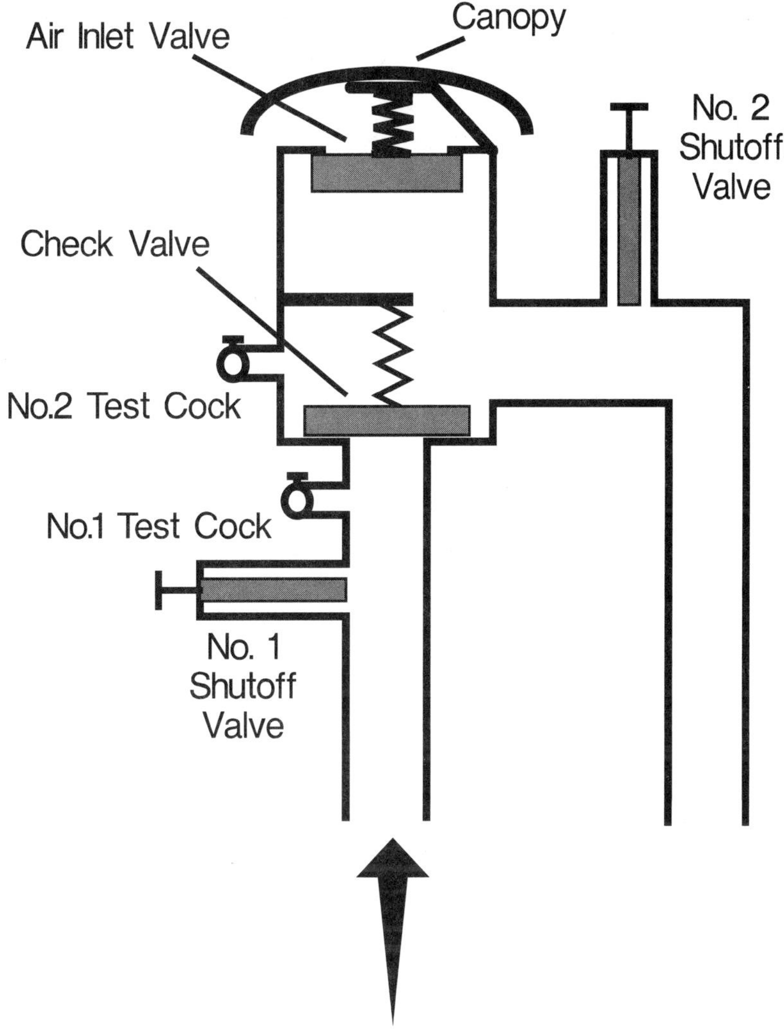

Fig. 10.5

Pressure Vacuum Breaker Backsiphonage Prevention Assembly Showing Location of Shutoff Valves, Check Valve, Air Inlet Valve and Test Cocks

10.2.4 DESIGN, OPERATIONAL AND EVALUATION SPECIFICATIONS FOR PRESSURE VACUUM BREAKER BACKSIPHONAGE PREVENTION ASSEMBLIES (PVB)

10.2.4.1 Design and Operational Specifications

a. This assembly shall include an approved internally loaded check valve and a loaded air inlet valve opening to atmosphere on the discharge side of the check valve between two tightly closing resilient seated shutoff valves; and, will include two properly located resilient seated test cocks. (See Fig. 10.5).

b. The air inlet valve of the pressure vacuum breaker shall open when the internal pressure is a minimum of 1 psi (6.89 KPa). It shall be fully open when the water drains from the body.

c. The maximum allowable pressure drop across the assembly, from the upstream face of the No. 1 shutoff valve to the downstream face of the No. 2 shutoff valve, shall not exceed 10 psi (68.94 KPa) for any rate of flow up to and including the rated flow listed in Table 10-5.

d. The check valve shall be internally loaded and shall at all times be drip-tight in the normal direction of flow with the inlet pressure at 1 psi (6.89 KPa) and the outlet under atmospheric pressure.

e. The effective size of the air inlet port(s) of the assembly shall be governed by the vacuum dissipation test. If an air inlet port shield or canopy is used, it shall extend down around the body of the assembly to the lowest portion of the port(s). To reduce potential for fouling, the minimum clearance between the air inlet port(s) and the shield or canopy shall be 3/16 inch (5 mm).

10.2.4.2 Laboratory Evaluation

One assembly shall be inspected and evaluated in the following order. Unless otherwise noted, tests shall be performed at ambient temperature.

a. Conformance to general, design and material requirements outlined in Section 10.1.

b. Conformance to the operational requirements outlined in Section 10.2.

c. Conformance to the working drawings and materials specifications.

d. Hydrostatic tests — the test assembly shall be subjected to the conditions of Section 10.1.1.5.

e. Pressure loss characteristics for flow rates up to the rated conditions set forth in Table 10-5.

f. Conformance to the air inlet requirements.

g. Static closing point of the check valve.

h. Thermal Test — the assembly shall be evaluated at the greater of 140°F (60°C) or the maximum working water temperature (MWWT), maximum working water pressure (MWWP), and specified flow rate.

i. Life cycle test.

10.2.4.3 Evaluation Procedure

10.2.4.3.1

Purpose: To determine the capability of the assembly to withstand the required hydrostatic test pressure.

Requirement: All components of the assembly shall withstand a hydrostatic pressure of twice (2×) the maximum working water pressure (MWWP), as stated by the manufacturer, for a minimum period of ten (10) minutes without any damage, permanent deformation or impairment of operation.

Steps:

a. With both shutoff valves closed and all of the air bled from the assembly, a hydrostatic pressure of twice (2×) the rated working water pressure shall be supplied through the No. 1 test cock for a minimum period of ten (10) minutes. Any evidence of leakage shall be cause for rejection.

b. The required hydrostatic pressure shall be supplied through test cock No. 2 with test cock No. 1 open to atmosphere for a minimum period of ten (10) minutes. Any evidence of leakage shall be cause for rejection.

c. The assembly shall be disassembled and inspected for any damage, permanent deformation, or impairment of operation. Evidence of such shall be cause for rejection.

10.2.4.3.2

Purpose: To determine the overall pressure loss of the complete assembly as a function of the rate of flow.

Requirement: At any rate of flow from zero (static) up to and including the rated flow the overall pressure loss shall not exceed the values shown in Table 10-5. The assembly shall withstand a minimum of 200 percent of the rated flow without damage, permanent deformation, or impairment of operation.

Steps:

a. Install the assembly in a suitable hydraulic test line having a flow capacity in excess of the rated flow conditions (see Table 10-5) and an available line pressure of at least 150 psi (1034 KPa).

b. Install in the supply line at a suitable location upstream from the assembly a piezometer ring; and, install a similar piezometer ring at a suitable location downstream from the assembly (located as indicated in ANSI/ISA Standard S75.02). Then install a mercury manometer, or other suitable means of measuring pressure differentials between the two piezometer rings to measure the observed overall pressure loss at various rates of flow.

c. Record differential pressures between piezometer rings at static condition. Increase flow rate and record steady state differential pressure at a sufficient number of increments to fully define flow curve characteristic. Data shall be taken for both increasing and decreasing flow conditions. The flow rate shall then be increased to 200 percent of the rated flow for five (5) minutes, then returned to static.

d. Remove the assembly from the line and couple the downstream piping directly to the supply piping. Then observe data for a friction head correction curve for the pipe fittings required to adapt the test assembly to the supply line between the existing piezometer rings. The data of this correction curve are then subtracted from the observed overall pressure loss data of the assembly from the upstream face of the upstream shutoff valve to the downstream face of the downstream shutoff valve.

e. Exceeding the maximum allowable pressure loss listed in Table 10-5 for any rate of flow up to and including rated flow, shall be cause for rejection. The assembly shall be disassembled and inspected for any damage, permanent deformation, or impairment of operation. Evidence of such shall be cause for rejection.

10.2.4.3.3

Purpose: To test the opening pressure differential of the air inlet valve.

Requirement: The air inlet valve shall open when the pressure in the body is a minimum of 1.0 psi (6.89 KPa) above atmospheric pressure. The air inlet valve shall be fully open when the water drains from the body.

Steps:

a. Install the assembly in a suitable hydraulic test line with a minimum pressure of 150 psi (1034 KPa). Remove air inlet valve canopy.

b. Install the high side hose of the differential pressure gage, or other suitable means of measuring pressure differentials, to test cock No. 2 and bleed air from the hose and gage.

c. Close shutoff valve No. 2; and then close shutoff valve No. 1.

d. Slowly (0.2 psid/ second maximum) open the high side bleed needle valve on the gage. Record the pressure differential at which the air inlet valve opens. Failure to open at a value of 1.0 psi (6.89KPa) or greater shall be cause for rejection.

e. Allow water to drain out of the body through test cock No. 2 to determine if the air inlet valve opens fully. Failure to open fully shall be cause for rejection.

10.2.4.3.4

Purpose: To determine the static pressure drop across the check valve.

Requirement: The check valve shall be drip-tight in the normal direction of flow when the inlet pressure is at least 1 psi (6.89 KPa) and the outlet pressure is atmospheric.

Steps:

a. Install a transparent tube approximately 6 feet (180 cm) long on test cock No. 1. Open test cock No. 1 to fill the tube with water, then close test cock No. 1.

b. Close shutoff valve No. 2, then close shutoff valve No. 1.

c. Open test cock No. 2 ,then open test cock No. 1. The air inlet valve will open; and, the water in the tube will fall until closure of the check valve is attained. When no further fall of the water in the tube is observed, record the height of water in the transparent tube above the centerline of test cock No. 2.

d. Failure to maintain a minimum level of 27 3/4 inches (705 mm) shall be cause for rejection.

10.2.4.3.5

Purpose: To compare the air flow capacity of the air inlet valve to that of the effective water through-way of the assembly.

Requirement: The time required to dissipate a vacuum (25 to 5 inches of mercury vacuum) through the air inlet valve shall be less than the time required to dissipate the same vacuum through the water through-way.

Steps:

a. Install the assembly in a normal operating position, with the outlet of the assembly attached to a vacuum source.

b. Close No. 1 shutoff valve to simulate the check valve being closed. (No. 2 shutoff valve remains fully open during entire test procedure.)

c. With the air inlet valve open, record the time required to dissipate the vacuum from 25 to 5 inches of mercury vacuum when this flow is controlled by means of a quick opening valve. Repeat this step a minimum of three (3) times, then average the recorded times.

d. Open No. 1 shutoff valve and attach a 12 inch (30.5 cm) long pipe nipple, the same nominal pipe diameter as the assembly, to the inlet of the No. 1 shutoff valve.

e. Hold the air inlet valve closed; then record the time required to dissipate the vacuum from 25 to 5 inches of mercury vacuum when this flow is controlled by means of a quick opening valve. Repeat this step a minimum of three (3) times, then average the recorded times.

f. Based upon the average of at least three (3) tests, the time required for step ***c*** shall be less than the time required for step ***e***. Failure to comply with this requirement shall be cause for rejection.

10.2.4.3.6

Purpose: To test that a backsiphonage condition together with a fouled check valve shall not allow a carry-over of downstream water back into the inlet.

Requirement: Water shall not be backsiphoned more than three (3) inches (76 mm) above the free surface of water in the discharge pipe standing six (6) inches (152 mm) below the bottom or critical level line of the assembly.

Steps:

a. Install the assembly in a normal operating position, with the inlet of the assembly connected to the vacuum source and the discharge side equipped with a transparent tube that extends down into a vessel of water where the free surface of the water is six (6) inches (152 mm) below the bottom or critical level line of the assembly.

b A wire of the size shown in Table 10-6 shall be placed across the seating surface in the lower quadrant of a hinged or horizontally moving poppet check valve, or at a single point of a vertical moving check valve. The wire shall be placed so that the entire width of the seating contact surface is covered.

c. By the application of vacuum conditions ranging from zero to 25 inches of mercury under both steady flow and instantaneous conditions, the water in the discharge tube shall not rise more than three (3) inches (76 mm) above the original free surface of the water in the vessel. Evidence of such shall be cause for rejection.

TABLE 10-6

SIZE OF FOULING WIRE FOR BACKSIPHONAGE TESTS

Size of Assembly (inches)	Size of Assembly (mm)	Diameter of Wire (inches)	Diameter of Wire (mm)
1/2 and below	12	0.032	0.81
5/8×3/4	19	0.040	1.01
3/4	19	0.040	1.01
1	25	0.048	1.22
1 1/4	31	0.056	1.42
1 1/2	38	0.064	1.62
2	50	0.080	2.03
2 1/2	63	0.096	2.44
3	75	0.112	2.84
4	100	0.144	3.66
6	150	0.210	5.33
8	200	0.275	6.98
10	254	0.340	8.64
12	305	0.400	10.16
14	356	0.460	11.68
16	406	0.525	13.34

10.2.4.3.7

Purpose: To determine if all components of the assembly shall operate properly under rated temperature and pressure conditions, and flow rates as specified in Table 10-7.

Requirements: All components of the assembly shall operate at, and withstand a thermal test at the greater of 140°F (60°C) or at the maximum working water temperature (MWWT), maximum working water pressure (MWWP), and flow rate as specified in Table 10-7 for a minimum period of one hundred (100) hours without any damage, permanent deformation or impairment of operation.

Steps:

a. Install assembly into suitable test line which can generate the following minimum parameters:

   Maximum working water pressure *{minimum of 150 psi (1034 KPa)}*
   Maximum working water temperature *{minimum of 140°F (60°C)}*
   Flow requirements per Table 10-7

b. The dimensions and durometer hardness of all elastomer components shall be inspected and recorded.

c. Set controls to the rated temperature and pressure conditions of the assembly, as well as rate of flow as specified in Table 10-7.

d. The temperature and pressure conditions shall be continuously monitored and recorded.

e. During and after a minimum of one hundred (100) hours at rated temperature and pressure the assembly shall be tested to determine if the assembly operates satisfactorily per 10.2.4.3.3 and 10.2.4.3.4. Failure to comply with these tests shall be cause for rejection.

f. The assembly shall then be disassembled and inspected for any internal damage or permanent deformation. Evidence of such shall be cause for rejection.

g. Once the assembly returns to ambient temperature conditions, the assembly shall then be tested per Section 10.2.4.3.1. Failure to comply with this test shall be cause for rejection.

10.2.4.3.8

Purpose: To determine if the assembly sustains any damage, permanent deformation, or impairment of operation following the specified cycles.

Requirement: The assembly shall withstand the specified cycle without leakage, damage, permanent deformation, or impairment of operation.

Steps:

a. Install assembly into suitable test line which can generate the following parameters:

   Rated flow of assembly under test (per Table 10-5)
   Line pressure of 60 psi ± 10 psi (414 KPa ± 69 KPa)
   Temperature of 110°F $^{+30}/_{-0}$ °F (43°C. $^{+17}/_{-0}$ °C)
   Control valves capable of closing/opening within five (5) seconds *.
   *(*Resulting rise of pressure shall not exceed MWWP of the assembly.*)

b. With Valve #3 (Valve #3A closed and Valve #3B open) and Valve #4 (see Fig. 10.3) in the closed position, and Valve #1 and Valve #2 in the open position, establish flow through the assembly equal to 25 percent of the maximum rated flow per Table 10-5.

c. Maintain the flow through the assembly for a minimum of six (6) seconds. Close Valve #1.

d. After five (5) seconds, close Valve #2.

e. After three (3) seconds, open Valve #3.

f. After ten (10) seconds, close Valve #3.

g. After three (3) seconds, open Valve #2.

h. After three (3) seconds, open Valve #1.

i. Repeat steps ***c*** through ***h*** 1250 times.

j. Raise flow through the assembly to 50 percent of rated flow. Repeat steps ***c*** through ***h*** 1250 times.

k. Raise flow through the assembly to 75 percent of rated flow. Repeat steps ***c*** through ***h*** 1250 times.

l. Raise flow through the assembly to 100 percent of rated flow. Repeat steps ***c*** through ***h*** 1250 times.

m. After a minimum of 5000 total cycles the assembly shall be tested to determine if the assembly complies with 10.2.4.3.3 and 10.2.4.3.4. Failure to comply with these tests shall be cause for rejection.

n. The assembly shall then be disassembled and inspected for any damage or permanent deformation. Evidence of such shall be cause for rejection.

10.2.4.3.9

Purpose: To determine the overall compliance of the assembly with the Specifications as set forth in Section 10 of this Manual.

Requirement: That all Sections of the Specifications of this Manual which are not specifically covered by the above tests are satisfactorily met.

Steps:

a. For each pertinent Section of the Specifications inspect, test or otherwise be assured that the assembly meets the minimum conditions of these Specifications.

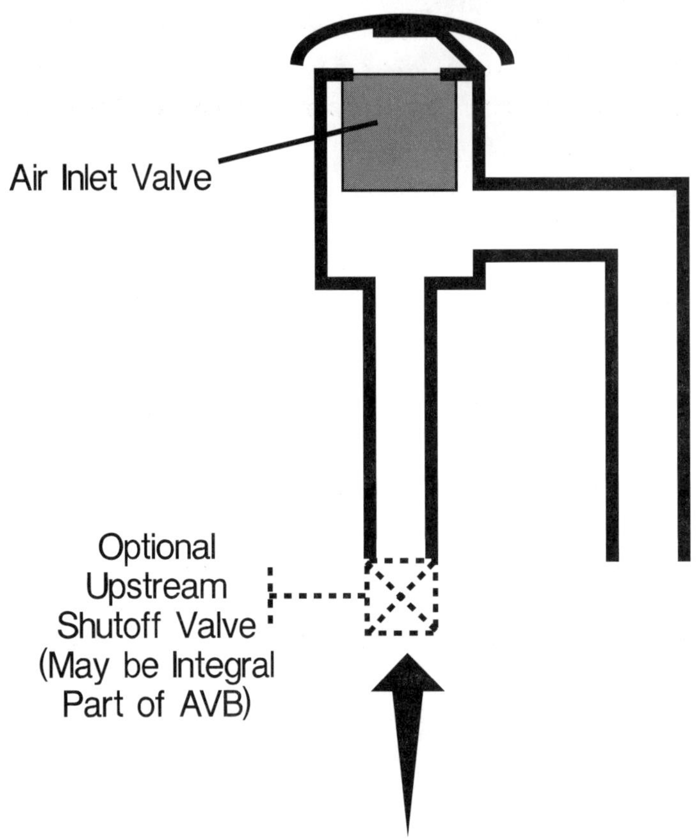

Fig. 10.6
Atmospheric Vacuum Breaker Backsiphonage Prevention Assembly — Showing Location of Air Inlet Valve and Optional Integral Shutoff Valve

10.2.5 DESIGN, OPERATIONAL AND EVALUATION SPECIFICATIONS FOR ATMOSPHERIC VACUUM BREAKER BACKSIPHONAGE PREVENTION ASSEMBLIES (AVB)

10.2.5.1 Design and Operational Specifications

a. This assembly shall include an air inlet valve or float check, a check seat, and air inlet port(s). A tightly closing integral shutoff valve may only be included on the upstream side of the air inlet valve. (See Fig. 10.6.)

b. The air inlet valve of the atmospheric vacuum breaker shall open when the internal pressure is atmospheric. It shall be fully open when the water drains from the body.

c. The maximum allowable pressure drop across the assembly, from the upstream face of the assembly to the downstream face of the assembly, for any rate of flow up to and including the rated flow shall not exceed the values given in Table 10-5.

d. The effective size of the air inlet port(s) of the assembly shall be governed by the vacuum dissipation test. If an air inlet port shield or canopy is used, it shall extend down around the body of the assembly to the lowest portion of the port(s). To reduce the potential for fouling, the minimum clearance between the air inlet port(s) and the shield or canopy shall be 3/16 inch (5 mm).

10.2.5.2 Laboratory Evaluation

One assembly shall be inspected and evaluated in the following order. Unless otherwise noted, tests shall be performed at ambient temperature.

a. Conformance to general, design and material requirements outlined in Section 10.1.

b. Conformance to the operational requirements outlined in Section 10.2.

c. Conformance to the working drawings and materials specifications.

d. Hydrostatic tests — the test assembly shall be isolated and subjected to the conditions of Section 10.1.1.5.

e. Pressure loss characteristics for flow rates up to the rated conditions.

f. Conformance to the air inlet requirements.

g. Thermal Test — the assembly shall be evaluated at the greater of 140°F (60°C) or the maximum working water temperature (MWWT), maximum working water pressure (MWWP), and specified flow rate.

h. Life cycle test.

10.2.5.3 Evaluation Procedure

10.2.5.3.1

Purpose: To determine the capability of the assembly to withstand the required hydrostatic test pressure.

Requirement: All components of the assembly shall withstand a hydrostatic pressure of twice (2×) the maximum working water pressure (MWWP), as stated by the manufacturer, for a minimum period of ten (10) minutes without any damage, permanent deformation or impairment of operation.

Steps:

a. With the assembly isolated both upstream and downstream and all of the air bled from the assembly, a hydrostatic pressure of twice (2×) the maximum working water pressure shall be supplied through the inlet of the assembly for a minimum period of ten (10) minutes. Any evidence of leakage shall be cause for rejection.

b. The assembly shall be disassembled and inspected for any internal damage as well as damage to the external cover plates or other components. Evidence of such shall be cause for rejection.

10.2.5.3.2

Purpose: To determine the overall pressure loss of the complete assembly as a function of the rate of flow.

Requirement: At any rate of flow from zero (static) up to and including the rated flow the overall pressure loss shall not exceed the values shown in Table 10-5. The assembly shall withstand a minimum of 200 percent of the rated flow without any damage, permanent deformation, or impairment of operation.

Steps:

a. Install the assembly in a suitable hydraulic test line having a flow capacity in excess of the rated flow conditions (see Table 10-5) and an available line pressure of at least 150 psi (1034 KPa).

b. Install in the supply line at a suitable location upstream from the assembly a piezometer ring;, and, install a similar piezometer ring at a suitable location downstream from the assembly (located as indicated in ANSI/ISA Standard S75.02). Then install a mercury manometer, or other suitable means of measuring pressure differentials between the two piezometer rings to measure the observed overall pressure loss at various rates of flow.

c. Record differential pressure between piezometer rings at static condition. Increase flow rate and record steady state differential at a sufficient number of increments to fully define flow curve characteristics. Data shall be taken for both increasing and decreasing flow conditions The flow rate shall then be increased to 200% of the rated flow for five (5) minutes, then returned to static.

d. Remove the assembly from the line and couple the downstream piping directly to the supply piping. Then observe data for a friction head correction curve for the pipe fittings required to adapt the test assembly to the supply line between the existing piezometer rings. The data of this correction curve are then subtracted from the observed overall pressure loss data of the assembly from the upstream face of the assembly to the downstream face of the assembly.

e. Exceeding the maximum allowable pressure loss listed in Table 10-5 for any rate of flow up to and including rated flow, shall be cause for rejection. The assembly shall be disassembled and inspected for any damage, permanent deformation, or impairment of operation. Evidence of such shall be cause for rejection.

10.2.5.3.3

Purpose: To test the opening of the air inlet valve.

Requirement: The air inlet valve shall open when the pressure in the body is atmospheric. And, the air inlet valve shall be fully open when the water drains from the body.

Steps:

a. Remove air inlet valve canopy.

b. Install the assembly in a suitable hydraulic test line with a minimum pressure of 150 psi (1034 KPa).

c. Isolate the assembly in the test line both upstream and downstream.

d. Lower the pressure in the body to atmospheric, the air inlet valve shall begin to open. Continue to drain the water from the body. Once all of the water has drained out, the air inlet valve shall be fully open.

e. Failure to fully open shall be cause for rejection.

10.2.5.3.4

Purpose: To compare the air flow capacity of the air inlet valve to that of the effective water throughway of the assembly.

Requirement: The time required to dissipate a vacuum 25 to 5 inches of mercury vacuum through the air inlet valve shall be less than the time required to dissipate the same vacuum through the water throughway.

Steps:

a. Install the assembly in the normal operating position, with the discharge outlet of the assembly attached to a vacuum source.

b. Close the inlet by plugging or other suitable means. (If integral upstream shutoff valve is present, close shutoff valve) With the air inlet valve open, record the time required to dissipate the vacuum from 25 to 5 inches of mercury vacuum when this flow is controlled by means of a quick opening valve. Repeat this step a minimum of three (3) times, then average the recorded values.

c. Open the inlet (if integral upstream shutoff valve is present, open shutoff valve) and attach a 12 inch (30.5 cm) long pipe nipple, the same nominal pipe diameter as the inlet connection of the assembly.

d. Hold the air inlet valve closed; then record the time required to dissipate the vacuum from 25 to 5 inches of mercury vacuum when this flow is controlled by means of a quick opening valve. Repeat this step a minimum of three (3) times, then average the recorded times.

e. Based upon the average of at least three (3) tests, the time required for step ***b*** shall be less than the time required for step ***d***. Failure to comply with this requirement shall be cause for failure.

10.2.5.3.5

Purpose: To test that a backsiphonage condition together with a fouled check valve shall not allow a carry-over of downstream water back into the inlet.

Requirement: Water shall not be backsiphoned more than three (3) inches (76 mm) above the free surface of water in the discharge pipe standing six (6) inches (152 mm) below the bottom or critical level line of the assembly.

Steps:

a. Install the assembly in a normal operating position, with the inlet of the assembly connected to the vacuum source and the discharge side equipped with a transparent tube that extends down into a vessel of water where the free surface of the water is six (6) inches (152 mm) below the bottom or critical level line of the assembly.

b. A wire of the size shown in Table 10-6 shall be placed across the seating surface in the lower quadrant of a hinged or horizontally moving poppet check valve, or at a single point of a vertical moving check valve. The wire shall be placed so that the entire width of the seating contact surface is covered.

c. By the application of vacuum ranging from zero to 25 inches of mercury under both steady flow and instantaneous conditions, the water in the discharge tube shall not rise more than three (3) inches above the original free surface of the water in the vessel. Evidence of such shall be cause for rejection.

10.2.5.3.6

Purpose: To determine if all components of the assembly shall operate properly under rated temperature and pressure conditions, and flow rates as specified in Table 10-7.

Requirements: All components of the assembly shall operate at, and withstand a thermal test at 140°F (60°C) or at the maximum working water temperature (MWWT) whichever is greater, maximum working water pressure (MWWP), and flow rate as specified in Table 10-7, for a minimum period of one hundred (100) hours without any damage, permanent deformation or impairment of operation.

Steps:

a. Install assembly into suitable test line which can generate the following minimum parameters:

   Maximum working water pressure *{minimum of 150 psi (1034 KPa}*
   Maximum working water temperature *{minimum of 140°F (60°C)}*
   Flow requirements per Table 10-7

b. The dimensions and durometer hardness of all elastomer components shall be inspected and recorded.

c. Set controls to the rated temperature and pressure conditions of the assembly as well as rate of flow as specified in Table 10-7.

d. The temperature and pressure conditions shall be continuously monitored and recorded.

e. During and after a minimum of one hundred (100) hours at rated temperature and pressure the assembly shall be tested to determine if the assembly operates satisfactorily per 10.2.5.3.3. Failure to comply with this test shall be cause for rejection.

f. The assembly shall then be disassembled and inspected for any internal damage or permanent deformation. Evidence of such shall be cause for rejection.

g. Once the assembly returns to ambient temperature conditions, the assembly shall then be tested per Section 10.2.5.3.1. Failure to comply with this test shall be cause for rejection.

10.2.5.3.7

Purpose: To determine if the assembly sustains any damage, permanent deformation, or impairment of operation following the specified cycles.

Requirement: The assembly shall withstand the specified cycle without leakage, damage, permanent deformation, or impairment of operation.

Steps:

a. Install assembly into suitable test line which can generate the following parameters:

Rated flow of assembly under test (per Table 10-5)
Line pressure of 60 psi ± 10 psi (414 KPa ± 69 KPa)
Temperature of 110°F $^{+30}/_{-0}$ °F (43°C $^{+17}/_{-0}$ °C)
Control valves capable of closing/opening within five (5) seconds *.
*(*Resulting rise of pressure shall not exceed MWWP of the assembly.*)

b. With Valve #3 (Valve 3A closed and Valve 3B open) and Valve #4 (see Fig. 10.3) in the closed position, and Valve #1 and Valve #2 in the open position, establish flow through the assembly equal to 25 percent of the maximum rated flow per Table 10-5.

c. Maintain the flow through the assembly for a minimum of six (6) seconds. Close Valve #1.

d. After five (5) seconds, close Valve #2

e. After three (3) seconds, open Valve #3.

f. After ten (10) seconds, close Valve #3.

g. After three (3) seconds, open Valve #2.

h. After three (3) seconds, open Valve #1.

i. Repeat steps ***c*** through ***h*** 5000 times.

j. After a minimum of 5000 total cycles the assembly shall be tested to determine if the assembly complies with 10.2.5.3.3. Failure to comply with this test shall be cause for rejection.

k. The assembly shall then be disassembled and inspected for any damage or permanent deformation. Evidence of such shall be cause for rejection.

10.2.5.3.8

Purpose: To determine the overall compliance of the assembly with the Specifications as set forth in Section 10 of this Manual.

Requirement: All Sections of the Specifications of this Manual which are not specifically covered by the above tests shall be satisfactorily met.

Steps:

a. For each pertinent Section of the Specifications inspect, test or otherwise be assured that the assembly meets the minimum conditions of these Specifications.

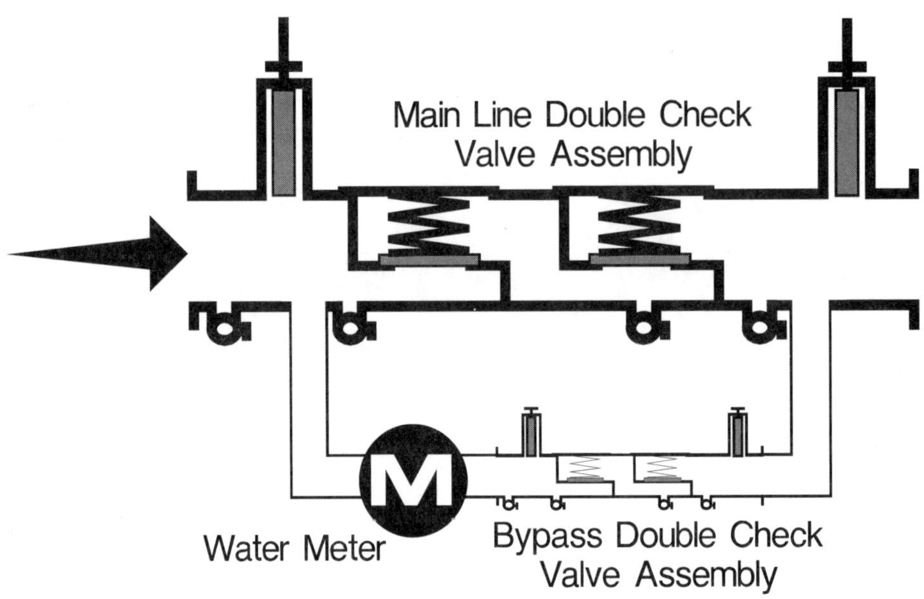

Fig. 10.7
Double Check-Detector Backflow Prevention Assembly (DCDA)

10.2.6 DESIGN, OPERATIONAL AND EVALUATION SPECIFICATIONS FOR DOUBLE CHECK — DETECTOR BACKFLOW PREVENTION ASSEMBLIES (DCDA)

10.2.6.1 Design and Operational Specifications

a. This assembly shall include a line-size approved double check valve assembly with a parallel water meter and parallel approved double check valve assembly (see Fig. 10.7). The purpose of this assembly is to provide double check valve protection for the distribution system and at the same time provide a telltale of the fire sprinkler system showing any system leakage or unauthorized use of water.

b. All low flow demands up to a minimum of 3 gpm (0.189 L/s) shall pass only through the bypass water meter and bypass double check valve assembly and be accurately recorded.

c. All flows above that of paragraph b) above, shall be permitted to pass through both the line-size double check valve assembly and bypass without accurate registration by or damage to the water meter.

d. The double check valve assemblies used for this complete assembly shall meet all of the specifications for double check valve assemblies as set forth in Section 10.2.3 of this Manual.

e. The bypass piping must attach to the line-size assembly body downstream of the No. 1 shutoff valve and upstream of the No. 2 shutoff valve. This is to allow for testing and maintenance of the bypass assembly.

f. The test cock No. 4 of the line-size assembly shall not be located on the bypass piping, and test cock No. 4 shall not be attached to the main-line body at the same location as the bypass piping.

10.2.6.2 Laboratory Evaluation

One assembly shall be inspected and evaluated in the following order. Unless otherwise noted, tests shall be performed at ambient temperature.

a. Conformance to the general design and material requirements outlined in Section 10.1.

b. Conformance of each individual double check valve assembly to the operational requirements outlined in Section 10.2.3.

c. Conformance to the working drawings and materials specifications.

d. Hydrostatic tests — the test assembly shall be subjected to the conditions of Section 10.1.1.5.

e. Pressure loss characteristics for flow rates up to the rated conditions.

f. Cross-over point from bypass to line-size assembly.

g. Thermal Test — the assembly shall be evaluated at the greater of 140°F (60°C) or the maximum working water temperature (MWWT), maximum working water pressure (MWWP), and specified flow rate.

10.2.6.3 Evaluation Procedure

10.2.6.3.1

Purpose: To determine the capability of the assembly to withstand the required hydrostatic test pressure.

Requirement: All components of the assembly shall withstand a hydrostatic pressure of twice (2×) the maximum working water pressure (MWWP), as stated by the manufacturer, for a minimum period of ten (10) minutes without any damage, permanent deformation or impairment of operation.

Steps:

a. With both shutoff valves closed on the line-size assembly and all of the air bled from the assembly, a hydrostatic pressure of twice (2×) the maximum working water pressure shall be supplied through the No. 2 test cock of the line-size assembly for a minimum period of ten (10) minutes. Any evidence of leakage shall be cause for rejection.

b. The assemblies shall be disassembled and inspected for any damage, permanent deformation, or impairment of operation. Evidence of such shall be cause for rejection.

10.2.6.3.2

Purpose: To determine the overall pressure loss of the complete assembly as a function of the rate of flow.

Requirement: At any rate of flow from zero (static) up to and including the rated flow the overall pressure loss shall not exceed the values shown in Table 10-1 for double check valve assemblies. The assembly shall withstand a minimum of 200 percent of the rated flow without damage, permanent deformation, or impairment of operation.

Steps:

a. Install the assembly in a suitable hydraulic test line having a flow capacity in excess of the rated flow conditions (see Table 10-1) and an available line pressure of at least 150 psi (1034 KPa).

b. Install in the supply line at a suitable location upstream from the assembly a piezometer ring; and, install a similar piezometer ring at a suitable location downstream from the assembly (located as indicated in ANSI/ISA Standard S75.02). Then install a mercury manometer, or other suitable means of measuring pressure differentials between the two piezometer rings to measure the observed overall pressure loss at various rates of flow.

c. Record differential pressure between piezometer rings at static condition. Increase flow rate and record steady state differential pressure at a sufficient number of increments to fully define the flow curve characteristic. Data shall be taken for both increasing and decreasing flow conditions. The flow rate shall then be increased to 200% of the rated flow for five (5) minutes, then returned to static.

Manifold Assembly (Section 10.1.2.19) Only: Close all shutoff valves so that only one of the individual line size assemblies of the manifold assembly is open. Gradually increase the flow of water through the individual assembly until the rated flow of the manifold assembly is reached, then return to static condition. Repeat flow test for each individual line size assembly contained in the manifold assembly.

d. Remove the assembly from the line and couple the downstream piping directly to the supply piping. Then observe data for a friction head correction curve for the pipe fittings required to adapt the test assembly to the supply line between the existing piezometer rings. The data of this correction curve are then subtracted from the observed overall pressure loss data of the assembly from the upstream face of the upstream shutoff valve to the downstream face of the downstream shutoff valve.

e. Exceeding the maximum allowable pressure loss listed in Table 10-1 for double check valve assemblies for any rate of flow up to and including rated flow, shall be cause for rejection. The assembly shall be disassembled and inspected for any damage, permanent deformation, or impairment of operation. Evidence of such shall be cause for rejection.

10.2.6.3.3

Purpose: To determine the flow rate at which the line-size double check valve assembly begins to open.

Requirement: All low flow demands up to a minimum of 3 gpm (0.189 L/s) shall pass only through the bypass meter and bypass double check valve assembly.

Steps:

a. Install the assembly in a suitable hydraulic test line with an available line pressure of at least 150 psi (1034 KPa).

b. Attach a control valve to the discharge end of the double check-detector assembly, so that it discharges into a weigh tank or other suitable means of measuring flow rate.

c. Open the control valve until approximately 1 gpm (0.06 L/s) is discharging into the weigh tank, or other means of measuring flow rate. Determine the flow rate through the bypass meter and the flow rate through the entire assembly.

d. Increase the flow rate through the control valve at small increments, and at each increase determine what the flow rates are through the bypass meter and the entire assembly. When the total flow rate into the weigh tank, or other means of measuring flow rate, exceeds the flow rate through the bypass meter, then the line-size assembly has begun to open. This *cross-over point* must be equal to or greater than 3 gpm (0.189 L/s).

e. Failure to cross-over at 3 gpm or above shall be cause for rejection.

10.2.6.3.4

Purpose: To determine if all components of the assembly shall operate properly under rated temperature and pressure conditions, and flow rates as specified in Table 10-7.

Requirements: All components of the assembly shall operate at, and withstand a thermal test at the greater of 140°F (60°C) or at the maximum working water temperature (MWWT), maximum working water pressure MWWP, and the flow rate as specified in Table 10-7, for a minimum period of one hundred (100) hours without any damage, permanent deformation or impairment of operation.

Steps:

a. Install assembly into suitable test line which can generate the following minimum parameters:

   Maximum working water pressure *{minimum of 150 psi (1034 KPa)}*
   Maximum working water temperature *{minimum of 140°F (60°C)}*
   Flow requirements per Table 10-7

b. The dimensions and durometer hardness of all elastomer components shall be inspected and recorded.

c. Set controls to the rated temperature and pressure conditions of the assembly as well as rate of flow as specified in Table 10-7.

d. The temperature and pressure conditions shall be continuously monitored and recorded.

e. During and after a minimum of one hundred (100) hours at rated temperature and pressure the assembly shall be tested to determine if the assembly operates satisfactorily per 10.2.3.3.3 and 10.2.3.3.4. Failure to comply with these tests shall be cause for rejection

f. The assembly shall then be disassembled and inspected for any damage or permanent deformation. Evidence of such shall be cause for rejection.

g. After reaching ambient temperature, the assembly shall then be tested per Section 10.2.3.3.1. Failure to comply with this test shall be cause for rejection.

10.2.6.3.5

Purpose: To determine the overall compliance of the assembly with the Specifications as set forth in Section 10 of this Manual.

Requirement: All Sections of the Specifications of this Manual which are not specifically covered by the above tests shall be satisfactorily met.

Steps:

For each pertinent Section of the Specifications inspect, test or otherwise be assured that the assembly meets the minimum conditions of these Specifications.

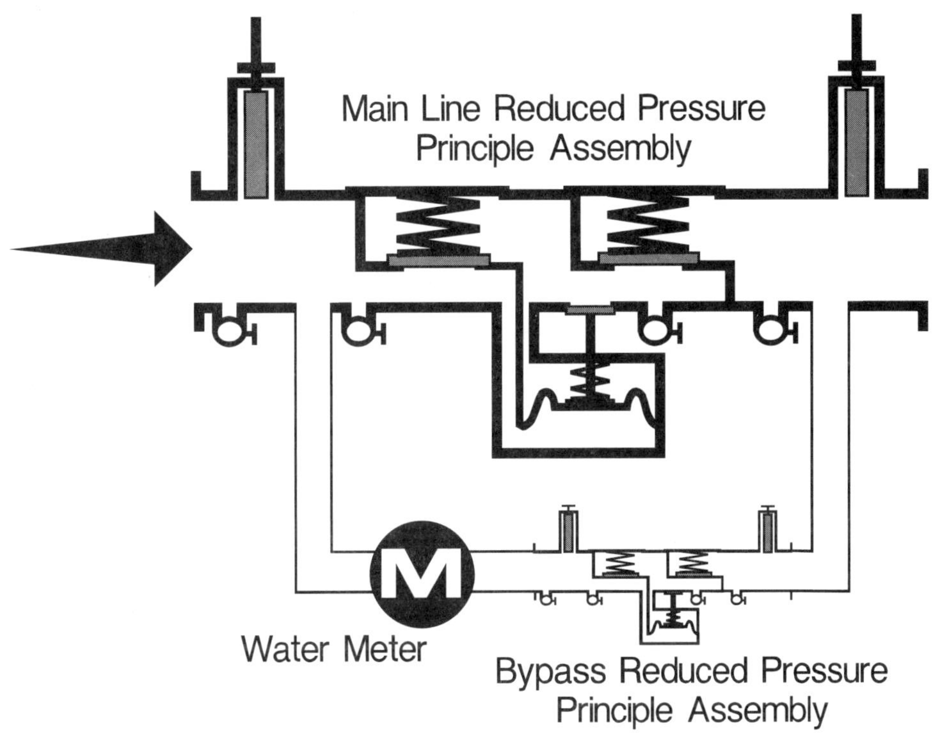

Fig. 10.8

Reduced Pressure Principle-Detector Backflow Prevention Assembly (RPDA)

10.2.7 DESIGN, OPERATIONAL AND EVALUATION SPECIFICATIONS FOR REDUCED PRESSURE PRINCIPLE-DETECTOR BACKFLOW PREVENTION ASSEMBLIES (RPDA)

10.2.7.1 Design and Operational Specifications

a. This assembly shall include a line-size approved reduced pressure principle backflow prevention assembly with a parallel water meter and parallel approved reduced pressure principle backflow prevention assembly (see Fig. 10.8). The purpose of this assembly is to provide reduced pressure principle protection for the distribution system and at the same time provide a telltale of the fire sprinkler system showing any system leakage or unauthorized use of water.

b. All low flow demands up to a minimum of 3 gpm (0.189 L/s) shall pass only through the bypass water meter and bypass reduced pressure principle backflow prevention assembly and be accurately recorded.

c. All flows above that of paragraph b) above, shall be permitted to pass through both the line-size reduced pressure principle backflow prevention assembly and bypass without accurate registration by or damage to the water meter.

d. The reduced pressure principle backflow prevention assemblies used for this complete assembly shall meet all of the specifications for reduced pressure principle backflow prevention assemblies as set forth in Section 10.2.2 of this Manual. The maximum allowable pressure loss through the RPDA shall be two (2) psi (13.78 KPa) above the values stated for the reduced pressure principle backflow prevention assemblies in Table 10-1 for each specific size.

e. The bypass piping shall attach to the line-size assembly body downstream of the No. 1 shutoff valve and upstream of the No. 2 shutoff valve. This is to allow for testing and maintenance of the bypass assembly.

f. Test cock No. 4 of the line-size assembly shall not be located on the bypass piping, and test cock No. 4 shall not be attached to the main-line body at the same location as the bypass piping.

10.2.7.2 Laboratory Evaluation

One assembly shall be inspected and evaluated in the following order. Unless otherwise noted, tests shall be performed at ambient temperature.

a. Conformance to the general design and material requirements outlined in Section 10.1.

b. Conformance of each individual reduced pressure principle assembly to the operational requirements outlined in Section 10.2.2.

c. Conformance to the working drawings and materials specifications.

d. Hydrostatic tests — the test assembly shall be subjected to the conditions of Section 10.1.1.5.

e. Pressure loss characteristics for flow rates up to the rated conditions.

f. Cross-over point from bypass to line-size assembly.

g. Thermal test — the assembly shall be evaluated at the greater of 140°F (60°C) or the maximum working water temperature (MWWT), maximum working water pressure (MWWP), and specified flow rate.

10.2.7.3 Evaluation Procedure

10.2.7.3.1

Purpose: To determine the capability of the assembly to withstand the required hydrostatic test pressure.

Requirement: All components of the assembly shall withstand a hydrostatic pressure of twice (2×) the maximum working water pressure (MWWP), as stated by the manufacturer, for a minimum period of ten (10) minutes without any damage, permanent deformation or impairment of operation.

Steps:

a. With both shutoff valves closed on the line-size assembly and all of the air bled from the assembly, a hydrostatic pressure of twice (2×) the maximum working water pressure shall be supplied through the No. 2 test cock for a minimum period of ten (10) minutes. Any evidence of leakage shall be cause for rejection.

b. The assembly shall be disassembled and inspected for any damage, permanent deformation, or impairment of operation. Evidence of such shall be cause for rejection.

10.2.7.3.2

Purpose: To determine the overall pressure loss of the complete assembly as a function of the rate of flow.

Requirement: At any rate of flow from zero (static) up to and including the rated flow the overall pressure loss shall not exceed the values detailed in Section 10.2.7.1d. The assembly shall withstand a minimum of 200 percent of the rated flow without damage, permanent deformation, or impairment of operation.

Steps:

a. Install the assembly in a suitable hydraulic test line having a flow capacity in excess of the rated flow conditions (see Table 10-1) and an available line pressure of at least 150 psi (1034 KPa).

b. Install in the supply line at a suitable location upstream from the assembly a piezometer ring; and, install a similar piezometer ring at a suitable location downstream from the assembly (located as indicated in ANSI/ISA Standard S75.02). Then install a mercury manometer, or other suitable means of measuring pressure differentials between the two piezometer rings to measure the observed overall pressure loss at various rates of flow.

c. Record differential pressure between piezometer rings at static condition. Increase flow rate and record steady state differential pressure at a sufficient number of increments to fully define flow curve characteristic. Data shall be taken for both increasing and decreasing flow conditions. The flow rate shall then be increased to 200% of the rated flow for five (5) minutes, then returned to static.

Manifold Assembly (Section 10.1.2.19) Only: Close all shutoff valves so that only one of the individual line-size assemblies of the manifold assembly is open. Gradually increase the flow of water through the individual assembly until the rated flow of the manifold assembly is reached, then return to static condition. Repeat flow test for each individual line-size assembly contained in the manifold assembly.

d. Remove the assembly from the line and couple the downstream piping directly to the supply piping. Then observe data for a friction head correction curve for the pipe fittings required to adapt the test assembly to the supply line between the existing piezometer rings. The data of this correction curve are then subtracted from the observed overall pressure loss data of the assembly from the upstream face of the upstream shutoff valve to the downstream face of the downstream shutoff valve.

e. Exceeding the maximum allowable pressure loss detailed in Section 10.2.7.1d for any rate of flow up to and including rated flow, or discharge from either differential pressure relief valve, shall be cause for rejection. The assembly shall be disassembled and inspected for any damage, permanent deformation, or impairment of operation. Evidence of such shall be cause for rejection.

10.2.7.3.3

Purpose: To determine the flow rate at which the line-size reduced pressure principle backflow prevention assembly begins to open.

Requirement: All low flow demands up to a minimum of 3 gpm (0.189 L/s) shall pass only through the bypass meter and bypass reduced pressure principle backflow prevention assembly.

Steps:

a. Install the assembly in a suitable hydraulic test line with an available line pressure of at least 150 psi (1034 KPa).

b. Attach a control valve to the discharge end of the reduced pressure principle-detector assembly, so that it discharges into a weigh tank or other suitable means of measuring flow rate.

c. Open the control valve until approximately 1 gpm (0.06 L/s) is discharging into the weigh tank, or other means of measuring flow rate. Determine the flow rate through the bypass meter and the flow rate through the entire assembly.

d. Increase the flow rate through the control valve at small increments, and at each increase determine what the flow rates are through the bypass meter and the entire assembly. When the total flow rate into the weigh tank, or other suitable means of measuring flow rate, exceeds the flow rate through the bypass meter, then the line-size assembly has begun to open. This cross-over point must be equal to or greater than 3 gpm (0.189 L/s).

e. Failure to cross-over above 3 gpm shall be cause for rejection.

10.2.7.3.4

Purpose: To determine if all components of the assembly shall operate properly under rated temperature and pressure conditions, and flow rates as specified in Table 10-7.

Requirements: All components of the assembly shall operate at, and withstand a thermal test at the greater of 140°F (60°C) or at the maximum working water temperature (MWWT), maximum working water pressure (MWWP), and the flow rate as specified in Table 10-7, for a minimum period of one hundred (100) hours without any leakage, damage, permanent deformation or impairment of operation.

Steps:

a. Install assembly into suitable test line which can generate the following minimum parameters:

   Maximum working water pressure *{minimum of 150 psi (1034 KPa)}*
   Maximum working water temperature *{minimum of 140°F (60°C)}*
   Flow requirements per Table 10-7

b. The dimensions and durometer hardness of all elastomer components shall be inspected and recorded.

c. Set controls to the rated temperature and pressure conditions of the assembly as well as rate of flow as specified in Table 10-7.

d. The temperature and pressure conditions shall be continuously monitored and recorded.

e. During and after a minimum of one hundred (100) hours at rated temperature and pressure the assembly shall be tested to determine if the assembly operates satisfactorily per 10.2.2.3.3, 10.2.2.3.5, and 10.2.2.3.6. Failure to comply with these tests shall be cause for rejection.

f. The assembly shall then be disassembled and inspected for any damage or permanent deformation. Evidence of such shall be cause for rejection.

g. Once the assembly returns to ambient temperature conditions, the assembly shall then be tested per Section 10.2.2.3.1. Failure to comply with this test shall be cause for rejection.

10.2.7.3.5

Purpose: To determine the overall compliance of the assembly with the Specifications as set forth in Section 10 of this Manual.

Requirement: All Sections of the Specifications of this Manual which are not specifically covered by the above tests shall be satisfactorily met.

Steps:

For each pertinent Section of the Specifications inspect, test or otherwise be assured that the assembly meets the minimum conditions of these Specifications.

TABLE 10-7

THERMAL TEST MINIMUM FLOW RATES

Size of Assembly		Minimum Flow Rate	
Inch(es)	(mm)	GPM	(L/s)
1/4	(6)	1	(0.06)
3/8	(9)	3	(0.19)
1/2	(12)	7.5	(0.47)
5/8×3/4	(16×20)	20	(1.26)
3/4 - 1	(20 - 25)	30	(1.89)
1-1/4 - 2	(31-50)	40	(2.52)
2-1/2 - 16	(63 - 406)	50	(3.15)

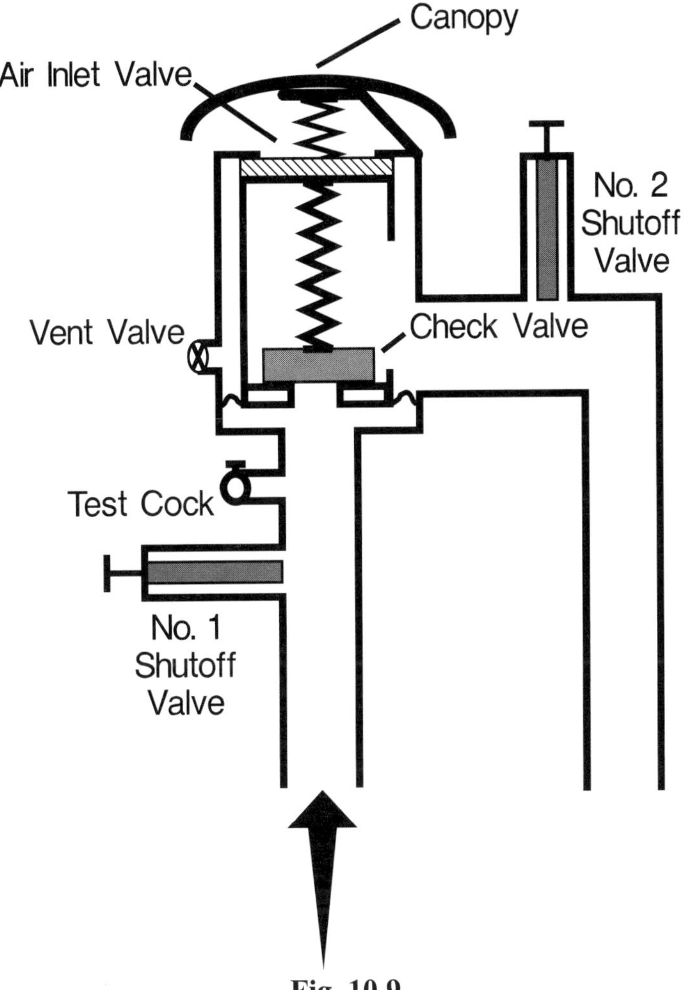

Fig. 10.9
Spill-Resistant Pressure Vacuum Breaker Backsiphonage Prevention Assembly (SVB)

10.2.8 DESIGN, OPERATIONAL AND EVALUATION SPECIFICATIONS FOR SPILL-RESISTANT PRESSURE VACUUM BREAKER BACKSIPHONAGE PREVENTION ASSEMBLIES (SVB)

10.2.8.1 Design and Operational Specifications

a. This assembly shall include an internally loaded check valve and a loaded air inlet valve opening to atmosphere on the discharge side of the check valve between two tightly closing resilient seated shutoff valves; and, will include one properly located resilient seated test cock, and one vent valve. (See Fig. 10.9.)

b. The air inlet valve of the spill-resistant vacuum breaker shall open when the inlet pressure is 1 psi (6.89 KPa) or greater, and the outlet pressure is atmospheric. The air inlet valve shall be fully open when the inlet pressure is atmospheric and the outlet pressure is atmospheric.

c. The maximum allowable pressure drop across the assembly, from the upstream face of the No. 1 shutoff valve to the downstream face of the No. 2 shutoff valve, shall not exceed 10 psi (68.94 KPa) for any rate of flow up to and including the rated flow listed in Table 10-5.

d. The check valve shall be internally loaded and shall at all times be drip-tight in the normal direction of flow with the inlet pressure at 1 psi (6.89 KPa) and the outlet under atmospheric pressure.

e. The effective size of the air inlet port(s) of the assembly shall be governed by the vacuum dissipation test. If an air inlet port shield or canopy is used, it shall extend down around the body of the assembly to the lowest portion of the port(s). To reduce potential for fouling, the minimum clearance between the air inlet port(s) and the shield or canopy shall be 3/16 inch (5 mm).

f. There shall be no water leakage from the air inlet port(s) when the assembly is pressurized from atmospheric to working pressure.

10.2.8.2 Laboratory Evaluation

One assembly shall be inspected and evaluated in the following order. Unless otherwise noted, tests shall be performed at ambient temperatures.

a. Conformance to general, design and material requirements outlined in Section 10.1.

b. Conformance to the operational requirements outlined in Section 10.2.

c. Conformance to the working drawings and materials specifications.

d. Hydrostatic tests — the test assembly shall be subjected to the conditions of Section 10.1.1.5.

e. Pressure loss characteristics for flow rates up to the rated conditions set forth in Table 10-5.

f. Conformance to the air inlet requirements.

g. Static closing point of the check valve.

h. Thermal Test — the assembly shall be evaluated at the greater of 140°F (60°C), or the maximum working water temperature (MWWT), maximum working water pressure (MWWP), and specified flow rate.

i. Life cycle test.

10.2.8.3 Evaluation Procedure

10.2.8.3.1

Purpose: To determine the capability of the assembly to withstand the required hydrostatic test pressure.

Requirement: All components of the assembly shall withstand a hydrostatic pressure of twice (2×) the maximum working water pressure (MWWP), as stated by the manufacturer, for a minimum period of ten (10) minutes without any damage, permanent deformation or impairment of operation.

Steps:

a. With both shutoff valves closed and all of the air bled from the assembly, a hydrostatic pressure of twice (2×) the maximum working water pressure shall be supplied through the No. 1 test cock for a minimum period of ten (10) minutes. Any evidence of leakage shall be cause for rejection.

b. The required hydrostatic pressure shall be supplied through shutoff valve No. 2 with test cock No. 1 open to atmosphere for a minimum period of ten (10) minutes. Any evidence of leakage shall be cause for rejection.

c. The assembly shall be disassembled and inspected for any damage, permanent deformation, or impairment of operation. Evidence of such shall be cause for rejection.

10.2.8.3.2

Purpose: To determine the overall pressure loss of the complete assembly as a function of the rate of flow.

Requirement: At any rate of flow from zero (static) up to and including the rated flow the overall pressure loss shall not exceed the values shown in Table 10-5. The assembly shall withstand a minimum of 200 percent of the rated flow without damage, permanent deformation, or impairment of operation.

Steps:

a. Install the assembly in a suitable hydraulic test line having a flow capacity in excess of the rated flow conditions (see Table 10-5) and an available line pressure of at least 150 psi (1034 KPa).

b. Install, in the supply line at a suitable location upstream from the assembly, a piezometer ring; and, install a similar piezometer ring at a suitable location downstream from the assembly (located as indicated in ANSI/ISA Standard S75.02). Then install a mercury manometer, or other suitable means of measuring pressure differentials between the two piezometer rings to measure the observed overall pressure loss at various rates of flow.

c. Record differential pressures between piezometer rings at static condition. Increase flow rate and record steady state differential pressure at a sufficient number of increments to fully define flow curve characteristic. Data shall be taken for both increasing and decreasing flow conditions. The flow rate shall then be increased to 200 percent of the rated flow for five (5) minutes, then returned to static.

d. Remove the assembly from the line and couple the downstream piping directly to the supply piping. Then observe data for a friction head correction curve for the pipe fittings required to adapt the test assembly to the supply line between the existing piezometer rings. The data of this correction curve are then subtracted from the observed overall pressure loss data of the assembly from the upstream face of the upstream shutoff valve to the downstream face of the downstream shutoff valve.

e. Exceeding the maximum allowable pressure loss listed in Table 10-5 for any rate of flow up to and including rated flow, shall be cause for rejection. The assembly shall be disassembled and inspected for any damage, permanent deformation, or impairment of operation. Evidence of such shall be cause for rejection.

10.2.8.3.3

Purpose: To test the opening pressure differential of the air inlet valve.

Requirement: The air inlet valve shall open when the inlet pressure is a minimum of 1.0 psi (6.89 KPa) above atmospheric pressure, and the outlet is atmospheric. The air inlet valve shall be fully open when inlet pressure is at atmospheric, and the outlet is atmospheric.

Steps:

a. Install the assembly in a suitable hydraulic test line with a minimum pressure of 150 psi (1034 KPa). Remove air inlet valve canopy.

b. Install the high side hose of the differential pressure gage, or other suitable means of measuring pressure differentials, to the test cock, open the test cock, and bleed air from the hose and gage.

c. Close shutoff valve No. 2, and then close shutoff valve No. 1.

d. Open the vent valve to reduce the outlet pressure to atmospheric.

e. Slowly (0.2 psid/ second maximum) open the high side bleed needle valve on the gage. Record the pressure differential at which the air inlet valve opens. Failure to open at a value of 1.0 psi (6.89KPa) or greater shall be cause for rejection.

f. Allow water to drain out of the test cock until the inlet pressure is atmospheric, to determine if the air inlet valve opens fully. Failure to open fully shall be cause for rejection.

10.2.8.3.4

Purpose: To determine the static pressure drop across the check valve.

Requirement: The check valve shall be drip-tight in the normal direction of flow when the inlet pressure is at least 1 psi (6.89 KPa) and the outlet pressure is atmospheric.

Steps:

a. Install a transparent tube approximately 6 feet (180 cm) long on the test cock. Open the test cock to fill the tube with water, then close the test cock.

b. Close shutoff valve No. 2, then close shutoff valve No. 1.

c. Open vent valve, then open the test cock. The water in the tube will fall until closure of the check valve is attained. When no further fall of the water in the tube is observed, record the height of water in the transparent tube above the centerline of the vent valve .

d. Failure to maintain a minimum level of 27 3/4 inches (705 mm) shall be cause for rejection.

10.2.8.3.5

Purpose: To compare the air flow capacity of the air inlet valve to that of the effective water through-way of the assembly.

Requirement: The time required to dissipate a vacuum 25 to 5 inches of mercury vacuum through the air inlet valve shall be less than the time required to dissipate the same vacuum through the water through-way.

Steps:

a. Install the assembly in a normal operating position, with the outlet of the assembly attached to a vacuum source.

b. Close No. 1 shutoff valve to simulate the check valve being closed. (No. 2 shutoff valve remains fully open during entire test procedure.)

c. With the air inlet valve open, record the time required to dissipate the vacuum from 25 to 5 inches of mercury vacuum when this flow is controlled by means of a quick opening valve. Repeat this step a minimum of three (3) times, then average the recorded times.

d. Open No. 1 shutoff valve and attach a 12 inch (30.5 cm) long pipe nipple, the same nominal pipe diameter as the assembly, to the inlet of the No. 1 shutoff valve.

e. Hold the air inlet valve closed, then record the time required to dissipate the vacuum from 25 to 5 inches of mercury vacuum when this flow is controlled by means of a quick opening valve. Repeat this step a minimum of three (3) times, then average the recorded times.

f. Based upon the average of at least three (3) tests, the time required for step ***c*** shall be less than the time required for step ***e***. Failure to comply with this requirement shall be cause for rejection.

10.2.8.3.6

Purpose: To test that a backsiphonage condition together with a fouled check valve shall not allow a carry-over of downstream water back into the inlet.

Requirement: Water shall not be backsiphoned more than three (3) inches (76 mm) above the free surface of water in the discharge pipe standing six (6) inches (152 mm) below the bottom or critical level line of the assembly.

Steps:

a. Install the assembly in a normal operating position, with the inlet of the assembly connected to the vacuum source and the discharge side equipped with a transparent tube that extends down into a vessel of water where the free surface of the water is six (6) inches (152 mm) below the bottom or critical level line of the assembly.

b. A wire of the size shown in Table 10-6 shall be placed across the seating surface in the lower quadrant of a hinged or horizontally moving poppet check valve, or at a single point of a vertical moving check valve. The wire shall be placed so that the entire width of the seating contact surface is covered.

c. By the application of vacuum conditions ranging from zero to 25 inches of mercury under both steady flow and instantaneous conditions, the water in the discharge tube shall not rise more than three (3) inches (76 mm) above the original free surface of the water in the vessel. Evidence of such shall be cause for rejection.

10.2.8.3.7

Purpose: To determine if all components of the assembly shall operate properly under rated temperature and pressure conditions, and flow rates as specified in Table 10-7.

Requirements: All components of the assembly shall operate at, and withstand a thermal test at the greater of 140°F (60°C) or at the maximum working water temperature (MWWT), maximum working water pressure (MWWP), and flow rate as specified in Table 10-7 for a minimum period of one hundred (100) hours without any damage, permanent deformation or impairment of operation.

Steps:

a. Install assembly into suitable test line which can generate the following minimum parameters:

   Maximum working water pressure *{minimum of 150 psi (1034 KPa)}*
   Maximum working water temperature *{minimum of 140°F (60°C)}*
   Flow requirements per Table 10-7

b. The dimensions and durometer hardness of all elastomer components shall be inspected and recorded.

c. Set controls to the rated temperature and pressure conditions of the assembly, as well as rate of flow as specified in Table 10-7.

d. The temperature and pressure conditions shall be continuously monitored and recorded.

e. During and after a minimum of one hundred (100) hours at rated temperature and pressure the assembly shall be tested to determine if the assembly operates satisfactorily per 10.2.8.3.3 and 10.2.8.3.4. Failure to comply with these tests shall be cause for rejection.

f. The assembly shall then be disassembled and inspected for any internal damage or permanent deformation. Evidence of such shall be cause for rejection.

g. Once the assembly returns to ambient temperature conditions. The assembly shall then be tested per Section 10.2.8.3.1. Failure to comply with this test shall be cause for rejection.

10.2.8.3.8

Purpose: To determine if the assembly sustains any damage, permanent deformation, or impairment of operation following the specified cycles.

Requirement: The assembly shall withstand the specified cycle without leakage, damage, permanent deformation, or impairment of operation.

Steps:

a. Install assembly into suitable test line which can generate the following parameters:

   Rated flow of assembly under test (per Table 10-5)
   Line pressure of 60 psi ± 10 psi (414 KPa ± 69 KPa)
   Temperature of 110°F $^{+30}/_{-0}$ °F (43°C. $^{+17}/_{-0}$ °C)
   Control valves capable of closing/opening within five (5) seconds. *
   *(*Resulting rise of pressure shall not exceed MWWP of the assembly.*)

b. With Valve #3 (Valve #3A closed and Valve #3B open) and Valve #4 (see Fig. 10.3) in the closed position, and Valve #1 and Valve #2 in the open position, establish flow through the assembly equal to 25 percent of the maximum rated flow per Table 10-5.

c. Maintain the flow through the assembly for a minimum of six (6) seconds. Close Valve #1.

d. After five (5) seconds, close Valve #2.

e. After three (3) seconds, open Valve #3.

f. After three (3) seconds, open Valve #4. Maintain Valve #4 open for six (6) seconds. Close Valve #4.

g. After two (2) seconds, close Valve #3.

h. After three (3) seconds, open Valve #2.

i. After two (2) seconds open Valve #1.

j. Repeat steps ***c*** through ***i*** 1250 times.

k. Raise flow through the assembly to 50 percent of rated flow. Repeat steps ***c*** through ***i*** 1250 times.

l. Raise flow through the assembly to 75 percent of rated flow. Repeat steps ***c*** through ***i*** 1250 times.

m. Raise flow through the assembly to 100 percent of rated flow. Repeat steps ***c*** through ***i*** 1250 times.

n. After a minimum of 5000 total cycles the assembly shall be tested to determine if the assembly complies with 10.2.8.3.3, and 10.2.8.3.4. Failure to comply with these tests shall be cause for rejection.

o. The assembly shall then be disassembled and inspected for any damage or permanent deformation. Evidence of such shall be cause for rejection.

10.2.8.3.9

Purpose: To determine the overall compliance of the assembly with the Specifications as set forth in Section 10 of this Manual.

Requirement: That all Sections of the Specifications of this Manual which are not specifically covered by the above tests are satisfactorily met.

Steps:

a. For each pertinent Section of the Specifications inspect, test or otherwise be assured that the assembly meets the minimum conditions of these Specifications.

SECTION 11

Summary of Case Histories

These figures represent only a small percentage of the total number of incidents that have actually occurred. Only those incidents which have been submitted for inclusion in this Section with documentation are included.

Date	City & State	Cause of Contamination	Location	Organization	Type of Service	Source of Data
—	Eau Claire, WI	Cross-Connections	—	—	Not given	—
—	Nashua, NH	Cross-Connections	—	—	Not given	—
—	Piedmont, WV	Cross-Connections	—	—	Not given	—
1903	Lawrence, MA	Cross-Connections	—	—	Not given	—
1903	Lowell, MA	Cross-Connections	—	—	Not given	—
1907	Hyde Park, MA	Cross-Connections	—	—	Not given	—
1908	Auburn, NY	Cross-Connections	—	—	Not given	*1
1910	Rochester, NY	Cross-Connections	—	—	Not given	—
1911	Springfield, OH	Cross-Connections	—	—	Not given	—
1912	Milwaukee, WI	Cross-Connections	—	—	Not given	—
1913	Philadelphia, PA	Cross-Connections	—	—	Not given	—
1914	Circleville, OH	Cross-Connections	—	—	Not given	—
1914	St. Paul, MN	Cross-Connections	—	—	Not given	—
1916	Bluff Points, NY	Cross-Connections	—	—	Not given	—
1916	Elgin, WI	Cross-Connections	—	—	Not given	--
1916	Mannington, WV	Cross-Connections	—	—	Not given	—
1918	Hartford, CT	Cross-Connections	—	—	Not given	—
1918	Minneapolis, MN	Cross-Connections	—	—	Not given	—
1918	Newburgh, NY	Cross-Connections	—	—	Not given	—
1920	Bloomington, IL	Cross-Connections	—	—	Not given	—
1920	Maywood, IL	Cross-Connections	—	—	Not given	—
1920	Philadelphia Navy Yd	Cross-Connections	—	—	Not given	—
Sept 1920	Neenah, WI	Cross-Connection in Industrial Plant	Factory	—	Deep Well	*2
Oct 1920	Amsterdam, NY	X-Conn to polluted creek	County	--	River	*3
1922	Franklin, NJ	Single Check Valve failed	Town	—	River	*5
1923	Binghamton, NY	Cross-Connections	—	—	Not given	--
May 1923	Rockaway, NJ	X-Conn with raw river water	City	—	Deep Well	*5
July 1923	Everett, WA	X-Conn, City with Mill fire system	Factory	City Water	River	*6
Apr 1923	Oakland, CA	Fire Dept pumped salt water into	Factory	Calif Cotton	Public	
1923	Wausau, WI	X-Conn with contaminated on-site well	Factory	—	Deep Well	*2
Dec 1923	Ft. Wayne, IN	X-Conn with river supply	City	—	Deep Well	—
1924	Rockford, IL	Cross-Connections	—	—	Not Given	—

Date	City & State	Cause of Contamination	Location	Organization	Type of Service	Source of Data
1924	Sterling, IL and Rockfalls, IL	Cross-Connections	—	—	Not Given	—
Jan 1924	Bloomington, IL	X-Conn with contaminated industrial on-site supply	Town	—	Deep Well	*9
June 1924	Wausau, WI	X-Conn with contaminated industrial on-site supply	Factory	—	Deep Well	*2
Sept 1924	Newport, VT	Cross-Connection	Town	—	Lake	*10
1924	Oakland, CA	X-Conn in plant pumped molasses into city mains	Factory	Albers Milling	Public	*7
1925	Niagara Falls, NY	Cross-Connection	—	—	Not given	—
1925	Sheboygan, WI	Cross-Connection	—	—	Not given	—
Jan 1925	Yonkers, NY	Cross-Connection	Town	—	River	*11
Feb 1925	Sterling, IL	X-Conn with contaminated factory on-site supply	Factory	—	Rock well	*9
July 1925	Winona Lake, IN	X-Conn between Town mains and contaminated lake	Town	—	Deep Well	*8
Oct 1925	Eco City, MI	Temporary X-Conn with contaminated stream	Town	—	Well	*12
Jan 1926	Woodbury & 2 adj townships	X-Conn with contaminated surface waters	City	—	Deep Well	*3
Mar 1926	Ada, OK	X-Conn between city and industrial line from contaminated lake	Plant	—	3 wells	*13
May 1926	Tiverton, NJ	Cross-Connections	Town	—	no info	*14
Aug 1926	Richmond, CA	High pressure salt water forced into city mains	Oil Company	Std.Oil Cal	Public	*7
Dec 1926	Rochester, NY	X-Conn in Department store with fire sprinkler supply	Store	—	Lake	*15
1927	Western PA	Cross-Connection	Town	—	River	*16
Feb 1927	Beaverton, OR	Cross-Connection	Town	—	River	*13
Aug 1927	Oswego, NY	X-Conn to aux. river supply	Factory	—	Lake	*17
Sept 1927	Cohoes, NY	X-Conn to aux. river supply	Factory	—	River	*15
Oct 1927	Albany, NY	X-Conn to aux. river supply	Factory	—	River	*15
Jan 1929	Two Rivers, WI	X-Conn in plant	Industrial Plant	—	Lake	*2
Mar 1929	Ft. Wayne, IN	X-Conn with contaminated on-site reservoir	Convent	—	Deep Well	*18

Date	City & State	Cause of Contamination	Location	Organization	Type of Service	Source of Data
June 1929	So. St. Paul, MN	Cross-Connection	Plant	—	River	*19
1929	Ft. Wayne, PA	X-Conn to contaminated Railroad	City	—	City	*20
1929	Oakland, CA	X-Conn with high pressure triplex pumps forced soap mix into mains	Garage	Stadium	Public	*7
Jan 1930	Oakland, CA	X-Conn with on-site high pressure well supply	Dry Dock	Moore Co.	Public	*7
Feb 1930	Oakland, CA	X-Conn with liquified pigment under high pressure	Chemical	—	Public	*7
Aug 1930	Milford, DE	X-Conn to contaminated stream	Town	—	Deep Well	*21
Dec 1930	Lawrence, MA	X-Conn to contaminated, private fire sprinkler supply	City	—	River	*12
June 1932	Lapeer, MI	X-Conn with contaminated creek	Town	—	Rock well	*12
June 1933	Chicago, IL	X-Conn with contaminated fire sprinkler system	Plant	—	Lake Michigan	*22
June 1933	Chicago, IL	X-Conn with sewer system	Hotel	—	Lake Michigan	*22
July 1933	Marseilles, IL	Cross-Connection within camp	Camp	—	Rock Well	*22
Dec 1933	Chicago, IL	X-Conn with sewer in hotel	Hotel	McCormak	Lake Michigan	*22
Dec 1933	Chicago, IL	X-Conn with contaminated fire sprinkler system	Factory	—	Lake Michigan	*22
Jan 1934	Winooski, VT	8" X-Conn within woolen mill	Factory	—	Spring	*10
May 1934	Brentwood, CA	Backsiphonage	Home	Mrs. Briggs	City	*20
Dec 1934	Burbank, CA	X-Conn to boiler feed water	Winery	McClure	City	*20
May 1935	Worcester, MA	Cross-Connection	City	—	Shallow reservoir	*12
June 1935	North Adams, MA	X-Conn with contaminated private supply	Town	—	Shallow reservoir	*12
July 1935	Indianapolis, IN	Check valve on pump discharge failed	Town	—	Dug well	*23
1936	Pomona, CA	Several Cross-Connections on campus	College	Pomona	City	*20
Jan 1936	Oakland, CA	Ship pumped salt water into potable water system	Railroad yard	Santa Fe	Public	*7
Juky 1936	India Lake, NY	X-Conn with contaminated lake	Town	—	Spring, Lake	*17
Aug 1936	Schroon, NY	X-Conn within Hotel	Hotel	—	Spring	*24
Sept 1936	Los Angeles, CA	Backsiphonage	Apartment House	Nirvana	City	*20
Dec 1936	Cleveland, MS	X-Conn on construction project	Road Building	Carey-Reed Const. Co.	City	*20

Date	City & State	Cause of Contamination	Location	Organization	Type of Service	Source of Data
June 1938	Camarillo, CA	Backsiphonage from irrigation ditch	Building Project	Phillip Murphy Corp.	City	*20
Aug 1938	Mamakating, NY	Backsiphonage	Summer Hotel	—	Private Spring	*25
Nov 1938	Middleboro, MA	Cross-Connection	Industrial Plant	—	Private well	*25/pg 46
Jan 1939	Lansing, MI	Backsiphonage	College Lab	Michigan St. College	City	*26
Jan 1939	Palmer, MA	X-Conn with contaminated river	Town	—	Semi-public	*25/pg 51
Mar 1939	Los Angeles, CA	Backsiphonage at nurses quarters	Hospital	—	City	*27
Aug 1939	Boston, MA	Faulty Plumbing	Coop work rooms	—	City	*25/pg 51
June 1939	Hopkinsville, KY	Backsiphonage	Town	—	Public	*25/pg 53
June 1939	Fullerton, CA	High pressure from oil well forced	Oil Well	Bartholomae	Public	*28
Sept 1939	Cleveland, OH	Liquid parafin pumped into city water mains	Oil Refinery	Standard Oil	Public	*29
Oct 1939	Oakland, CA	X-Conn in car wash pumped soap into city mains	Service Station	—	Public	*7
Jan 1940	Joliet, IL	Backpressure from private well	Soda Factory	—	Public	*30
Jan 1940	Joliet, IL	Backpressure forced beer into city mains	Brewery	—	Public	*28
Apr 1940	So. Hadley, MA	X-Conn with contaminated fire sprinkler system	College	Mt. Holyoke	Public	*25/pg 56
May 1940	Albany, NY	Numerous internal X-Conns	Hotel	Eayck	Public	*27
May 1940	Canastota, NY	Backflow of sewage thru faulty valve	Village	—	Public	*25/pg 56
July 1940	Coxsackie, NY	X-Conn with contaminated well	Children's Camp	—	Private well	*25/pg 56
Aug 1940	Vallejo, CA	Salt water backpressured into potable water mains on Base	Navy Base	US Navy	Public	*31
Sept 1940	Louisville, KY	Cross-Connection	Apt House	—	Public	*25/pg 56
Nov 1940	Clymer, NY	Backsiphonage from toilets	School	—	Well	*25/pg 56
Dec 1940	Rochester, NY	X-Conn with sewage	City	—	Public	*25/pg 56
June 1941	Fall River, MA	X-Conn with contaminated river	Mill	—	Public	*25/pg 63
Aug 1941	Oro Grande, NM	X-Conn with contaminated source	Town	—	Public	*25/pg 62
Sept 1941	Berkeley, CA	Lake water pumped into public supply	Park	Tilden Park	Public	*7
Oct 1941	High Point, NC	X-Conn with sewage	City	—	Public	*25/pg 64
— 1942	Los Angeles, CA	Cross-Connections	Aircraft Factory	North Amer.	Public	*26
Feb 1942	Los Angeles, CA	Several Cross-Connections	Labor Camp	Hopkins	City	*20
Mar 1942	Baltimore, MD	Cross-Connection	Dept. Store	—	Public	*25/pg 71

Date	City & State	Cause of Contamination	Location	Organization	Type of Service	Source of Data
Mar 1942	St. Mary's OH	X-Conn with contaminated canal	City	—	Municipal	*25/pg 72
May 1942	San Francisco,CA	Ship pumped harbor water ashore	Pier 16	—	City	*33
July 1942	Lee County, IL	Open valve between contaminated on-site supply & public supply	Ordnance Plant	Green River	Public	*25/pg 70
July 1942	Pittsburgh, PA	X-Conn of raw and treated waters	Industrial Plant	—	Public	*25/pg 70
July 1942	Fallsburg, NY	X-Conn with contaminated supply	Summer Hotel	—	Drilled well	*25/pg 71
Aug 1942	Laurel Hills, NY	X-Conn with contaminated creek	——	——	Municipal	*25/pg 72
Aug 1942	Superior, WI	X-Conn, broken main at yard	Shipyard	——	Public	*25/pg 70
Aug 1942	Los Angeles, CA	Partial vacuum in drinking water syphoned toilet contents into water line	School	Aero. Indus.	Public	*27
Aug 1942	Los Angeles, CA	Heavy use caused venturi effect resulting in backflow from toilets	Army Post Outpost	Canoga Park	Public	*27
Aug 1942	Los Angeles, CA	Ship pumped harbor water into domestic system	Shipyard building	L.A. Ship-	City	*33
Sept 1942	Newton, KS	X-Conn with sewer system	City	——	Public	*25/pg 70
Oct 1942	Camden, NJ	X-Conn with contaminated river	Shipyard	——	Public	*25/pg 71
Oct 1942	Los Angeles, CA	Local contamination, source not determined	Port of Embarkation	——	City	*34
Nov 1942	Los Angeles, CA	Ship pumped harbor water into both ship & Base domestic system	Navy Base	U.S.Navy	City	*33
Nov 1942	San Francisco,CA	Ship pumped harbor water from fire system into City system	Pier 25	Port Auth.	City	*33
Dec 1942	Los Angeles, CA	Ship pumped harbor water thru its fresh water tanks into City system	Shipyard	Bethlehem	City	*33
Dec 1942	San Francisco,CA	Ship fire system backpressured salt water into City system	Pier 20	Port Auth.	City	*33
Dec 1942	Los Angeles, CA	X-Conn to high pressure process water in plant	Plant Mach. Shop	Frazier Wright	Public	*27
— 1942	Los Angeles, CA	Harbor water backpressured into domestic and City mains	Shipyard	Consolidated Shipbuilding	City	*33
Feb 1943	San Pedro, CA	985 X-Conn found during inspection	Army Post	Ft. MacArthur	City	*35
— 1943	Alameda, CA	Testing of pump backpressured salt water into plant's system	Engr. Co.	United Engr.	Public	*7
Feb 1943	Van Nuys, CA	X-Conn to vacuum pump in plant	Indust. Plant	Aircraft Components	Public	*27

Date	City & State	Cause of Contamination	Location	Organization	Type of Service	Source of Data
Apr 1943	Seattle, WA	X-Conn with untreated supply	Shipyard	——	Private	*25/pg 77
Aug 1943	Fall River, MA	X-Conn with raw river water for fire sprinkler system test	Factory	——	Public	*28
Aug 1943	San Francisco,CA	X-Conn with ship fire system	Pier 23, 25, 29	Port Auth.	City	*33
Sept 1943	Los Angeles, CA	X-Conn backsiphoned toilet contents into domestic system	Army Field HDQ	Brazil St.	Public	*36
Sept 1943	Orinda, CA	X-Conn between on-site spring and public supply	Estate	Delaveaga	Public	*7
Sept 1943	Long Beach, CA	Ship pumped harbor water into city mains	Naval Base	U. S. Navy	City	*33
Oct 1943	San Pedro, CA	Ship pumped harbor water through fire line into city mains	S.S. Avalon Terminal	Catalina	City	*33
Oct 1943	San Francisco,CA	Ship pumped harbor water into both on-board tanks and city mains	Pier 46	——	City	*33
Oct 1943	Portland, OR	X-Conn with contaminated river	Shipyard	——	Public	*25/pg 77
— 1944	Los Angeles, CA	618 X-Conn found during campus inspection	College	Occidental	City	*20
Jan 1944	Forest City, NC	Backsiphonage of waste water into city storage tank	City	——	Public	*25/pg 82
Feb 1944	Los Angeles, CA	Ship pumped harbor water into both on-board tanks and city mains	Pier 232	L.A. Harbor	City	*33
Feb 1944	San Diego, CA	Ship pumped harbor water into city mains	Destroyer Base	U.S. Navy	City	*33
Feb 1944	Terminal Is., CA	Ship pumped harbor water through fire lines to city mains	Berth 232	U.S. Navy	City	*33
Mar 1944	Oakland, CA	Ship pumped harbor water through fire lines into city mains	Grove St. pier	——	City	*33
Mar 1944	Elizabethtown,KY	Backsiphonage through frost-proof toilets	Town	——	Public	*25/pg 81
Mar 1944	Bay City, TX	Possible backsiphonage during fire or contamination during flood	City	——	Public wells	*25/pg 82
Apr 1944	Oklahoma	Cross-Connections	School	——	not identified	*25/pg 82
May 1944	Alameda, CA	Ship pumped harbor water thru fire system into city mains	Pier Packers	Alaska	City	*33

Date	City & State	Cause of Contamination	Location	Organization	Type of Service	Source of Data
June 1944	San Francisco,CA	Ship pumped harbor water thru fire system into city mains	Pier 23	——	City	*33
July 1944	Oakland, CA	Ship pumped harbor water thru fire system into city mains	Pier 17	——	City	*33
July 1944	San Pedro, CA	X-Conn to fire sprinkler system	Port of Embarkation	U.S. Army	City	*33
Aug 1944	Blooming Grive, NY	X-Conn with unapproved spring	Resort Lodge	——	Private well	*25/pg 81
Aug 1944	Southern MI	X-Conn permitted backsiphonage	College	——	Raw	*25/pg 81
Dec 1944	Los Angeles, CA	Pumps on fire line forced harbor water into domestic lines	Berth 195-197 Terminal Co.	Inner Harbor	Public	*37
Feb 1945	Green Bayou,TX	X-Conn with flush water	Navy Freighter	——	Private wells	*25/pg 86
Feb 1945	Los Angeles, CA	Many X-Conn found during inspection	Hospital	Cedars of Lebanon	City	*33
Mar 1945	Los Angeles, CA	Ship pumped harbor water thru fire lines into city water mains	Pier Shipbuilders	Standard	City	*33
June 1945	Compton, CA	X-Conn between gas and water mains allowed gas to enter water mains	Flushing Conn.	——	Public	*28
Oct 1945	Terminal Is., CA	Ship fire pumps forced harbor water into city mains	Pier	Bethlehem	City	*20
Oct 1945	Chicago, IL	Backflow from toilet connection	School	——	City	*25/pg 86
Jan 1946	North Central IN	X-Conn with only one gate valve	Industrial Plant	——	Well	*38
Oct 1947	Sawtelle, CA	X-Conn with unapproved well	Vet. Adm.	U.S. Govt.	City	*33
Aug 1948	New York, NY	Backsiphonage thru low inlets into city mains	Office Bldg.	——	Public	*28
Dec 1948	Van Nuys, CA	Backsiphonage of Sodium Arsenite thru unprotected hose	Construction site Paving Co.	Security	City	*35
June 1949	Clinton, IA	X-Conn - single check removed	Factory	——	Public	*30
Aug 1949	El Segundo, CA	Zinc Sulphate pumped back into city mains	Chemical Plant	——	Public	*39
Dec 1950	San Diego, CA	7200 ppm chloride backsiphoned into city mains	Residential	——	Public	*28
July 1952	Venice, CA	Tret-o-lite solution pumped back into city mains	Oil well	——	Public	*29
July 1955	San Pedro, CA	Ship pumped harbor water thru five obsolete check valves into city mains	Shipyard	Todd Shipyard, Inc	Public	*40

Date	City & State	Cause of Contamination	Location	Organization	Type of Service	Source of Data
Aug 1956	Shreveport, LA	X-Conn between contaminated bayou and city mains through lawn system	Private Res.	——	Public	*41
Oct 1956	Renton, WA	X-Conn between petroleum product transmission line @130psi and city mains @70psi	Pumping Station	Olympic Pipeline Co.	Public	*42
Oct 1958	Sunland, CA	Weedicide pumped back into city main	Orchard	——	Public	*40
— 1960	Crestview, FL	Gas from 250 Gal. butane tank back-pressured into city mains	Cafe	——	Public	*41
— 1960	Atlanta, GA	Blood backsiphoned from water operated aspirator into drinking fountain	Mortuary	——	Public	*41
— 1960	Center Point AL	Sludge from septic tank pumped into drinking water system through pump priming line. Existing check valve failed	School	——	Public	*41
April 1963	Bay City, MI	Shipboard fire pumps backpressured harbor water into city mains	Shipyard	Dafoe Shipbuilding	Public	*76
Dec 1964	Michigan	Bachsiphonage from unprotected autopsy table contaminated hospital water	Hospital	——	Public	*44
— 1966	Wyoming, MI	Plating tanks backsiphoned into city mains	Industrial plant	Plating Plant	Public	*76
Oct 1967	Renton, WA	X-conn of gasoline pipeline with city water system causing 2000 gallons of gasoline to enter the water system.	pipeline company		Public	*107
April 1969	Berkeley, CA	Backsiphonage from treated hot water heating system into potable system	8-story building	Univ of California	Public	*45
Aug 1969	Elmhurst, IL	Fire hose used to flush out gasoline pipeline to airport	Pipeline	City Water Dept.	Public	*46
Aug 1969	Worcester, MA	83 football team members/coaching staff stricken with infectious hepatitis via sunken hose-bibs backsiphonage due to local fire	Football field	Holy Cross	Public	*47
Sept 1969	Peoria, AZ	Sodium Arsenate backsiphoned from tank of weed killer	Farm	——	Public	*158
Nov 1969	Atlanta, GA	Chromate treated cooling water X-Conn to potable system	Commercial bldg.	——	Public	*48

Date	City & State	Cause of Contamination	Location	Organization	Type of Service	Source of Data
Dec 1969	Silverton, OH	Wine flushed back into city mains	Wine cellar	Commercial	Public	*49
— 1970	New York, N.Y.	X-Conn of chromate treated cooling tower system with domestic system	College	Skidmore College	Public	*50
Jan 1970	Atlanta, GA	X-Conn of chromate treated cooling system with potable water system	Bank	Fulton Nat. Bank	Public	*43
Jan 1970	Vanvouver, B.C.	X-Conn on board ship	Cruise ship	SS Oronsay	Shipboard	*51
June 1970		Backpressure of radioactive water into plant potable water system	Nuclear plant	Dresden 2 BWR	Internal	*52
June 1970	Mattoon, IL	Hot wash water from asphalt plant backpressured into city mains coupled with flow testing of fire hydrants in the area	Asphalt plant	——	Public	*53
July 1970	San Bruno, CA	Strong detergent water from car wash pumped thru single, faulty check into city water mains	Car wash	——	Public	*54
Sept 1970	New Jersey	Soft drink vending machine X-Conn to building heating system in which hexavalent chromium was added	Caddy house	Golf club	Public	*53
Nov 1970	Atlanta, GA	Backsiphonage from air conditioning system treated with bichromate of soda	Commercial bldg.	—	Public	*48
— 1972	Boston, MA	Sodium Chromate/caustic soda backpressured from cooling tower	Office Building	——		*99
— 1972	Huskerville, NE	Backsiphonage from toilets into potable, looped system	65 unit Housing	——	Public	*56
— 1972	Lyndon, OR	Liquid alum backsiphoned into city mains	Factory	not specified	Public	*57
Mar 1972	Huachuca City, AZ	Backsiphonage of pesticide thru garden hose submerged in mixing tank	Exterminating Co		Public	*157
April 1972	Bakersfield, CA	Aldrite (termite pesticide) back-siphoned thru X-Conn	200 home area	Clark Pest	Public	*58
July 1972	San Antonio, TX	Private well pumping into city system Meter found running backwards	Packing Company	--	Public	*140
Sept 1972	Bydgoszcz, Poland	Single, faulty valve permitted beer to backflow into city mains	Brewery	State	Public	*59

Date	City & State	Cause of Contamination	Location	Organization	Type of Service	Source of Data
Sept 1972	Salt Lake City,UT	Boiler compound backpressured into potable system thru leaking check valve	Apartment Bldg		Public	*159
— 1973	Klamath Falls,OR	Contaminated irrigation water backpressured into city	Farm	not specified	Public	*57
Mar 1973	Seattle, WA	X-conn between hydrant drain and the sewer.	city	city water	Public	*108
May 1973	Morganville and Wickatunk, NJ	Submerged hose in chlordane mix backsiphoned during main break	Residence	Marlboro Twsp Muni Water Dist	Public	*77
May 1973	Salt Lake City,UT	Backsiphonage of slag from submerged inlets in plating process	Plating Company		Public	*159
Feb 1973	Roebling, NJ	X-Conn between fire line using contaminated river and service connection to building	U.S. Post Office	City	Public	*60
— 1974	Boston, MA	Backpressure of harbor water from ship into city main	Harbor	——	——	*99
Feb 1974	San Antonio, TX	Salt water from water softener equipment backsiphoned into internal system	Pharmacy	Oakdell Pharmacy	Public	*61
— 1974	Woburn, MA	Fungicide backpress from checmical application system through single check valve	Greenhouse	——	——	*99
Jun 1974	Soldiers Grove, IL	Backsiphonage of herbicide Balan to the water system.	village	village water	Public	*109
Jul 1974	Seattle, WA	Backsiphonage of a chemical De-Germ and other pollutants into water system.	airport		Public	*110
Sep 1974	New York, NY	Backpressure from chromate treated air conditioning system into bldg. potable water system	Office bldg.	Internal	Public	*77
Dec 1974	No. Carolina	Chemically treated boiler feed backpressured into city mains	Fertilizer plant	——	Public	*62
Dec 1974	San Antonio, TX	Chemically treated steam backpressured into potable system through faulty check valve	Food Manufacturer		Public	*140
— 1975	Chelmsford, MA	Ethyline glycol from solar heating system	Commercial Bldg.	——	——	*99

Date	City & State	Cause of Contamination	Location	Organization	Type of Service	Source of Data
Mar 1975	Redwood City,CA	Detergent laden wash water back-pressured into city mains	Car Wash	City Water	Public	*63
Oct 1975	Langley, SC	Ethylacroglate backsiphoned from industrial plant into city mains	Residences	unknown	Public	*64
Nov 1975	East Monitro,WA	Endrin treated water (pesticide) backsiphoned into 21 homes	Fire Hydrant	Town Water Dept.	Public	*65
— 1975	Salt Lake City,UT	Backsiphonage of scale from water heater plugging water users systems	Beverage Company		Public	*159
Mar 1976	Chattanooga, TN	Backsiphonage of Chlorine following a break in a city water main.	city	city water	Public	*111
Mar 1976	San Antonio, TX	Contaminated private lake back-pressured thru lawn sprinkler system into city mains	Private grounds	U.A.A.A.	Public	*61
May 1976	San Jose, CA	Direct connection between automatic water softener and sewer line back-siphoned into camp water system	Private camp	Getting in	Private	*63
June 1976	San Antonio, TX	6" and 4" services looped thru hospital causing backflow	Hospital	Santa Rosa	Public	*61
Oct 1976	Washington, PA	Pesticide pump discharge hose connected to house hose bib	140 homes	Western PA	Public	*66
— 1977	North Dakota	Backsiphonage of DDT from two homes into city mains.	residences	City mains	Public	*80
Jan 1977	San Antonio, TX	Resin of water softener backsiphoned into house system due to main break	Private home	Binney res.	Public	*61
Feb 1977	Seattle, WA	X-conn of closed hot water heating system and city water resulted in backflow of Borate-Nitrite from the heating boiler.	Hospital		Public	*112
Mar 1977	Burlington, VT	CO2 and carbonated water backflowed into copper piping of hospital causing copper poisoning	Hospital	Medical Center	Public	*75
July 1977	San Antonio, TX	Failure of control switch pumped scale inhibitor into domestic lines	Industrial bldg	Paul Anderson	Public	*61
July 1977	Kulm, ND	Backsiphonage from garden sprayers contaminated city mains with DDT	Several parts of the town	City Water Dept.	Public	*67

Date	City & State	Cause of Contamination	Location	Organization	Type of Service	Source of Data
Aug 1977	Chico, CA	Detergent ladened car wash water pumped back into city mains	Car wash	——	Public	*68
Oct 1977	San Antonio, TX	"Black water" from fire sprinkler system polluted major portion of hospital complex	Hospital	Santa Rosa Hospital	Public	*61
Oct 1977	Chicago, IL	X-Conn in apartment house caused Shigella in one case and acute symptoms in others	Apt. house	——	Public	*69
-- 1978	Salt Lake City,UT	Backpressure of carbon dioxide from beverage dispenser	Restaurant		Public	*159
Jan 1978	Vancouver, B.C.	Backflow Carbon Dioxide from a beverage machine resulting in copper contamination.	theater		Public	*114
Jan 1978	Hinsdale, IL	High pressure air from hospital escaped thru "stuck valve" to mains	Hospital	Hinsdale Memorial	Public	*70
Feb 1978	Kelly AFB, TX	X-Conn of chromate laden waste water with potable mains	Air Force Base	Kelly AFB	Internal	*71
Mar 1978	Vancouver, B.C.	Backflow of corrosion inhibitor BRAMCO 750 from apartment's heating boiler to water system.	high rise apartment	domestic water	Public	*115
May 1978	Pierce County, WA	Backsiphonage of septic tank sewage to school's water system resulting in 33% student absenteeism.	school	school	Public	*113
July 1978	San Antonio, TX	Backsiphonage of 'blue water' from toilet without vacuum breaker	residence		Public	*140
Aug 1978	South Carolina	X-Conn backsiphoned Chlordane from exterminator tank during meter repair	Church	Church	Public	*77
Feb 1979	Seattle, WA	X-Conn allowed backpressuring of industrial detergent into 100 blocks	Car wash	City Water	Public	*79
Feb 1979	San Antonio, TX	Steam from chemically treated boiler backpressured into city mains	Plastics plant	Alamo Foam	Public	*61
Mar 1979	Kulm, N.Dakota	Backsiphonage of DDT from an aspirator sprayer causing contamination of municipal water system.	city		Public	*116

Date	City & State	Cause of Contamination	Location	Organization	Type of Service	Source of Data
— 1979	Suffolk Co., NY	Suspected backsiphonage from slop sink	Public bldg	——	Public	*72
April 1979	Marshalltown, IA	X-Conn of waste water line and new on-site well caused gross contamination of approx $2,000,000 of pork	Packing Plant	Swift Co.	Private	*78
June 1979	Meridian, ID	Backsiphonage of "stagnant water" containing high bacterial count through leaking check valve.	city		Public	*118
June 1979	Meridian, ID	Leaky alarm check on fire line system	Supermarket	——	Public	*106
June 1979	Adair Co., MO	Backsiphonage thru submerged hose in herbicide tank	Farm	Adair Co. Public Water Dist.	Public	*73
June 1979	Callaway Co., MO	Backsiphonage thru submerged hose in herbicide tank	Farm	Private	Private	*73
June 1979	Lake Havasu, AZ	Backpressure effluent tree bubbler system into potable water supply	Marina	——	Public	*105
Aug 1979	Los Alamos, NM	Backsiphonage of treated boiler water into school potable water system	Pueblo Jr. Hi. School	Los Alamos Sch. Dist.	Public	*74
Aug 1979	San Antonio, TX	Backsiphonage of "blue water" from several newly installed toilets	Apartment		Public	*140
Sep 1979	Portland, OR	Backflow of detergent water through faulty R P device.	dairy		Public	*117
Oct 1979	Roanoke, VA	X-Conn backsiphoned Chlordane into City mains. Court ruled both City & exterminator company negligent	Domestic	Roanoke Water Dept	Public	*93
Oct 1979	San Antonio, TX	"Blue water" from commode tank back siphoned into potable water supply	Residence		Public	#137
Oct 1979	Seattle, WA	Backflow of cooling solution into building's water system due to faulty plant control valve.	Industrial		Public	*119
Nov 1979	San Antonio, TX	Backsiphonage of "blue water" from unprotected toilet	Residence		Public	*140
— 1980	Boston, MA	Chromates backpressured from boiler	High school			*99
Mar 1980	Manchester, NH	X-Conn backsiphoned chemical in hot water system into potable supply during heavy draft for fire	Office Bldg.	Public Utility	Public	*81

Date	City & State	Cause of Contamination	Location	Organization	Type of Service	Source of Data
Apr 1980	San Antonio, TX	Carbonated water backpressured thru faulty check valve from soda dispensing equipment	Restaurant		Public	*140
June 1980	San Antonio, TX	Swiming pool water backsiphoned into potable water supply	Residence		public	*140
Aug 1980	Salt Lake City, UT	Backpressure through unprotected make up line to heating/cooling system on 18th floor	Highrise		Public	*159
Sep 1980	Unalaska, AL	Backflow of sewage contaminated water due to X-conn aboard a crab processing ship.	ship	crab processing ship	private	*121
Oct 1980	Tacoma, WA	Oil is pumped into city water system due to leaks in a heat exchanger.	pumping station		Public	*120
Nov 1980	San Antonio, TX	Main break caused backsiphonage of "blue water" from toilets	residence		public	*140
Nov 1980	Salt Lake City,UT	Backsiphonage of detergent from hose left in large vat	Warehouse complex		Public	*159
Dec 1980	Beechview, PA	X-Conn backsiphoned Chlordane from exterminator truck into mains. 25,000 out of water for weeks due contamination of mains	Residential Water Co.	West PA	Public	*82
Dec 1980	Salt Lake City,UT	Backpressure of chemicals from solar heating system thru hose	Office Building		Public	*159
— 1981	South Bend, IN	X-conn between gas line and water causing water in the gas lines		gas company	Public	*122
Jan 1981	Norfolk, VA	X-Conn backpressured harbor water from fire system into city mains	Shipyard & fast-food restaurant	Norfolk Water Dept	Public	*77
Feb 1981	San Antonio, TX	Backsiphonage of filter material into hospital's internal system	Hospital		internal	*140
Mar 1981	Salt Lake City,UT	Backpressure of corrosion inhibitor thru bypass around RP on make up line to chiller	Office building		Public	*159
Mar 1981	Salt Lake City,UT	Contractor broke water main which then washed out sewer line. Sewage backsiphoned into water main when it was shut down.	Construction area		Public	*159
May 1981	San Antonio, TX	Aspiration of chrome plating solution into customer system	Plating Plant		Public	*140

Date	City & State	Cause of Contamination	Location	Organization	Type of Service	Source of Data
June 1981	Chester, VA	Polyvinyl garden hose with pistol type nozzle left in sun permitted gas from hose to enter residence producing taste & odor problem	Residence	Mobile Home	Master Meter	*83
July 1981	Salt Lake City,UT	Backpressure of onsite recycled water	Garage		Public	*159
Sept 1981	San Antonio, TX	Backpressure of chemically treated water from heat exchanger through faulty gage valve	Hospital		Public	*140
Nov 1981	Chicago, IL	X-Conn apparently allowed contaminated fluids from industrial plant into city mains. Officials warned residents not to drink the water even after boiling	Industrial Plant	City Water	Public	*84
Nov 1981	Bangor, ME	X-Conn backsiphoned home hot water heating system into mains during main repairs	Residence	Bangor Water Dist.	Public	*85
Dec 1981	Bailey, CO	X-Conn allowed antifreeze in solar heating system to contaminate school potable water system	School	City Water Dept.	Public	*86
Jan 1982	San Antonio, TX	Carbonated water backpressured into water supply through leaking check valve on soda dispenser	convenience store	Stop & Go	Public	*140
Mar 1982	Tacoma, WA	A leaking check valve permitted soapy water from a car wash to be pumped into city water system.	car wash facility		Public	*125
Mar 1982	San Antonio, TX	Carbonated water from soda dispenser backpressured into water system. Several illnesses reported.	Restaurant			*136
May 1982	Bancroft, MI	X-Conn backsiphoned Malathion from home hose-end applicator during main shutdown to replace valve. Village "out of water" for 2 days.	Residence	Municipal Water Dept	Public	*87
Jun 1982	Buffalo, NY	X-Conn backpressured from industrial plant into 10-block Lovejoy area producing taste & odor suggesting contamination	Industrial	City Water Division	Public	*88

Date	City & State	Cause of Contamination	Location	Organization	Type of Service	Source of Data
June 1982	San Antonio, TX	Carbonated water backpressured into water supply through leaking check valve on soda dispenser	restaurant		Public	*140
Jun 1982	Springfield, OR	Backflow from irrigation hose resulted in an insect larvae entering a food prcessing vat.	food processing plant		Public	*123
July 1982	North Andover MA	X-Conn allowed backpressure of Hexavalent Chromium into potable water system	Mfg. plant	Western Electric	Public	*89
July 1982	Jersey City, NJ	Rupture of 72" main with inoperable isolation valves forced 300,000 customers "out of water"	City	Jersey City	Public	*91
July 1982	Laguna Hills, CA	Hydroseed mulch backpressured into fire hydrant by contractor	Fire Hydrant	Church	Public	*162
Sept 1982	Alice, TX	X-Conn backpressured petroleum products contaminating city mains for several days	Petro-Chem Allegra	Rancho	Public	*90
Oct 1982	Springfield, OR	X-conn between potable water system and process water & fire system.	lumbermill	lumber mill	Public	*124
Oct 1982	Salt Lake City, UT	Backpressure of CO_2 from beverage dispenser	Restaurant		Public	*159
Dec 1982	Salt Lake City, UT	Backpressure of CO_2 from beverage dispenser	Hospital cafeteria		Public	*159
Feb 1983	McKeesport, PA	Giardia Lambia probably from open reservoir coupled with X-Conn in filtration plant. 45,000 req'd to boil water	Inlet	City Water Dept	Public	*92
May 1983	San Antonio, TX	Blue water from commode tank back siphoned into potable water supply	Residential		Public	*137
Aug 1983	Akron, OH	Numerous X-Conn in major industrial plants permitted contamination of several blocks of city	Industrial & Residential	City Water Dept	Public	*94
Sept 1983	Belleview, FL	Gasoline from undetermined source	Residential	City Water Dept	Public	*95
Oct 1983	Vancouver, WA	X-conn in dental office pumped air into the building's water system.	dental building		Public	*126

Date	City & State	Cause of Contamination	Location	Organization	Type of Service	Source of Data
Dec 1983	San Antonio, TX	Effluent from Rilling Road Treatment Plant backpressured into plant lines during plant maintenance, Incident isolated to in-plant system by RP at service connection.	Sewage Treatment	San Antonio Wastwater	Industrial	*96
Dec 1983	Philadelphia, PA	Direct connection to tank containing car wash product backsiphoned into adjacent buildings	Laboratory		Public	*154
Jan 1984	Philadelphia, PA	Backsiphonage of antifreeze from heating system	Clothing company		Public	*154
Feb 1984	Riverbend, OR	X-conn between a solar hot water system and domestic hot water system.	mobile home park		Public	*127
Feb 1984	Seattle, WA	Air conidtioning water containing Nitrate-Borate was pumped into a high rise office building's & city's water system	high rise office building		Public	*128
Apr 1984	Willets, CA	Internal X-Conn of private well with domestic water backpressured into company's distribution system	Residential	Private	Public	*97
Apr 1984	San Antonio, TX	Chromates backpressured into internal system thru faulty check valves an gate valve on heat exchanger. City system protected by RP	High School		Public	*140
Aug 1984	Bellevue, WA	Backsiphonage of Sodium Silicate into the building's potable water system.	nursing home	city water	Public	*129
Oct 1984	Philadelphia, PA	Backpressure from air compressors into 12" service main	Office buildings		Public	*154
Nov 1984	Camarillo, CA	CO_2 backpressured from soft drink vending machine into copper piping contaminated water supply	High School	Camarillo High School	Public	*98
Feb 1985	Salt Lake City, UT	Backsiphonage from circuit board washer when system was shut down for repairs	Manufacturing plant		Public	*159
Mar 1985	New York, NY	Hospitalized woman died after being exposed to Ethylene Glycol while undergoing Hemodialysis. Backpressure from air conditioning system.	Hospital			*102

Date	City & State	Cause of Contamination	Location	Organization	Type of Service	Source of Data
Mar 1985	San Antonio, TX	Backsiphonage of "blue water" from toilet	residence		Public	*140
Mar 1985	Salt Lake City,UT	Backpressure of CO_2 from beverage dispenser	Post office		Public	*159
— 1985	Chelmsford, MA	Backpressure of wash water from car wash.	car wash		Public	*99
— 1985	Boston, MA	Hexavalent Chromium backflow from cooling tower into potable system.	Condominium Complex		Public	*99
— 1985	Boston, MA	500-1000 gallons of water treated with ethylene glycol and hydrazline backflowed through temporary hose on chilled water system.	Hospital		Public	*99
Apr 1985	San Antonio, TX	Carbonated water backpressured into water supply through leaking check valve on soda dispenser	restaurant		Public	*140
Apr 1985	Salt Lake City,UT	Air backpressure through dental equipment on three separate dates	Dental office		Public	*159
Jun 1985	Yakima, WA	X-conn of an irrigation system resulted in pesticide contamination of a well supplying from residence.	private well		Private	*130
Jul 1985	Arpelan, OK	Break in waterline caused backsiphonage of Chlorine, Malathion, Sevin Diazanon into the public water system.	residential	city water	Public	*131
Sep 1985	Sacramento,CA	Malathion backpressured through faulty check valves into potable system.	Port of Sacramento	Cargill, Inc.	Public	*101
Jan 1986	Bonnev Springs	Pesticide Malathion was fed into the water system by the aspirator casuing grain mill employee to become ill.	grain mill plumbing system		Public	*132
Feb 1986	San Antonio, TX	Backsiphonage of "blue water" from toilet into potable water supply	residential		Public	*140
Apr 1986	Withrow, WA	Herbicide 2, 4-D was backsiphoned into the community's water system causing residents to go without water for four days.	residential	city water	Public	*133

Date	City & State	Cause of Contamination	Location	Organization	Type of Service	Source of Data
Apr 1986	Salt Lake City,UT	Backsiphonage from ornamental ponds when repairs performed. 40+ units affected with blue dyed water.	Condominium		Public	*159
Apr 1986	San Rafael, CA	Air from a single jacketed air compressor cooler backpressure into potable system.	Auto Painting Co.			*104
Apr 1986	Philadelphia, PA	Backpressure thru unprotected makeup line to chilled water system	Office Bldg	Fidelity Bank Building	Public	*154
Apr 1986	Salt Lake City,UT	Carbonated water backpressured into water supply through leaking check valve on soda dispenser	Golf course cafe		Public	*159
Jun 1986	San Luis Obispo CA	Surface water was siphoned into the city water due to defective operating valves on a lawn sprinkler system.	lawn sprinkler system	city water	Public	*134
June 1986	San Antonio, TX	Oil slurry from asphalt plant backpressured into internal water system. City supply protected by RP and not affected	Asphalt Manufacturing Plant		Public	*140
July 1986	San Antonio, TX	Carbonated water backpressured into water supply through leaking check valve on soda dispenser	convenience store		Public	*140
Aug 1986	San Antonio, TX	Recirculated wash water backpressured into internal water system. City supply protected by RP and not affected	car wash		Public	*140
Nov 1986	Salt Lake City,UT	Backsiphonage of "blue water" from toilet on 2nd floor when building was shut off.	Condominium		Publlic	*159
Feb 1987	San Antonio, TX	Carbonated water backpressured into water supply through leaking check valve on soda dispenser	restaurant		Public	*140
Mar 1987	Fountain Valley CA	Secondary effluent backpressured into domestic water used within the plant	Treatment plant		Internal	*153
Mar 1987	Santa Ana, CA	Domestic water use interconnected to industrial water line	Modular office		Public	*153
Apr 1987	North Dakota	Heating system backpressure ethylene glycol into potable water which was used to mix a powdered beverage at a picnic.	Firehall			*102

Date	City & State	Cause of Contamination	Location	Organization	Type of Service	Source of Data
May 1987	Evanston, IL	Water cooled air conditioner backflowed through leaky valve	Hospital	Evanston Hospital	Public	*163
Jun 1987	Gridley, KS	Herbicide backsiphoned from tanker truck when main was broken during excavation.	Grain and Feed Store			*103
July 1987	San Antonio, TX	Swimming pool water backpressured into into potable water thru make up connection to pool circulation system	Recreation Center		Public	*140
July 1987	Salt Lake City,UT	Backsiphonage of liquid fertilizer from fertigation system with no protection	Residence		Public	*159
Aug 1987	Salt Lake City,UT	Backpressure of sodium silica thru water flush line.	Chemical Co		Public	*159
Sept 1987	San Antonio, TX	Non potable water from heat exchanger backpressured thru faulty check valve in ten story building. Numerous illnesses reported	Office Bldg.		Public	*140
Oct 1987	Gainesville, FL	Improperly connected pipes allowed backflow of chemically treated water from an air conditioning unit.	University of Florida	University		*100
Oct 1987	San Antonio,TX	Backsiphonage of "blue water" from toilet	Apartment	Royal Apt.	Public	*140
Oct 1987	San Antonio, TX	Carbonated water backpressured into water supply through leaking check valve on soda dispenser	restaurant		Public	*140
Dec 1987	Costa Mesa, CA	Chromates from chilled water system backpressured into cafeteria	Newspaper		Public	*153
Mar 1988	San Antonio,TX	Backsiphonage of "blue water" from toilet into apartment's water supply	Apartment Complex		Public	*140
Mar 1988	San Antonio,TX	Backsiphonage of "green water" from toilet into apartment's water supply	Apartment Complex		Public	*140
Apr 1988	Edgewater, FL	Propylene glycol backflowed due to valve malfunction.	Paint Factory Paint Co.	Coronado	Public	*150
Apr 1988	Salt Lake City, UT	Backpressure of boiler compound thru malfunctioning unapproved RP	Office Bldg		Public	*159
May 1988	San Antonio,TX	Backsiphonage of soapy water from toilet into apartment's water supply	Apartment Complex		Public	*140

Date	City & State	Cause of Contamination	Location	Organization	Type of Service	Source of Data
June 1988	San Antonio, TX	Acid backpressured throughout internal system from air injected degreaser	Maintenance	NEISD	Public	*140
June 1988	Glendo, WY	Backsiphonage of synthetic detergent from truck tank filling in railroad yard	Railroad yard		Public	*151
Aug 1988	San Antonio, TX	Carbonated water backpressured into water supply through leaking check valve on soda dispenser	restaurant		Public	*140
Aug 1988	Gilbert, AZ	Interconnection between irrigation system and decorative pond	Industrial Park		Public	*155
Aug 1988	Fresno, CA	Backflow of chemically treated water from chiller into schools potable system	School		Public	*152
Sept 1988	Edmonton Canada	A faulty wafer check valve permitted the backflow of water from a fire sprinkler system into buildings potable supply	Office Building		Public	*141
Dec 1988	Newport Beach CA	Backwash from sand filters backpressured into restaurant's beverage dispensers.	Restaurant		Public	*149
Mar 1989	San Antonio, TX	Diesel fuel backpressured into internal system. Service connection protected by RP.	Asphalt manuf. plant		Public	*135
Mar 1989	Tucson, AZ	Diazinon backflowed through garden hose to next door neighbor	Residence		Public	*156
May 1989	Edmonton Canada	Backflow through an alarm check valve on a fire sprinkler system into department stores water system.	Department Store		Public	*141
June 1989	San Antonio, TX	Carbonated water backpressured through faulty check valve in soda dispenser.	Restaurant		Public	*135
Jun 1989	Salt Lake City,UT	Propylene glycol backpressured thru leaking check valve on fire sprinkler system during shutdowns to replace sprinkler heads.	Office bldg		Public	*159
June 1989	San Antonio, TX	Condensate from cooling system backpressured into internal water system. City supply protected by RP	Retirement Center		Public	*140
July 1989	Cincinnati, OH	Algae retardant backflowed into water from air conditioners under repair. About 50 people reported nausea.	Office building	Hamilton Co. Dept of Human Res.	Public	*161
July 1989	San Antonio, TX	Backsiphonage of "blue water" from toilet	Residence		Public	*135

Date	City & State	Cause of Contamination	Location	Organization	Type of Service	Source of Data
July 1989	Edmonton Canada	Leakage thru three wafer check valves on a fire system allowed backflow of oily water into station's restrooms	Rail Transit Station		Public	*141
Aug 1989	Memphis, TN	Backsiphonage of "blue water" from toilet into apartment's water supply	Apartment Complex		Public	*143
Aug 1989	Edmonton Canada	Leakage thru wafer check on fire sprinkler system allowed backflow of brown water into printing shop restroom	Printing Shop		Public	*141
Sept 1989	San Antonio,TX	Backsiphonage of "blue water" from toilet	Residence		Public	*140
Sept 1989	San Antonio,TX	Backsiphonage of "blue water" from toilet into apartment's water supply	Apartment Complex		Public	*140
Oct 1989	Edmonton Canada	Water quality complaint lead to discovery of drinking fountain piped from fire hose cabinet	Building		Public	*141
Dec 1989	San Antonio,TX	Backsiphonage of "blue water" from toilet into apartment's water supply. Four illnesses report by health department	Apartment Complex		Public	*140
Dec 1989	San Antonio,TX	Backsiphonage of "blue water" from toilet into apartment's water supply	Apartment Complex		Public	*140
Jan 1990	Brighton, CO	Antifreeze in heating system backpressured into school water system. Nine students treated	School	Overland	Public	*148
May 1990	Webster, NY	Solvent backpressured through hole in heat exchanger. Bypass around backflow preventer on water makeup line	Manufacturing plant	Xerox Corp.	Public	*147
May 1990	Tucson, AZ	Backflow of cooling system water into potable water system	Police station	Tucson police Station	Public	*160
Jul 1990	Salt Lake City,UT	Backpressure of boiler treatment compound into potable water sytem.	Elderly housing		Public	*159
Aug 1990	Bellevue, WA	Carbonated water backpressured into water supply through leaking check valve on soda dispenser	restaurant		Public	*145
Nov 1990	Memphis, TN	Backsiphonage of "blue water" from toilet into apartment's water supply	Apartment Complex		Public	*143
Jan 1991	San Antonio, TX	Water from heat exchanger backpressured through unprotected make up connection. City system protected by RP	Office Bldg.		Public	*140

Date	City & State	Cause of Contamination	Location	Organization	Type of Service	Source of Data
Feb 1991	Fountain Valley CA	Compressed air used for dental equipment backpressured into building water supply	Medical Office	Medical Arts	Public	*146
Feb 1991	Memphis, TN	Backsiphonage of "blue water" from toilet into apartment's water supply	Apartment Complex		Public	*142
Apr 1991	Kelly AFB, TX	Backpressure from chilled water system thru two leaking check valves			Internal	*144
Apr 1991	Edmonton Canada	Leaking check valve on fire sprinkler system backflowed into washrooms	Rail Transit Station		Public	*141
May 1991	San Antonio, TX	Swiming pool water backpressured from direct make-up connection to recirculation system	Motel		Public	*140
June 1991	San Antonio, TX	Carbonated water backpressured into water supply through leaking check valve on soda dispenser	lounge		Public	*140
July 1991	San Antonio, TX	Stagnate water from looped fire sprinkler system backflow into city main	Motel		Public	*140
Aug 1991	San Antonio,TX	Backsiphonage of "blue water" from toilet	Residence		Public	*140
Sept 1991	Mission Viejo,CA	Direct connection between potable water and reclaimed water line irrigating slope next to house.	Residence		Public	*162
Nov 1991	San Antonio,TX	Backsiphonage of "blue water" from toilet into apartment's water supply. Affected several apartments.	Apartment Complex		Public	*140
Oct 1992	San Antonio,TX	Backsiphonage of "blue water" from toilet into apartment's water supply	Apartment Complex		Public	*140
Nov 1992	San Antonio, TX	Soap backsiphoned into potable water of adjacent lease space. City water not affected.	Soap Mfg. Plant		Public	*140
May 1993	Fullerton, CA	Backpressure of industrial water loop from the tissue manufacturing process.	Paper mill		Public	*162
Aug 1993	Superior, AZ	Backflow of antifreeze solution thru fouled single check valve on fire sprinkler system	Arboretum	State Park	Private	*139

Footnotes for Source of Data

Footnote Number

1	Wallace & Tiernan - Technical Bulliten #99
2	Journal, American WaterWorks Association, Vol 31, No 2, Feb 1939 - pg 365
3	Ibid, Pg 346
4	DWP Public Health News, Feb 1938, pg 38. File Water Pollution Instances
5	Journal, American Water Works Association, Vol 31, No 2, Feb 1939 - pg 345
6	Ibid, Pg 362
7	East Bay Municipal Utility District, Cross-Connections, July 1944
8	Journal, American Water Works Association, Vol 31, No 2, Feb 1939 - Pg 330
9	Ibid, Pg 328
10	Ibid, Pg 361
11	Ibid, Pg 347
12	Ibid, Pg 339
13	Ibid, Pg 353
14	Ibid, Pg 358
15	Ibid, Pg 348
16	Ibid, Pg 355
17	Ibid, Pg 349
18	Ibid, Pg 331
19	Ibid, Pg 341
20	DWP File on Cross-Connections
21	Journal, American Water Works Association, Vol 31, No 2, Feb 1939 - Pg 327
22	Ibid, Pg 329
23	Ibid, Pg 332
24	Ibid, Pg 350
25	Government Printing Office, No.62370 - Hearing Before Senate Committee on Stream Pollution Control
26	American Journal of Public Health, #7, July 1939
27	Los Angeles Times, March 25, 1939
28	Letter from R. E. Dodson to R. L. Derby, Jan 17, 1951 - Photostat in LADW&P Files
29	Don Short, Supt., Engineering News Record, Sept. 28, 1939 - Pg 389
30	University of Michigan Bulletin, 1934 - Pg 84
31	Journal, American Water Works Association, Vol 33, No 3, march 1941 - Pg 406
32	DWP File on Cross-Connections; Information on number effected not available due to war time
33	Regional Water Wroks Engineer; W. E. Shaw Report, 1944; DWP File on Pollution Case History
34	Ibid
35	DWP File on Cross-Connections; Information on number effected not available due to war time
36	DWP File on National Research Council; Chart prepared by Goudey
37	DWP Public Health News, February 1948, Pg. 38; File, Water Pollution Instances
38	Public Works Magazine, 1946, Pg. 33
39	File; El Segundo Water Company
40	DWP File on Cross-Connections
41	Public Health Reports, July 1958
42	Fire Engineering, March 1968
43	Atlanta Journal, Jan 15, 1970
44	Journal of Environmental Health, Jan-Feb 1965
45	Journal of Environmental Health, July-Aug 1970
46	Chicago Daily News, Aug 20, 1969
47	Journal of American Medical Association, Feb 7, 1972 also, Journal of American Water Works Association, April 1972
48	Atlanta Journal, Jan 15, 1970
49	Cincinnati, Ohio newspaper, Dec 22, 1969
50	Chicago Daily News
51	Los Angeles Hearld Examinor, Jan 1970
52	The Silent Bomb by Faulkner, 394-72270-1
53	Willing Water, Feb 1973
54	"The Times", San Mateo, CA., Dec 27, 1972
55	Contractor, Oct 15, 1970
56	World Herald, Omaha, NB, Mar 25, 1956
57	Oregon State Health Department

Footnotes for Source of Data

Footnote Number	
58	"Mercury" San Jose, CA.,May 4, 1972
59	Los Angeles Times, Sept 23, 1972
60	National Inquirer
61	San Antonio Water Board
62	Watts News
63	San Mateo County Health Department
64	Los Angeles Times, Oct 14, 1975
65	Washington's Water, January 1976
66	Analytical Control, Vol 2, No 1,1977
67	Offical Bulletin, North Dakota Water & Pollution Conference, June 1978
68	City Activity Report, Sept 1977
69	Chicago Tribune, Dec 11, 1977
70	Hinsdale "Doings", Jan 16, 1978
71	"The Light", Feb 14, 1978
72	Journal, American Water Works Association, May 1972
73	Missouri Water & Wastewater Digest, June 1979
74	Los Alamos "Monitor", Aug 30, 1979
75	Morbidity & Mortality Weekly Report, July 8, 1977
76	Michigan State Health Dept Memo, April 22, 1963 and OpFlow (AWWA) Feb 1977
77	Aquarus and Watts News, Oct 1974
78	Food Chemical News, Apr 30, 1979
79	Pacific Northwest Section AWWA Conference, May 11, 1979
80	OpFlow (AWWA) June 1980
81	Manchester Water Department
82	Pittsburgh "Press", Dec 16, 1980
83	OpFlow (AWWA), Sept 1981
84	Chicago "Tribune", Nov 22, 1981
85	Bangor "Daily News", Nov 21, 1981
86	Denver "Post", May 5, 1982
87	Rocky Mountain News, May 6, 1982
88	Buffalo "Currier Express", June 6, 1982
89	Sunday "Eagle Tribune", Lawrence, MA
90	Pasadena (CA) Star News, Sept 8, 1982
91	Mainstream (AWWA), Aug 1982
92	Los Angeles "Times", Feb 24, 1983
94	Akron "Beacon Journal", Aug 21, 1983
95	Sarasota (FL) "Herald Tribune"
96	San Antonio "Express", Dec 23, 1983
97	"The Willets News", April 14, 1984
98	Ventura County Environmental Health Department internal report, Nov 16, 1984
99	Massachusett Department of Environmental Quality Engineering - CCC Manual, Nov 1987
100	University of Florida "Alligator", Oct 1987, Vol. 81 No. 28
101	"Daily Democrat", Woodland, CA, September 6, 1985
102	Center for Disease Control, Mortality and Morbidity Weekly Report, Vol. 36, No. 36
103	The Gridley Gleam, July 2, 1987, Vol. 5 Issue #6
104	Backflow Incident Report, April 30, 1986, Marin Municipal Water District
105	Investigation Chronology of the Sand Point Marina Water Borne Disease Outbreak, December 5, 1979
106	State of Idaho, Division of Environmental, Meridian Case Report, December 2, 1980
107	The Record-Chronicle Newspaper, Vol LXVII, No. 42
108	Seattle Water Department & Pacific Northwest Section of AWWA (PNWS-AWWA)
109	"Water & Sewage Work" & PNWS-AWWA
110	Seattle Water Department & PNWS-AWWA
111	Morbidity & Mortality Weekly Report & PNWS-AWWA
112	Department of Public Health, Seattle and King Company
113	Department of Social and Health Service, State of Washington
114	Vancouver Water Department & PNWS-AWWA
115	Vancouver Water Department & PNWS-AWWA
116	AWWA "Opflow", May 1979

Footnotes for Source of Data

Footnote Number	
117	Portland Water Bureau & PNWS-AWWA
118	Department of Health & Welfare, State of Idaho
119	Seattle Water Department & PNWS-AWWA
120	Tacoma Water Department
121	Seattle Post Intelligencer, October 18, 1980
122	"News Leaks" Indiana Section of AWWA
123	Springfield Utility Board & PNWS-AWWA
124	Lane County Health Division & PNWS-AWWA
125	Tacoma Water Department & PNWS-AWWA
126	Vancouver Water Department & PNWS-AWWA
127	Oregon Health Division & PNWS-AWWA
128	Seattle Water Department & PNWS-AWWA
129	Bellevue Public Works/Utilities Department & PNWS-AWWA
130	Department of Social and Health Services, State of Washington
131	"Rural Water Magazine", 1986, Volume 6, Number 2
132	"Backflow Prevention", 1986, Volume 3, Number 4
133	"Wenatchee World" Newspaper & PNWS-AWWA
134	County Engineering Department, San Luis Obispo County
135	The Direct Connection, San Antonio Chapter American Backflow Prevention Association (ABPA), Jan 1993, Vol.5, Issue 1
136	The Direct Connection, San Antonio ABPA, January 1991, Vol. 3, Issue 1
137	The Direct Connection, San Antonio ABPA, July 1991, Vol. 3, Issue 7
138	The Direct Connection, San Antonio ABPA, August 1991, Vol. 3, Issue 8
139	Interim Backflow Incident Report, Boyce Thompson Arboretum PWS No. 11-067, Arizona Department of Environmental Quality, August 27, 1993
140	San Antonio, TX, City Water Board, Case History Summary, February 1993
141	Western Canada Water and Wastewater Association, Bulletin, December 1991, Volume 43, No. 4
142	"Crossfire", Newsletter of the Memphis Light, Gas and Water Cross Connection Dept., Spring 1991, Volume 1, Number 2
143	"Crossfire", Newsletter of the Memphis Light, Gas and Water Cross Connection Dept., Winter 1990, Volume 1, Number 1
144	"Express-News", San Antonio, TX, April 20, 1991.
145	Incident Report, King County Water District Number 107, Renton, WA, 3 January 1991
146	Incident Report, County of Orange, Santa Ana, CA, 13 February 1991
147	"Democrat & Chronicle", Rochester, NY, May 10, 1990
148	Rocky Mountain News, Denver, CO, 30 January 1990
149	Incident Report, County of Orange, Santa Ana, CA, 5 January 1989
150	Gainesville Sun, Gainesville, FL, 28 April 1988
151	"Backflow Prevention", August 1988
152	Incident Report, State of California, Department of Health Services, 15 August 1988
153	Incident Reports, County of Orange, Santa Ana, CA
154	Incident Reports, City of Philadelphia, PA
155	"Tribune", Mesa, AZ, 25 August 1988
156	"Tucson Citizen", Tucson, AZ, 30 March 1989
157	Incident Report, Huachuca City, AZ, March 23, 1972
158	Incident Report, Peoria, AZ, September 25, 1969
159	Incident Reports, Salt Lake City Public Utilities, Incidents 1966 to 1991
160	"The Arizona Daily Star", Tucson, AZ, May 1, 1990
161	"Cincinnati Enquirer", Cincinnati, OH, 29 July 1989
162	Incident Reports, County of Orange, Santa Ana, CA
163	"Chicago Tribune", Chicago, IL 17 May 1987

*** Data not available due to war time restrictions

APPENDIX A
Optional Field Tests and Reporting

A.1 General Information

A.1.1 Line Pressure

Some jurisdictions require that the line pressure be recorded during the test of a backflow prevention assembly. To accomplish this a suitable pressure gage should be attached to the No. 1 or No. 2 test cock, and the value recorded.

A.2 Reduced Pressure Principle Backflow Prevention Assembly

A.2.1 3.0 psi Buffer

The 3.0 psi buffer *(i.e., difference between the readings for the check valve No. 1 and relief valve opening point)* is designed into the assembly so that small fluctuations of line pressure do not cause the differential pressure relief valve to discharge. The continual discharge of water may be considered a possible nuisance due to wet conditions around the assembly, or cause erosion due to runoff. Therefore, an assembly which maintains a 3.0 psi buffer or greater is less likely to discharge due to pressure fluctuations.

An assembly which maintains less than a 3.0 psi buffer will provide the same level of backflow protection as an assembly greater than 3.0.

A.2.2 Optional: Check Valve No. 2 Test - *Direction-of- flow*

Purpose: To determine the static pressure drop across check valve No. 2.

Requirement: The static pressure drop across check valve No. 2 shall be at least 1.0 psid. If shutoff valve No. 2 is found to be leaking this test cannot be performed accurately.

Steps:

a. With gage attached as in Section 9.2.2 Test No. 3, *Step **a**,* close No. 2 test cock. If the gage reading drops to zero, or rises above the actual pressure drop across check valve No. 1, this implies that the shutoff valve No. 2 is leaking. This test can not be completed accurately. Proceed to *Step **b** of Test No. 3.* If gage reading remains steady, then proceed to *step **b** below.*

b. Close test cocks No. 3 and No. 4. Remove test equipment.

c. Attach hose from the high side of the differential pressure gage to the No. 3 test cock.

d. Attach hose from the low side of the differential pressure gage to the No. 4 test cock.

e. Open test cocks No. 3 and No. 4, and bleed all air from the hoses and gage by opening the high side bleed needle valve and the low side bleed needle valve.

f. Close the high side bleed needle valve, then *slowly* close the low side bleed needle valve.

g. After the gage reading stabilizes, the steady state differential pressure reading indicated on the gage is the static pressure drop across check valve No. 2 and is to be recorded as such. This reading must be 1.0 psi or greater.

h. Close all test cocks, slowly open shutoff valve No. 2, and remove all test equipment.

A.2 Reduced Pressure Principle Assembly

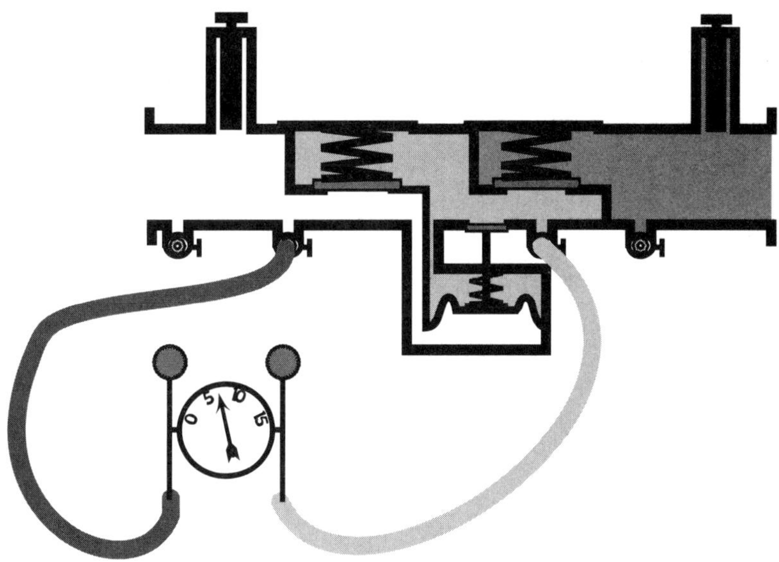

Fig. A.1
Hookup of Two needle valve gage configuration for Field Testing of Reduced Pressure Principle Assemblies

A.2.3 Field Test Procedure with 2-Needle Valve Gage Configuration (See Fig. A.1)

Test No. 1 - *Relief Valve Opening Point*

Purpose: To test the operation of the differential pressure relief valve.

Requirement: The differential pressure relief valve must operate to maintain the zone between the two check valves at least 2 psi less than the supply pressure.

NOTE: It is important that during this test the tester does not cause the relief valve to discharge before step ***i*** below. *(see Troubleshooting Section 9.2.3.1 - Exercising the Differential Pressure Relief Valve)*

Steps:

Follow all preliminary steps detailed in Section 9.1

a. To eliminate foreign material, open No. 4 test cock to establish flow through the unit, then flush water through test cocks No. 1, No. 2 (*open No. 2 test cock slowly*), and No. 3, by opening and closing each test cock one at a time. Be careful not to activate the relief valve during this process. Close test cock No. 4.

b. Install appropriate fittings to test cocks.

c. Attach hose from the high side of the differential pressure gage to the No. 2 test cock.

d. Attach hose from the low side of the differential pressure gage to the No. 3 test cock.

e. Open test cock No. 3 slowly and then bleed all air through the bypass hose by opening the low side bleed needle valve. Maintain the low side bleed needle valve in the open position while the test cock No. 2 is opened **slowly**. Open the high side bleed needle valve to bleed all air from the gage. Close the high side bleed needle valve, then close the low side bleed needle valve after the gage reading has reached the upper end of the scale.

f. Attach bypass hose from low side bleed needle valve to the high side bleed needle valve.

g. Close No. 2 shut-off valve.

h. Should the gage reading drop to the low end of the gage scale and the differential pressure relief valve discharges continuously, then the No. 1 check valve is leaking. Tests No. 1, No. 2 and No. 3 may not be completed. (*see Troubleshooting Section 9.2.3.2 - Leaking No. 1 Check Valve*) However, should the gage reading remain above the differential pressure relief valve opening point (i.e., relief valve does not discharge), then observe the gage reading. This is the apparent pressure drop across the No. 1 check valve. During Tests No. 1, No. 2, and No. 3 of this procedure the differential pressure gage is on line showing the pressure drop across the No. 1 check valve.

i. Open the high side bleed needle valve approximately one turn, and then open the low side bleed needle valve no more than one-quarter (1/4) turn to bypass water from the No. 2 test cock to the No. 3 test cock. If the low side bleed needle valve must be opened more than one-quarter (1/4) turn to lower the differential pressure reading to the relief valve opening point, then see *Troubleshooting Section 9.2.3.3 - Instructions for leaking No. 2 Shut-Off Valve.*

j. Observe the differential pressure reading as it **slowly** drops to the relief valve opening point. Record this opening point value when the first discharge of water is detected.

k. Close the needle valves. Detach the bypass hose from the low side bleed needle valve

Test No. 2 - *Tightness of No. 2 Check Valve*

Purpose: To test the No. 2 check valve for tightness against backpressure.

Requirement: The No. 2 check valve shall be tight against backpressure.

Steps:

a. Maintain the No. 2 shutoff valve in a closed position (from Test No. 1).

b. Attach the bypass hose from the high side bleed needle valve of the gage to the No. 4 test cock, then open the No. 4 test cock.

c. Bleed water from the zone by opening the low side bleed needle valve on the gage in order to re-establish the normal reduced pressure within the zone. Once the gage reading reaches a value above the apparent No. 1 check valve pressure drop, close the low side bleed needle valve.

d. Open the high side bleed needle valve. If the indicated differential pressure reading remains steady then the No. 2 check valve is reported as "closed tight". Go to Test No. 3. If the differential pressure - reading falls to the relief valve opening point, bleed water through the low side bleed needle valve until the gage reading reaches a value above the apparent No. 1 check valve pressure drop. If the differential pressure reading falls to the relief valve opening point again, then the No. 2 check valve is noted as "leaking", and Test No. 3 below cannot be completed. If the gage reading settles above the relief valve opening point, record the No. 2 check valve as "closed tight". *See the Troubleshooting Section 9.2.3.4 Disc Compression.* If the reading on the gage increases to the high end of the scale, *see Troubleshooting Section 9.2.3.5 - Backpressure Condition.*

Test No. 3 - *Tightness of No. 1 Check Valve*

Purpose: To determine the tightness of check valve No. 1, and to record the static pressure drop across the check valve No. 1.

Requirement: The static pressure drop across check valve No. 1 should be at least 3.0 psi greater than the relief valve opening point (Test No. 1). This 3.0 psi buffer will prevent the relief valve from discharging during small fluctuations in line pressure. A buffer of less than 3.0 psi does not imply a leaking check valve No. 1 (*i.e., allowing backflow to occur*), but rather is an indication of how well check valve No. 1 is sealing. *(See Appendix Section A.2.1 - 3.0 Psi Buffer)*

Steps:

a. With the bypass hose connected to test cock No. 4 as in step ***d*** of Test No. 2 (*high side bleed needle valve remaining open*), bleed water from the zone through the low side bleed needle valve on the gage until the gage reading exceeds the apparent No. 1 check valve pressure drop. Close the low side bleed needle valve. After the gage reading settles, the steady state differential pressure reading indicated (*reading is not falling on the gage*) is the actual static (*i.e., no flow*) pressure drop across check valve No. 1 and is to be recorded as such.

b. Close all test cocks, slowly open shut-off valve No. 2, and remove all test equipment

A.2.4 Field Test Procedure with 3-Needle Valve Gage Configuration (See Fig. A.2)

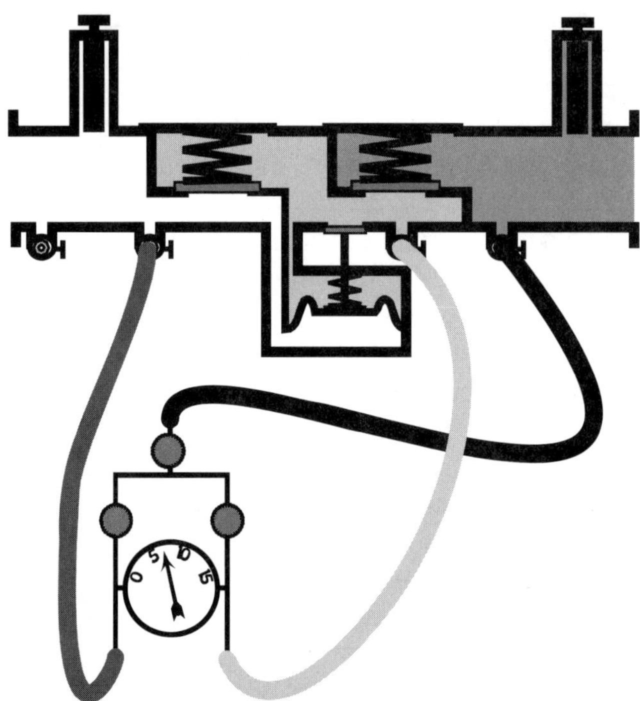

Fig. A.2
Hookup of Three needle valve gage configuration for Field Testing of Reduced Pressure Principle Assemblies

Test No. 1 - *Relief Valve Opening Point*

Purpose: To test the operation of the differential pressure relief valve.

Requirement: The differential pressure relief valve must operate to maintain the zone between the two check valves at least 2 psi less than the supply pressure.

NOTE: It is important that during this test the tester does not cause the relief valve to discharge before step ***i*** below. *(see Troubleshooting Section 9.2.3.1 - Exercising the Differential Pressure Relief Valve)*

Steps:

Follow all preliminary steps detailed in Section 9.1

a. To eliminate foreign material, open No. 4 test cock to establish flow through the unit, then flush water through test cocks No. 1, No. 2 (*open No. 2 test cock slowly*), and No. 3, by opening and closing each test cock one at a time. Be careful not to activate the relief valve during this process. Close test cock No. 4.

b. Install appropriate fittings to test cocks.

c. Attach hose from the high side of the differential pressure gage to the No. 2 test cock.

d. Attach hose from the low side of the differential pressure gage to the No. 3 test cock.

e. Open the bypass needle valve. Open test cock No. 3 slowly and then bleed all air through the bypass hose by opening the low side bleed needle valve. Maintain the low side bleed needle valve in the open position while the test cock No. 2 is opened **slowly**. Open the high side bleed needle valve to bleed all air from the gage. Close the high side bleed needle valve, then close the low side bleed needle valve after the gage reading has reached the upper end of the scale. Close the bypass needle valve.

f. Close #2 shut-off valve.

g. Should the gage reading drop to the low end of the gage scale and the differential pressure relief valve discharges continuously, then the No. 1 check valve is leaking. Tests No. 1, No. 2 and No. 3 may not be completed. (*see Troubleshooting Section 9.2.3.2 - Leaking No. 1 Check Valve*) However, should the gage reading remain above the differential pressure relief valve opening point (i.e., relief valve does not discharge), then observe the gage reading. This is the apparent pressure drop across the No. 1 check valve. During Tests No. 1, No. 2, and No. 3 of this procedure the differential pressure gage is on line showing the pressure drop across the No. 1 check valve.

h. Open the high side bleed needle valve approximately one turn, and then open the low side bleed needle valve no more than one-quarter (1/4) turn to bypass water from the No. 2 test cock to the No. 3 test cock. If the low side bleed needle valve must be opened more than one-quarter (1/4) turn to lower the differential pressure reading to the relief valve opening point, then see *Troubleshooting Section 9.2.3.3 - Instructions for Leaking No. 2 Shut-Off Valve.*

i. Observe the differential pressure reading as it **slowly** drops to the relief valve opening point. Record this opening point value when the first discharge of water is detected.

j. Close the low side bleed needle valve.

Test No. 2 - *Tightness of No. 2 Check Valve*

Purpose: To test the No. 2 check valve for tightness against backpressure.

Requirement: The No. 2 check valve shall be tight against backpressure.

Steps:

a. Maintain the No. 2 shut-off valve in a closed position (from Test No. 1).

b. Vent all of the air through the bypass hose by opening the bypass needle valve. Close the bypass needle valve. (*The high side bleed needle valve is to remain open*)

c. Attach the by-pass hose from the gage to the No. 4 test cock, then open the No. 4 test cock.

d. Bleed water from the zone by loosening the low side hose on the No. 3 test cock to re-establish the normal reduced pressure within the zone. Once the gage reading reaches a value above the apparent No. 1 check valve pressure drop, tighten the low side hose fitting.

e. Open the bypass needle valve. If the indicated differential pressure reading remains steady then the No. 2 check valve is reported as "closed tight". Go to Test No. 3. If the differential pressure reading falls to the relief valve opening point, bleed water through the low side bleed needle valve until the gage reading reaches a value above the apparent No. 1 check valve pressure drop. If the differential pressure reading falls to the relief valve opening point again, then the No. 2 check valve is noted as "leaking", and Test No. 3 below cannot be completed. If the gage reading settles above the relief valve opening point, record the No. 2 check valve as "closed tight". *See the Troubleshooting Section 9.2.3.4 - Disc Compression.* If the reading on the gage increases to the high end of the scale, *see Troubleshooting Section 9.2.3.5 - Backpressure Condition.*

Test No. 3 - *Tightness of No. 1 Check Valve*

Purpose: To determine the tightness of check valve No. 1, and to record the static pressure drop across the check valve No. 1.

Requirement: The static pressure drop across check valve No. 1 should be at least 3.0 psi greater than the relief valve opening point (Test No. 1). This 3.0 psi buffer will prevent the relief valve from discharging during small fluctuations in line pressure. A buffer of less than 3.0 psi does not imply a leaking check valve No. 1 (*i.e., allowing backflow to occur*), but rather is an indication of how well check valve No. 1 is sealing. *(See Appendix Section A.2.1 - 3.0 Psi Buffer)*

Steps:

a. With the bypass hose connected to test cock No. 4 as in step ***e*** of Test No. 2 (*high side bleed needle valve and the bypass needle valve remaining open*), bleed water from the zone by loosening the low side hose on the No. 3 test cock until the gage reading exceeds the apparent No. 1 check valve pressure drop. Tighten the low side hose. After the gage reading settles, the steady state differential pressure reading indicated (*reading is not falling on the gage*) is the actual static (*i.e., no flow*) pressure drop across check valve No. 1 and is to be recorded as such.

b. Close all test cocks, slowly open shut-off valve No. 2, and remove all test equipment

A.3 Pressure Vacuum Breaker (PVB & SVB)

A.3.1 Backpressure Evaluation

Purpose: To determine if the assembly (PVB or SVB) is subjected to continuous backpressure due to improper installation.

Requirement: Backpressure shall not be present. *(Test cannot be performed if the assembly is part of a parallel installation.)*

Steps:

After completing:
PVB: Section 9.4, Test No. 2, step ***f***
or SVB: Section 9.5, Test No. 2, step ***d***
proceed with the following:

a. Open shutoff valve No. 2. *(No. 1 shutoff valve is still closed.)*
b. *For PVB:* Open test cock No. 1, then open test cock No. 2.
For SVB: Open vent valve, then open test cock.
c. Observe and record if a continuous discharge of water occurs from test cock No. 2 (or vent valve)

NOTE: Discharge of water may occur for many minutes since the downstream must be depressurize through the test cock #2 (or vent valve). If the downstream piping contains any air in the system, then the depressuring may take a ***significant*** *amount of time.*

d. Close test cock No. 1 and No. 2 (test cock No. 1 and vent valve for SVB). Close shutoff valve No. 2.
e. Open shutoff valve No. 1, then slowly open shutoff valve No. 2.
f. Replace the air inlet valve canopy.

A.4 Double Check Valve Backflow Prevention Assembly

A.4.1 Bleed-off Valve Arrangement

The bleed-off valve used in Section 9.3 in the field test procedures for the double check valve assembly requires one each of:

1/4"Ø needle valve
1/4" nipple
1/4" tee
1/4" IPS × 45° SAE flare connector

constructed as shown in Fig. A.3.

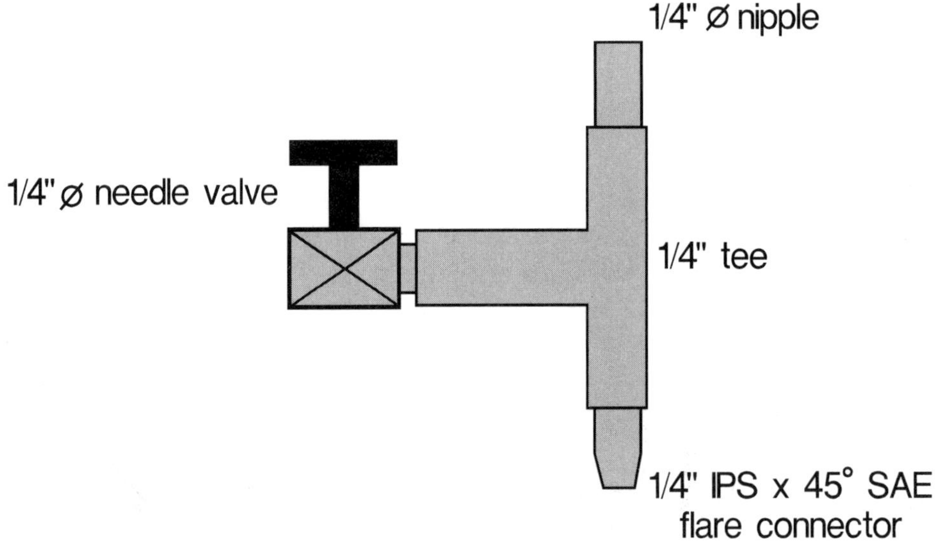

Fig. A.3
Bleed-Off Valve Arrangement

A.4.2 Limited Space

Below grade installations of the double check valve assembly (DC) may provide limited clearance around the assembly, such as in a meter-box or pit, and the gage may be prevented from being held at the proper elevation during the DC field test procedure. Should this occur, the DC field test procedure can still be performed accurately providing that the downstream reference point (See Section 9.3.3) is raised to an elevation where the gage is maintained. See Fig. A.4.

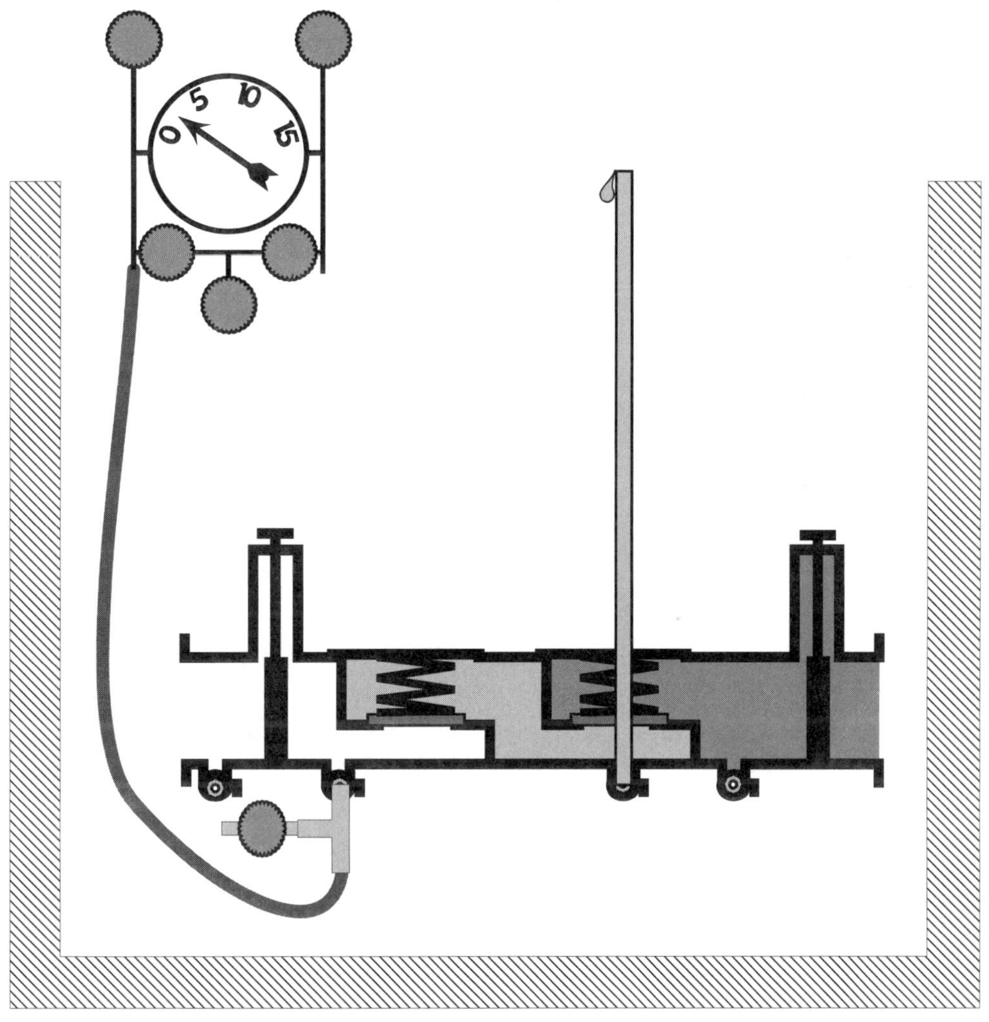

Fig. A.4

Testing the Double Check with Limited Clearance

Index

The Foundation for Cross-Connection control and Hydraulic Research at the University of Southern California offers several means to assist you in your cross-connection control program. In addition to the *Manual of Cross-Connection Control.* There are:

• Training Course for Testers,

• Training Courses for Cross-Connection Control Program Specialists,

• An Educational Video on Cross-Connection Control,

• A Membership Program which Entitles Provides the *List of Approved Backflow Preventers.*

For more information on each of these items, please complete one of the cards to the right and return it to the Foundation office or call the Foundation office at (213) 740-2032. By completing the attached card, you will automatically be informed when the next edition of the *Manual of Cross-Connection Control* is published.

Please send more information regarding the various tools available from the Foundation and send me information regarding the next edition of the *Manual of Cross-Connection Control* as soon as it is published.

Name:______________________________

Company Name:________________________

Street:______________________________

City, State, Zip________________________

Phone:______________________________

9th ed.

Please send more information regarding the various tools available from the Foundation and send me information regarding the next edition of the *Manual of Cross-Connection Control* as soon as it is published.

Name:______________________________

Company Name:________________________

Street:______________________________

City, State, Zip________________________

Phone:______________________________

9th ed.

Please send more information regarding the various tools available from the Foundation and send me information regarding the next edition of the *Manual of Cross-Connection Control* as soon as it is published.

Name:______________________________

Company Name:________________________

Street:______________________________

City, State, Zip________________________

Phone:______________________________

9th ed.

NO POSTAGE
NECESSARY
IF MAILED
IN THE
UNITED STATES

BUSINESS REPLY MAIL

FIRST CLASS MAIL PERMIT NO 55228 LOS ANGELES CA 90007

POSTAGE WILL BE PAID BY ADDRESSE

FOUNDATION FOR CROSS-CONNECTION CONTROL
AND HYDRAULIC RESEARCH
SCHOOL OF ENGINEERING MC-2531
UNIVERSITY OF SOUTHERN CALIFORNIA
P O BOX 77902
LOS ANGELES CA 90099-3334

NO POSTAGE
NECESSARY
IF MAILED
IN THE
UNITED STATES

BUSINESS REPLY MAIL

FIRST CLASS MAIL PERMIT NO 55228 LOS ANGELES CA 90007

POSTAGE WILL BE PAID BY ADDRESSE

FOUNDATION FOR CROSS-CONNECTION CONTROL
AND HYDRAULIC RESEARCH
SCHOOL OF ENGINEERING MC-2531
UNIVERSITY OF SOUTHERN CALIFORNIA
P O BOX 77902
LOS ANGELES CA 90099-3334

NO POSTAGE
NECESSARY
IF MAILED
IN THE
UNITED STATES

BUSINESS REPLY MAIL

FIRST CLASS MAIL PERMIT NO 55228 LOS ANGELES CA 90007

POSTAGE WILL BE PAID BY ADDRESSE

FOUNDATION FOR CROSS-CONNECTION CONTROL
AND HYDRAULIC RESEARCH
SCHOOL OF ENGINEERING MC-2531
UNIVERSITY OF SOUTHERN CALIFORNIA
P O BOX 77902
LOS ANGELES CA 90099-3334